Theory and Applications of Computability

In cooperation with the Association Computability in Europe

Books published in this series will be of interest to the research community and graduate students, with a unique focus on issues of computability. The perspective of the series is multidisciplinary, recapturing the spirit of Turing by linking theoretical and real-world concerns from computer science, mathematics, biology, physics, and the philosophy of science. The series includes research monographs, advanced and graduate texts, and books that offer an original and informative view of computability and computational paradigms.

Gleb Beliakov • Simon James • Jian-Zhang Wu

Choquet Capacities and Fuzzy Integrals

Springer

Gleb Beliakov
School of Information Technology
Deakin University
Geelong, VIC, Australia

Simon James
School of Information Technology
Deakin University
Geelong, VIC, Australia

Jian-Zhang Wu
Deakin University
Geelong, VIC, Australia

ISSN 2190-619X ISSN 2190-6203 (electronic)
Theory and Applications of Computability
ISBN 978-3-031-97069-6 ISBN 978-3-031-97070-2 (eBook)
https://doi.org/10.1007/978-3-031-97070-2

This Springer imprint is published by the registered company Springer Nature Switzerland AG
The registered company address is: Gewerbestrasse 11, 6330 Cham, Switzerland

To

Gelui Patricia, Chaquen and Sofia
G.B.
Rachel
S.J.
Fang Chen and Melinda
J.-Z. W.

Preface

As a computational model, the Choquet integral strikes a remarkable balance between complexity and interpretability. It is defined with respect to a Choquet capacity, which assigns value to each subset of input variables so that synergistic interactions, both positive and negative, can be modelled and accounted for in the aggregation process. While these interaction effects and resulting flexibility is what sets the Choquet integral apart from simpler (and largely additive) models, it is the emerging importance of transparency and model explainability that makes it an appealing alternative to other sophisticated methods currently used in machine learning and prediction. The Choquet and other fuzzy integrals have hence received a great deal of attention over the last few decades, particularly in fields related to artificial and computational intelligence, decision making, and data analysis. Along with various applications, extensions, and generalisations, considerable progress has also been made recently on simplifications and learning techniques that address the issue of tractability.

Choquet capacities are also referred to as *fuzzy measures* and *cooperative games* in different research areas, or simply *capacities*. They can be seen as a kind of generalised measure, relaxing the additivity requirement and replacing it with the monotonicity condition, i.e., that $\mathcal{A} \subseteq \mathcal{B} \rightarrow \mu(\mathcal{A}) \leqslant \mu(\mathcal{B})$. While they can be defined on continuous sets, the results of this book are usually contained to the discrete case, so that for a finite set of n inputs, denoted $\mathcal{N}$, a Choquet capacity assigns weight or value to every element of the power set $\mathcal{P}(\mathcal{N})$, requiring 2^n parameters in total. This exponential growth clearly presents a challenge when it comes to implementation, which either needs to be addressed through efficient approaches to eliciting values or simplification techniques that reduce the number of parameters that need to be specified explicitly.

Fuzzy or non-additive integrals aggregate sets of inputs according to the weight information provided by a capacity. The two most popular choices, the Choquet and Sugeno integrals, generalise the aggregation behaviour of the arithmetic mean and median, respectively, accounting not only for weight information but also for positive and negative interaction effects between variables. Interpretability of the underlying capacity is achieved via several value and interaction indices proposed by researchers, expressing in one way or another the extent to which sets of variables are complementary, redundant, or independent of one another. This explainability is one of the key features of the Choquet integral and other fuzzy integrals that makes them appealing for both prediction and analysis tasks in machine learning. The various indices can also be used to help construct Choquet capacities to meet application-specific requirements.

This book proceeds from the foundational theory, properties, and representations of capacities, to a focus on data-driven learning approaches that facilitate the application of Choquet capacities and fuzzy integrals to real-world problems. Chapter 1 provides the context which frames our study, giving a general overview of aggregation functions and the basic concepts related to Choquet capacities and fuzzy integrals. In Chap. 2 we focus on types of capacities and their properties, summarising different notions relating to interaction and additivity behaviours that can be used to define special families and subclasses. Indices based on value and interaction are then introduced in Chap. 3, many of which form the basis of alternative representations of capacities, which are covered in detail in Chap. 4. These representations, in turn, can be used to help define simplified and sparse capacities (which will be covered later in Chap. 7). Chapter 5 presents the key definitions and properties relating to fuzzy integrals, in particular the Choquet and Sugeno integrals. The Choquet integral with respect to symmetric capacities is equivalent to the ordered weighted averaging (OWA) operator, and so in Chap. 6 some specialised results for this case are summarised. Significant progress has been made in recent years in addressing the issue of tractability when applying Choquet capacities to large datasets. Chapters 7–9 combat this issue on multiple fronts. As already mentioned, Chap. 7 introduces a number of simplifying assumptions that result in sparse capacities in one or another representation. The tradeoff between complexity and the extent to which interactions are still accounted for is a key consideration, with the different approaches having different implications for interpretation and learning, the latter of which is covered in detail in Chap. 8. Here one can appreciate the extent to which these special capacity types, as well as some novel problem formulations, can help facilitate application of the Choquet integral to datasets where previously it would have been infeasible. Chapter 9 focuses on random sampling, summarising the challenges in effectively generating unbiased random capacities, both for applications and for experimental validation. Several approaches for generating different capacity types are presented, which can either be used by themselves or to augment the learning methods of the previous chapter. Chapter 10 then focuses on applications where the Choquet or other fuzzy integrals are used as objective functions in optimisation problems. Here too, we find that special capacity types can lead to significant reductions in complexity.

An appendix is provided giving details of a python package which implements many of the methods discussed in chapters of the book, with implementations in other languages also mentioned throughout. We make special mention here of RFMTool and Kappalab, which are available in R, as well as WOWA and AOTools available in C++.

It has been several years since *Discrete Fuzzy Measures: Computational Aspects* was published, with this new book aimed at bringing together some of the significant results obtained since then, particularly in addressing the problem of tractability when it comes to large datasets. Nonetheless, we are indebted to the great work of our colleagues and should mention several key monographs providing the theoretical foundations for what we have covered here and further detail on many of the concepts we discuss. Firstly is the book by M. Grabisch, *Set Functions, Games and Capacities in Decision Making*, 2016 (see the detailed references in the main text), where many theoretical aspects are addressed. We also mention the following (both as appreciation and a recommendation for further reading), grouped by topic:

Fuzzy integrals and capacities: V. Torra, Y. Narukawa and M. Sugeno's *Non-Additive Measures: Theory and Applications*, 2014; and Z. Wang and G.J. Klir's *Generalized Measure Theory*, 2009

Aggregation functions: G. Beliakov, H. Bustince and T. Calvo, A *Practical Guide to Averaging Functions*, 2016; M. Grabisch, J.-L. Marichal, R. Mesiar and E. Pap, *Aggregation Functions*, 2009; and M. Gagolewski, *Data Fusion: Theory, Methods, and Applications*, 2015.

Aggregation in engineering problems: J. Dujmovic, *Soft Computing Evaluation Logic: The LSP Decision Method and Its Applications*, 2018; J. Dujmovic, *Graded Logic: Modeling Human Commonsense Reasoning in Decision-Making and AI*, 2025.

G. Beliakov wishes to acknowledge support from his family he enjoyed during preparation of this book, as well as fruitful collaboration and friendship with his co-authors Simon and Wu. He also acknowledges fruitful discussions and collaboration with M. Grabisch, R. Mesiar, H. Bustince, and T. Calvo.

S. James would like to acknowledge the support of family and colleagues given throughout his career. For this book a special thanks goes out to valued friends and collaborators, Marek Gagolewski and Aoi Honda, as well as Gleb and Wu, with whom it is still a great privilege to collaborate.

J.-Z. Wu sincerely thanks his family for their strong support during the writing of this book. He is grateful for the great collaboration with Gleb and Simon. He also thanks Marek Gagolewski, Lei Pan, and Ye Zhu for their helpful advice and support throughout the project.

Melbourne, VIC, Australia
May 2025

Gleb Beliakov
Simon James
Jian-Zhang Wu

Declarations

Competing Interests The authors have no competing interests to declare that are relevant to the content of this manuscript.

Contents

Chapter 1
Introduction

Abstract Choquet capacities, which are also referred to as fuzzy measures and cooperative games, model contributions and interactions over elements within a given set. They are used to encode the weights of the Choquet integral, a widely studied averaging aggregation function that can be applied to analysis and prediction tasks. This chapter introduces Choquet capacities in the context of aggregation functions, providing an overview of the related definitions and properties, as well as providing an example to help guide intuition on their behaviour and how they can be constructed.

1.1 Rationale

Does $1 + 1 = 2$? This question seems odd as we have learned early in life to answer positively. But is it always the case? Bulk discounts provide fundamental incentives in any trade business. "Two for the price of one", "get the third item for free", and countless similar deals teach us that in basic economics we observe subadditive relationships in terms of prices, $1 + 1 < 2$.

Can we get the opposite inequality? We often hear the adage that "the whole is worth more than the sum of its parts". A can of tuna and a can opener on a hiking trip are more valuable together than individually. Two complementary building projects are more valuable if executed together. Thus, we also observe superadditivity in terms of value, $1 + 1 > 2$; see Fig. 1.1.

Decision sciences also provide countless examples of both synergies and redundancies among decision criteria. Several very common and individually insignificant symptoms taken together often provide strong pointers to a medical diagnosis. On the contrary, redundant criteria (e.g. opinions from mutually dependent experts) bring little value to the decision process and need to be downweighted.

This book is about formal mathematical modelling of the nonadditivity phenomena that surround us in decision sciences, economics, human reasoning, and by implication in artificial intelligence.

The default way of combining criteria, degrees of evidence, values, or costs is by sum or averaging (the arithmetic mean or its weighted counterparts). Being

G. Beliakov et al., *Choquet Capacities and Fuzzy Integrals*, Theory and Applications of Computability, https://doi.org/10.1007/978-3-031-97070-2_1

Fig. 1.1 Advertisements we often see in shopping centres

an additive aggregator, the arithmetic mean completely discards any interactions between its inputs. Each input only contributes its own value, potentially weighted by some factor. There is no room to account for contributions of groups of inputs combined together. In order to do so, we need to move away from additive aggregators.

The next least complex models after linear are quadratic and piecewise linear. Quadratic models account for pairwise interactions between inputs with pairwise product terms. In contrast, piecewise linear models in principle can account for all possible subsets of a set of inputs, which is the line we explore in this book.

In 1953, Gustave Choquet (1915–2006) introduced his theory of capacities, which we call *Choquet capacities*. In essence, Choquet capacities are functions defined on power sets of some set. That is, a value is assigned to all possible subsets of a set $\mathcal{N}$. This way we can quantify the value of every subset of inputs, often referred to as coalitions. By combining these values with the values of the inputs, one can aggregate the inputs in a sophisticated way, accounting for their synergies and redundancies. We shall see later in this book that mathematical functions to aggregate these inputs belong to the class of aggregation functions [BPC07, BBC16, Gra+09], and are often referred to as fuzzy integrals, among which the Choquet integral plays a prominent role.

Choquet capacities have been rediscovered several times and appear under the names of cooperative games [Gra16, PS07], nonadditive measures [Fuj14, Den94], and fuzzy measures [Sug74]. There are some minor distinctions between these terms, and we will establish their relations in the sequel. However, the commonality between the mentioned concepts lies in one fundamental property—monotonicity of the set functions. That is, the value of some set of inputs is no smaller than the values assigned to any of its subsets.

Monotonicity (with respect to set inclusion) extends the additivity of classical measures: addition of each input brings some extra value (which can also be 0), not necessarily equal to the input's individual value, but rather its marginal contribution to a particular coalition. One cannot diminish the value of a subset by bringing in new elements, which ensures a certain level of rationality in many contexts. For example, although we may see bundled item discounts, it would be quite irregular if the overall price is actually reduced by adding an additional item—it implies the seller is paying you to take it. Thus, a rational choice suggests that a set of items has a larger value than any of its subsets. In the language of cooperative game theory, addition of a player does not diminish the value of a coalition. Also, the value of a null set is taken as 0 by convention.

In this book, we focus on finite universal sets $\mathcal{N} = \{1, 2, \ldots, n\}$, as our interest lies in aggregating decision criteria or values of some items whose number is relatively small, even though the theories of set functions, capacities, cooperative games, and fuzzy measures extend to infinite sets. For this reason, one can always introduce a normalisation factor, so that the value of a monotone set function $\mu(\mathcal{N}) = 1$ (or any other positive number) without loss of generality. Although normalisation is always possible in our setting, we will not insist on it for reasons of interpretation, and hence define Choquet capacities as monotone grounded set functions, whereas we define fuzzy measures as normalised capacities to be consistent with current literature (see Fig. 1.2).

We start our formal presentation from aggregation functions in general, many of which are not designed to handle input dependencies, but provide a framework in which we view fuzzy or nonadditive integrals as a class of averaging aggregators. We also note that some classical aggregation functions are special cases of fuzzy integrals when the underlying capacity is additive, is symmetric, or has another

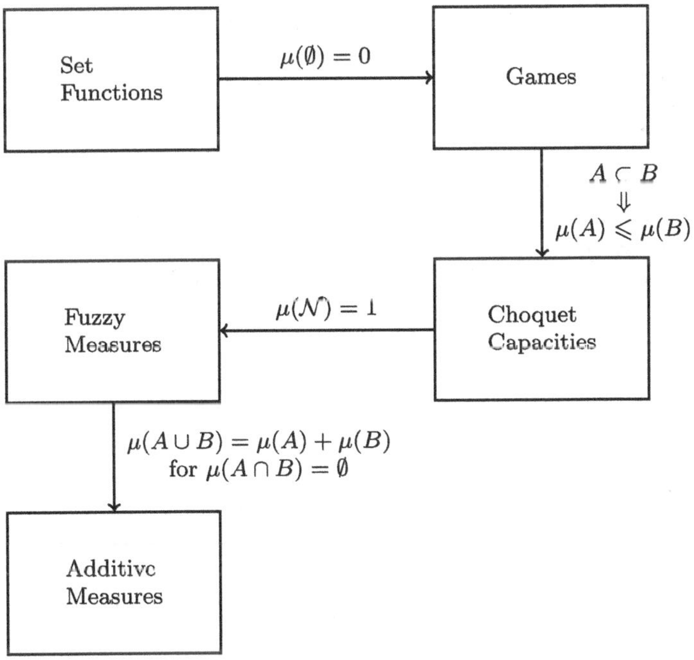

Fig. 1.2 Relationships between various concepts

property, which makes Choquet and other fuzzy integrals a generalisation of several commonly used means.

1.2 Review of Aggregation

The formal study of *aggregation functions* has largely been pursued in fields relating to fuzzy sets and decision-making, but also has clear connections to classical statistics and measures of centre, with many results arising independently in different mathematical contexts. An aggregation function $f : [0, 1]^n \to [0, 1]$ takes multiple inputs and combines them into a single output, either so that different options can be objectively compared or so as to give an overall evaluation leading to interpretations according to some predefined rules. Given the connection to fuzzy sets and the need to generalise logical operators, the two fundamental properties are the boundary conditions, $f(0, \ldots, 0) = 0$ and $f(1, \ldots, 1) = 1$, as well as monotonicity.

The two most widely used examples are the arithmetic mean, also referred to as the mean or simply "the average", and the median (along with their weighted counterparts). These functions are simply expressed and, for the most part, well understood. However, sometimes their behaviour can lead to misleading evaluations or they are misapplied in certain aggregation contexts. For example, if the price of household items increases by 4% 1 year due to inflation and 6% the next, the *average* yearly inflation, i.e. the rate that if constant would lead to the same final result, is not 5% as would be obtained by the arithmetic mean. Similarly, if I travel 40km/h on the way to work and 60km/h on the way home, it would not be correct to say my average speed is 50km/h. These typical examples, where one learns that we should actually apply the geometric and harmonic means, respectively, can provide that first insight into recognising that different and more sophisticated methods of aggregation are sometimes called for. Other common problems include the need to remove or downweight outliers, which is why the median is preferred over the arithmetic mean when summarising housing prices, and why the trimmed mean is used to measure inflation, e.g. by the Reserve Bank of Australia when determining whether to raise or lower interest rates. Such approaches may not only be about robustness against extreme observations but also about bias. The so-called Olympic average removes the lowest and highest scores, regardless of whether they are extreme or not.

As it turns out, aggregation with the Choquet integral includes the weighted arithmetic mean as well as the median, trimmed mean, and Olympic average as special cases, allowing one to model not only the importance of inputs but also taking into account the relative order of inputs. Often of most interest is the ability to account for correlations and other interacting behaviour between inputs; however, it is not the only aggregation function that fulfils this purpose. Before proceeding to some more formal definitions, we make note of the existing works summarising the many important developments and results in aggregation theory, including [BPC07, BBC16, Gra+09, BJW19, Gra16] as well as [Gag15, Jam16, Duj18, Bul03, Tor07].

1.2.1 Definitions

When aggregation functions such as the arithmetic mean or median are used to summarise data, whether it be for comparisons or providing some descriptive measure, an underlying implication is that the aggregated value somehow represents what is normal or typical when looking at the set of inputs as a whole. Many of the properties we associate with aggregation functions are geared towards practical interpretations or desirable behaviours in common applications, while others serve to make the theory more complete and consistent.

In this book, we will usually work with vectors of inputs $\mathbf{x} = (x_1, x_2, \ldots, x_n)$ with each x_i taking a value in the real interval $\mathbb{I} = [0, 1]$; however, with suitable transformations, it is also possible to use any interval $[a, b] \subset \mathbb{R}$. We use the vector inequality notation $\mathbf{x} \leqslant \mathbf{y}$ to indicate that $x_i \leqslant y_i$ for all $i \in \{1, \ldots, n\}$, which assumes that the vectors are comparable and of the same dimension. We can then express our formal definition of aggregation functions as follows.

Definition 1.1 (Aggregation Function) An aggregation function is a function of $n > 1$ arguments that maps the (n-dimensional) cube onto an interval $\mathbb{I} = [0, 1]$, $f : \mathbb{I}^n \to \mathbb{I}$, with the following properties:

(i) $f(\underbrace{0, 0, \ldots, 0}_{n-times}) = 0$ and $f(\underbrace{1, 1, \ldots, 1}_{n-times}) = 1$.

(ii) $\mathbf{x} \leqslant \mathbf{y}$ implies $f(\mathbf{x}) \leqslant f(\mathbf{y})$ for all $\mathbf{x}, \mathbf{y} \in \mathbb{I}^n$.

The first property is usually referred to as the boundary conditions, requiring that all zero inputs results in a zero and a vector of 1s results in a 1, or analogously for the minimum and maximum elements of $\mathbb{I}$ if a different interval is used. This is important for being able to reliably interpret the output as potentially taking values across the same domain as the inputs.

The second property prescribes the monotone behaviour of aggregation functions. It can be expressed equivalently in terms of the *univariate* functions $f_{\mathbf{x}}(t) = f(\mathbf{x})$, with $t = x_i$ and the rest of the components of $\mathbf{x}$ fixed. Monotonicity is perhaps the condition that most distinguishes approaches based on aggregation functions from other functions or input–output models used in similar applications. As an assumption, one should always consider whether this relationship between the inputs and output is justified or makes sense in the given context. When aggregating the evaluations of judges in sporting competitions, it makes sense that one judge increasing their score should not result in a lower overall score for the contestant. On the other hand, in predicting someone's salary based on the salaries of others in their department, we might not expect the relationship to be monotone, since sometimes one person earning more might come at the expense of another earning less.

Functions such as the arithmetic mean and median are defined consistently for any n and so allow input sets of varying size to be evaluated and compared. For some aggregation functions, however, especially weighted functions, there may not be any straightforward way of extending definitions for a particular value of n to higher or lower cardinalities. The term *extended aggregation function* [MC97] indicates that a family of functions is defined consistently for any value of n.

Definition 1.2 (Extended Aggregation Function) An extended aggregation function is a mapping

$$F : \bigcup_{n \in \{1,2,\ldots\}} \mathbb{I}^n \to \mathbb{I},$$

such that the restriction of this mapping to the domain $\mathbb{I}^n$ for a fixed n is an n-ary aggregation function f, with the convention $F(x) = x$ for $n = 1$.

Where there is a need to emphasise that an aggregation function has n arguments, we can use the notation f_n. The following example lists a number of well-known extended aggregation functions.

Example 1.1 The following aggregation functions are defined for any n.

Arithmetic mean	$\mathrm{AM}(\mathbf{x}) = \frac{1}{n}(x_1 + x_2 + \ldots + x_n).$
Geometric mean	$\mathrm{GM}(\mathbf{x}) = \sqrt[n]{x_1 x_2 \ldots x_n}.$
Harmonic mean	$\mathrm{HM}(\mathbf{x}) = \frac{n}{\frac{1}{x_1}+\frac{1}{x_2}+\ldots+\frac{1}{x_n}}.$
Minimum	$\min(\mathbf{x}) = \min\{x_1, x_2, \ldots, x_n\}$
Maximum	$\max(\mathbf{x}) = \max\{x_1, x_2, \ldots, x_n\}$
Product	$\mathrm{prod}(\mathbf{x}) = x_1 x_2 \ldots x_n$

Each of the means will provide an output that lies between the smallest and largest inputs, but the geometric mean and harmonic mean tend more toward the lower inputs. A well-established relationship between them is that for any $\mathbf{x} \in [0, 1]^n$, it holds that $\mathrm{HM}(\mathbf{x}) \leqslant \mathrm{GM}(\mathbf{x}) \leqslant \mathrm{AM}(\mathbf{x})$. For vectors over the unit hypercube, the product always results in values that are smaller or equal to the smallest input. We will see in the next section that such properties are used to define different classes of aggregation functions that often serve different purposes.

1.2.2 Main Classes of Aggregation Functions

The arithmetic mean and product given in Example 1.1 represent typical choices of averaging and conjunctive aggregation functions, respectively. Averaging behaviour would usually be desirable in situations where we want the aggregated value to be expressed in the same units or to range over the same values as the inputs, hence being easy to interpret in terms of its relative magnitude. Conjunctive behaviour, on the other hand, generalises the logical AND operation and so would usually reflect the degree to which all of the inputs are satisfied. There are also disjunctive aggregation functions, generalising the OR logical operation, and functions whose behaviour is mixed, i.e. with different behaviours over different parts of the domain. These four classes characterising the overall behaviour of aggregation functions are often studied separately as they are suited to different applications [DP85, Cal+02, DP80, DP00, DP04, BPC07]. We will outline their definitions below, before looking at the main families and prototypes in Sect. 1.2.4.

Definition 1.3 (Averaging Aggregation) An aggregation function f has averaging behaviour (or is averaging) if for every $\mathbf{x} \in \mathbb{I}^n$, it is bounded by

$$\min(\mathbf{x}) \leqslant f(\mathbf{x}) \leqslant \max(\mathbf{x}).$$

The arithmetic, geometric, and harmonic means from Example 1.1 are all averaging aggregation functions, while the minimum and maximum represent the border cases.

Definition 1.4 (Conjunctive Aggregation) An aggregation function f has conjunctive behaviour (or is conjunctive) if for every $\mathbf{x}$, it is bounded by

$$f(\mathbf{x}) \leqslant \min(\mathbf{x}) = \min(x_1, x_2, \ldots, x_n).$$

The minimum and product are common choices for modelling conjunctive behaviour.

Definition 1.5 (Disjunctive Aggregation) An aggregation function f has disjunctive behaviour (or is disjunctive) if for every $\mathbf{x}$, it is bounded by

$$f(\mathbf{x}) \geqslant \max(\mathbf{x}) = \max(x_1, x_2, \ldots, x_n).$$

The probabilistic sum $a + b - ab$ is an example of a disjunctive aggregation function, dual to the product. As we will see more formally later, the dual of an aggregation function can be obtained by taking $1 - f(1 - \mathbf{x})$. While the semantics and purposes for conjunctive and disjunctive functions are different, we can learn everything about one class and obtain dual results using this relationship.

Definition 1.6 (Mixed Aggregation) An aggregation function f is mixed if it does not belong to any of the above classes, i.e. it exhibits different types of behaviour on different parts of the domain.

Mixed behaviour is needed in some contexts so that the function behaves differently depending on the inputs, for example, we might want high or low values to reinforce each other.

1.2.3 Main Properties of Aggregation

While the class of an aggregation function should align with its intended purpose in the chosen application, there are a number of well-known properties that can be useful for ensuring desired behaviour and consistency or for understanding and interpreting results in context. We provide definitions here for those that are most commonly presented in aggregation research. These help to provide a background framework for understanding how Choquet capacities can be used in practice.

Idempotency, Symmetry, and Associativity

The desire for idempotency arises naturally whenever we want the output for an aggregation function to provide a faithful summary of input tendency. In short, if the inputs to an idempotent function are all the same, then the output is simply that value. Of course, if a function is averaging, then the upper and lower bounds ensure that this will be the case; however, examples of mixed aggregation functions that are idempotent are also common.

Definition 1.7 (Idempotency) An aggregation function f is called idempotent if for every input $\mathbf{x} = (t, t, \ldots, t), t \in \mathbb{I}$, the output is $f(t, t \ldots, t) = t$.

Note 1.1 Not only is it the case that all averaging functions are idempotent, but in fact, the monotonicity of aggregation functions ensures that averaging behaviour and idempotency are *equivalent* for aggregation functions.[1] The monotonicity requirement also ensures that the *minimum* and *maximum* are the only examples of conjunctive and disjunctive aggregation functions, respectively, that are idempotent.

A related but broader concept is local internality, which refers to functions whose arguments is always one of the inputs.

Definition 1.8 (Local Internality) An aggregation function f is called *locally internal* if for all $x_1, \ldots, x_n \in \mathbb{I}^n$, $f(x_1, \ldots, x_n) \in \{x_1, \ldots, x_n\}$.

The term “internal” sometimes is used synonymously with the definition we have given for averaging behaviour; however, we reserve its use according to the above. Clearly, all internal functions are idempotent (and hence averaging according to Note 1.1), with simple examples including order statistics, the minimum, the maximum, and the median when there is an odd number of inputs.

[1] Proof: Take any $\mathbf{x} \in \mathbb{I}^n$, and let p, q denote the values $p = \min(\mathbf{x}), q = \max(\mathbf{x})$. By monotonicity, $p = f(p, p, \ldots, p) \leqslant f(\mathbf{x}) \leqslant f(q, q, \ldots, q) = q$. Hence, $\min(\mathbf{x}) \leqslant f(\mathbf{x}) \leqslant \max(\mathbf{x})$. The converse: let $\min(\mathbf{x}) \leqslant f(\mathbf{x}) \leqslant \max(\mathbf{x})$. By taking $\mathbf{x} = (t, t, \ldots, t)$, $\min(\mathbf{x}) = \max(\mathbf{x}) = f(\mathbf{x}) = t$, hence idempotency.

Another property that is often considered essential when summarising sets of data evaluations is that of symmetry. For functions like the mean or median, consistency demands that the order in which the inputs are provided should not affect the results.

Definition 1.9 (Symmetry) An aggregation function f is called symmetric if its value does not depend on the permutation of the arguments, i.e.

$$f(x_1, x_2, \ldots, x_n) = f(x_{P(1)}, x_{P(2)}, \ldots, x_{P(n)}),$$

for every $\mathbf{x}$ and every permutation $P = (P(1), P(2), \ldots, P(n))$ of $(1, 2 \ldots, n)$.

This property amounts to commutativity in the bivariate case, e.g. addition and multiplication are both commutative, while division and subtraction are not. Symmetric functions can be interpreted as treating each of the inputs the same and sometimes is referred to as anonymity. Of course, there are situations where we would desire inputs to be weighted differently, e.g. if the input set represents a distribution of scores (with differing frequencies) or if the vote of a particular judge or stakeholder needs to carry more weight.

Our definition of symmetry makes use of ordering permutations, which are also useful for other concepts when it comes to aggregation and Choquet capacities. We will use some different forms of notation to represent ordering permutations throughout these chapters, aiming to ease the complexity of cumbersome notation where possible. A vector whose components are rearranged according to a permutation π can be given by $\mathbf{x}_\pi = (x_{\pi(1)}, x_{\pi(2)}, \ldots, x_{\pi(n)})$. For example, $(5, 3, 2, 1, 4)$ would be one permutation of $(1, 2 \ldots, 5)$, which would be represented as $\mathbf{x}_\pi = (x_5, x_3, x_2, x_1, x_4)$. The following definition summarises the shorthand we will often use for increasing/decreasing permutations.

Definition 1.10 (Increasing/Decreasing Permutations) We denote by $\mathbf{x}_\nearrow$ the vector obtained from $\mathbf{x}$ by arranging its components in increasing (non-decreasing) order, that is, $\mathbf{x}_\nearrow = \mathbf{x}_\pi$ where π is the permutation such that $x_{\pi(1)} \leqslant x_{\pi(2)} \leqslant \ldots \leqslant x_{\pi(n)}$.

Similarly, we denote by $\mathbf{x}_\searrow$ the vector obtained from $\mathbf{x}$ by arranging its components in decreasing order, that is, $\mathbf{x}_\searrow = \mathbf{x}_\pi$ where π is the permutation such that $x_{\pi(1)} \geqslant x_{\pi(2)} \geqslant \ldots \geqslant x_{\pi(n)}$.

Note 1.2 The notation $\mathbf{x}_{()} = (x_{(1)}, \ldots, x_{(n)})$ is often adopted in the literature to represent both $\mathbf{x}_\nearrow$ and $\mathbf{x}_\searrow$. For example, increasing permutations are often assumed for the Choquet integral, while decreasing permutations have traditionally been used for ordered weighted averaging functions.

Note 1.3 Rather than defining symmetry in terms of every possible permutation, one can use the equivalent condition:

$$f(\mathbf{x}) = f(\mathbf{x}_\nearrow) \ \ (\text{ or } f(\mathbf{x}) = f(\mathbf{x}_\searrow)), \ \text{ for all } \mathbf{x}.$$

Associativity allows bivariate functions to be extended to any number of arguments consistently.

Definition 1.11 (Associativity) A two-argument function f is associative if $f(f(x_1, x_2), x_3) = f(x_1, f(x_2, x_3))$ holds for all x_1, x_2, x_3 in its domain.

This is one way of constructing extended aggregation functions so that the aggregation rules are consistent for all n. For an associative bivariate aggregation function f_2, the corresponding extended aggregation function is given by

$$f_n(x_1, \dots, x_n) = f_2(f_2(\dots f_2(x_1, x_2), x_3), \dots, x_n).$$

The product, minimum, and maximum are associative aggregation functions; however, the arithmetic mean and statistical median are not.

The associativity property provides us with one example of how n-variate functions can be constructed from simpler functions. This is related to the idea of decomposability.

Definition 1.12 (Decomposable Extended Aggregation Function) An extended aggregation function F is decomposable if for all $m, n = 1, 2, \dots$ and for all $\mathbf{x} \in \mathbb{I}^m$, $\mathbf{y} \in \mathbb{I}^n$:

$$\begin{aligned} &f_{m+n}(x_1, \dots, x_m, y_1, \dots, y_n) = \\ &f_{m+n}(\underbrace{f_m(x_1, \dots, x_m), \dots, f_m(x_1, \dots, x_m)}_{m\ times}, y_1, \dots, y_n). \end{aligned} \tag{1.1}$$

This amounts to replacing m of the inputs with repetitions of their aggregated value. As such, continuous decomposable extended aggregation functions are always idempotent.

Note 1.4 In [MT15], the term used for this property is *barycentric associativity*.

The geometric mean is one example of a decomposable aggregation function. For example,

$$\sqrt[5]{x_1\ x_2\ x_3\ x_4\ x_5} = \sqrt[5]{\sqrt{x_1 x_2}\ \sqrt{x_1 x_2}\ x_3\ x_4\ x_5}.$$

The bisymmetry property for extended aggregation functions also provides conditions for when the aggregation process is broken down into multiple steps.

Definition 1.13 (Bisymmetry) An extended aggregation function f is bisymmetric if for all $m, n = 1, 2, \dots$ and for all $\mathbf{x} \in \mathbb{I}^{mn}$,

$$\begin{aligned} f_{mn}(\mathbf{x}) &= f_m(f_n(x_{11}, \dots, x_{1n}), \dots, f_n(x_{m1}, \dots, x_{mn})) \\ &= f_n(f_m(x_{11}, \dots, x_{m1}), \dots, f_m(x_{1n}, \dots, x_{mn})). \end{aligned} \tag{1.2}$$

Table 1.1 The table of scores to be aggregated by m judges with respect to n criteria

Judge\Criterion	1	2	3	...	n	Overall
1	x_{11}	x_{12}	x_{13}	...	x_{1n}	y_1
2	x_{21}	x_{22}	x_{23}	...	x_{2n}	y_2
3	x_{31}	x_{32}	x_{33}	...	x_{3n}	y_3
4	x_{41}	x_{42}	x_{43}	...	x_{4n}	y_4
$\vdots$	$\vdots$	$\vdots$	$\vdots$	$\vdots$	$\vdots$	$\vdots$
m	x_{m1}	x_{m2}	x_{m3}	...	x_{mn}	y_m
Overall	$\tilde{y}_1$	$\tilde{y}_2$	$\tilde{y}_3$	...	$\tilde{y}_n$	$\tilde{z}\backslash z$

The property specifies that the aggregation can be broken down into sub-aggregations of equal smaller groups and that the group size should not affect the result. An intuitive example is the task of aggregating the values contained in a spreadsheet with m rows and n columns. If we aggregate the values in each row to produce $y_1, \ldots, y_m$ and then aggregate these, we should arrive at the same result as when we aggregate the n columns to produce $\tilde{y}_1, \ldots, \tilde{y}_n$ in the first step, and both of these should be consistent with the aggregation of all mn values together. Table 1.1 provides an example of how the data would be arranged, where we assume that each of the rows correspond with a judge and each column with a criterion. We can either aggregate the judges' overall scores or aggregate the overall scores across the given criteria.

Note 1.5 An extended aggregation function that is both symmetric and associative will be bisymmetric; however, functions such as the arithmetic and geometric means are symmetric and bisymmetric but not associative. There are also examples of aggregation functions that are associative and bisymmetric functions but not symmetric, e.g. the projection to the first coordinate, $f_n(\mathbf{x}) = x_1$.

Further Properties

The monotonicity of aggregation functions ensures that in, for example, multicriteria decision-making, for two alternatives A and B, if A is equal to or better than B with respect to all criteria, then B should not be preferred to A. We note that this does not discount the possibility that A and B might be equally preferred. For example, taking the minimum of (3, 5) and (3, 10) would yield 3 in both cases, even though in a decision-making context, we might expect the latter to be the preferred option. Strict monotonicity requires that that an alternative that is better in at least one respect and no worse in all others should have a higher result.

Definition 1.14 (Strict Monotonicity) An aggregation function f is strictly monotone increasing if

$$\mathbf{x} \leqslant \mathbf{y} \text{ but } \mathbf{x} \neq \mathbf{y} \text{ implies } f(\mathbf{x}) < f(\mathbf{y}) \text{ for every } \mathbf{x}, \mathbf{y} \in \mathbb{I}^n. \tag{1.3}$$

Note 1.6 We draw attention to the difference between $[\mathbf{x} \leqslant \mathbf{y}, \mathbf{x} \neq \mathbf{y}]$ and $\mathbf{x} < \mathbf{y}$, with the latter implying that $x_i < y_i$ for *all* components. Strict monotonicity can be likened to the idea of Pareto dominance in economics.

Homogeneity and shift invariance are useful properties if we want to use aggregation functions on different scales, with homogeneity providing a kind of stability with respect to multiplication by a constant, while shift invariance allows for consistent results when a constant is added.

Definition 1.15 (Homogeneity) An aggregation function $f : \mathbb{I}^n \to \mathbb{I}$ is homogeneous of order 1 if for all λ and for all $(x_1, \ldots, x_n) \in \mathbb{I}^n$, it holds that

$$f(\lambda x_1, \ldots, \lambda x_n) = \lambda f(x_1, \ldots, x_n).$$

Definition 1.16 (Shift Invariance) An aggregation function $f : \mathbb{I}^n \to \mathbb{I}$ is shift-invariant (or stable for translations) if for all λ and for all $(x_1, \ldots, x_n) \in \mathbb{I}^n$, it holds that

$$f(x_1 + \lambda, \ldots, x_n + \lambda) = f(x_1, \ldots, x_n) + \lambda$$

whenever $(x_1 + \lambda, \ldots, x_n + \lambda) \in \mathbb{I}^n$.

Aggregation functions that satisfy both properties are referred to as *linear aggregation functions*, with the arithmetic mean being a prototypical example.

Note 1.7 From the boundary conditions, $f(a, \ldots, a) = a$ and $f(b, \ldots, b) = b$, aggregations that are either shift-invariant or homogeneous (or both) will also be idempotent.

For homogenous aggregation functions, it does not matter whether we use $\mathbb{I} = [0, 1]$ or $\mathbb{I} = [0, 100]$, while for a shift-invariant function, results on $[0, 1]$ would be equivalent to those on $[-1, 0]$. Such considerations can be essential when we apply aggregation functions in different contexts. For example, in averaging temperatures expressed in Fahrenheit and Celsius, or lengths in inches and centimetres, it would be important that the scale or units used do not have an impact on the results.

Figure 1.3 provides examples for each of the cases when we consider whether or not the aggregation function satisfies shift invariance and is homogeneous. Homogeneous aggregation functions, which include all power means, may be more familiar, especially widely used linear aggregation functions including the arithmetic mean and statistical median. However, we can see that even for averaging aggregation functions, there are cases that are not homogeneous and/or not shift-invariant. The Choquet integral that we focus on in later chapters satisfies both properties.

		Shift-Invariant	
		Yes	No
Homogeneous	Yes	$\frac{1}{n}\sum_{i=1}^{n} x_i$ Arithmetic Mean	$\left(\frac{1}{n}\sum_{i=1}^{n}\frac{1}{x_i}\right)^{-1}$ Harmonic Mean
	No	$\log\left(\frac{1}{n}\sum_{i=1}^{n}\exp(x_i)\right)$ Exponential Mean	$\left(\log\left(\frac{1}{n}\sum_{i=1}^{n}\exp(\frac{1}{x_i})\right)\right)^{-1}$ Radical Mean

Fig. 1.3 Examples of functions satisfying (or not satisfying) shift invariance and homogeneity

Duality

The conjunctive and disjunctive functions are dual to one another, so we can learn a lot about the behaviour of a disjunctive function by looking at its dual conjunctive counterpart and vice versa; however, it is also useful when working with averaging and mixed aggregation functions. Averaging aggregation functions are closed under duality, and so taking the dual results in another averaging function, however one that exhibits different behaviour. For example, the geometric mean has an absorbent element of 0 (any input vectors that include a zero will result in a zero output), while its dual function has an absorbent element of 1. We will often refer to duals when working with Choquet capacities and the Choquet integral.

We define dual aggregation functions using negations.

Definition 1.17 (Strict Negation) A univariate function N defined on $[0, 1]$ is called a strict negation if its range is also $[0, 1]$ and it is strictly monotone decreasing.[2]

Definition 1.18 (Strong Negation) A univariate function N defined on $[0, 1]$ is called a strong negation if it is strictly decreasing and involutive (i.e. $N(N(t)) = t$ for all $t \in [0, 1]$).

We usually make use of the standard negation, $N(t) = 1 - t$; however, other choices are possible.

Since a strictly monotone bijection is always continuous, negations that are both strict and strong are continuous.

[2] We can use the term *bijection*: a bijection is a function $f : A \to B$, such that for every $y \in B$, there is exactly one $x \in A$, such that $y = f(x)$, i.e. it defines a one-to-one correspondence between A and B. Because N is strictly monotone, it is a one-to-one function. Its range is [0,1]; hence, it is an *onto* mapping and therefore a bijection.

Once a negation is defined, the dual is then given as follows.

Definition 1.19 (Dual Aggregation Function) Let $N : [0, 1] \to [0, 1]$ be a strong negation and $f : [0, 1]^n \to [0, 1]$ an aggregation function. Then the aggregation function f^d given by

$$f^d(x_1, \ldots, x_n) = N(f(N(x_1), N(x_2), \ldots, N(x_n)))$$

is called the dual of f with respect to N, or, for short, the N-dual of f. When using the standard negation, f^d is given by

$$f^d(x_1, \ldots, x_n) = 1 - f(1 - x_1, \ldots, 1 - x_n),$$

and we will simply say that f^d is the dual of f.

While the averaging class of aggregation functions is closed under duality, it is also possible to have self-dual aggregation functions. The arithmetic mean is one such example.

Definition 1.20 (Self-Dual Aggregation Function) Given a strong negation N, an aggregation function f is self-dual with respect to N (for short, N-self-dual or N-invariant) if

$$f(\mathbf{x}) = N(f(N(\mathbf{x}))),$$

where $N(\mathbf{x}) = (N(x_1), \ldots, N(x_n))$. For the standard negation, we have

$$f(\mathbf{x}) = 1 - f(\mathbf{1} - \mathbf{x}) = f^d(\mathbf{x}),$$

and it is simply said that f is self-dual.

Comparability

From Example 1.1, we saw that sometimes inequalities hold between aggregation functions, i.e. it was noted that $\mathrm{GM}(\mathbf{x}) \leqslant \mathrm{AM}(\mathbf{x})$. Where this is the case, we can say that the aggregation functions are comparable, and we can remark on whether an aggregation function is stronger or weaker than another.

Definition 1.21 (Comparability of Aggregation Functions) An aggregation function f is stronger than another aggregation function of the same number of arguments g if for all $\mathbf{x} \in \mathbb{I}^n$:

$$g(\mathbf{x}) \leqslant f(\mathbf{x}).$$

It is expressed as $g \leqslant f$. When f is stronger than g, it is equivalently said that g is weaker than f.

For Choquet integrals, we may be able to compare different instances depending on the values of the capacity. However, for these, as well as pairs of aggregation functions in general, we may not always have comparability. A function can be stronger on one part of the domain but weaker on others, in which case we can say that the functions are *incomparable*.

Lipschitz Continuity and Stability

The useful interpretation of continuity[3] in the aggregation context is that it ensures, to some degree, that small changes in the input should not result in large changes to the output, so there is some level of stability or predictability. Lipschitz continuity is a stronger condition that limits the rate at which a function can increase to a constant M.

Definition 1.22 (Lipschitz Continuity) An aggregation function f is called Lipschitz continuous if there is a positive number M, such that for any two vectors $\mathbf{x}, \mathbf{y}$ in the domain of f,

$$|f(\mathbf{x}) - f(\mathbf{y})| \leqslant M\, d(\mathbf{x}, \mathbf{y}), \tag{1.4}$$

where $d(\mathbf{x}, \mathbf{y})$ is a distance[4] between $\mathbf{x}$ and $\mathbf{y}$. The smallest such number M is called the Lipschitz constant of f (in the distance d).

The Euclidean distance is usually used to define the distance between two vectors,

$$d(\mathbf{x}, \mathbf{y}) = \sqrt{(x_1 - y_1)^2 + (x_2 - y_2)^2 + \ldots + (x_n - y_n)^2}.$$

[3] A real function of n arguments is continuous if for any sequences $\{x_{ij}\}, i = 1, \ldots, n$ such that $\lim_{j\to\infty} x_{ij} = y_i$, it holds that $\lim_{j\to\infty} f(x_{1j}, \ldots x_{nj}) = f(y_1, \ldots, y_n)$. Because the domain $\mathbb{I}^n$ is a compact set, continuity is equivalent to its stronger version, uniform continuity. An aggregation function is uniformly continuous if and only if it is continuous in each argument (i.e. we can check continuity by fixing all variables but one and checking continuity of each univariate function. However, general non-monotone functions can be continuous in each variable without being continuous).

[4] A *distance* between objects x, y from a set $\mathcal{S}$ is a function defined on $\mathcal{S} \times \mathcal{S}$, whose values are non-negative real numbers, with the following properties: (1) $d(x, y) = 0$ if and only if $x = y$, (2) $d(x, y) = d(y, x)$ (symmetry), and (3) $d(x, z) \leqslant d(x, y) + d(y, z)$ (triangular inequality). Such a distance is called a *metric*.

However, we can choose any norm[5] $d(\mathbf{x}, \mathbf{y}) = ||\mathbf{x} - \mathbf{y}||$; typically a p-norm, which for $p \geqslant 1$ is a function $||\mathbf{x}||_p = \left(\sum_{i=1}^{n} |x_i|^p\right)^{1/p}$, for finite p, and $||\mathbf{x}||_\infty = \max_{i=1\ldots,n} |x_i|$.

When a function is Lipschitz-continuous, a change in the input of $\delta = ||\mathbf{x} - \mathbf{y}||$ will result in, at most, a change of $M\delta$ in the output. For differentiable functions, M denotes the upper bound on the norm of its gradient, and hence all differentiable functions are Lipschitz-continuous. On the other hand, Lipschitz-continuous functions are differentiable "almost everywhere", i.e. they are differentiable on the entire domain except for a subset with a measure of zero.

If an aggregation function is not Lipschitz-continuous, or if the Lipschitz constant is very high, then it could mean that a small inaccuracy (or noise) in the input has a disproportionate effect on the output.

The following definitions provide some further properties related to stability.

Definition 1.23 (p-Stable Aggregation Functions) Given $p \geqslant 1$, an aggregation function is called p-stable if its Lipschitz constant in the p-norm $|| \cdot ||_p$ is 1. An extended aggregation function is p-stable if it can be represented as a family of p-stable aggregation functions.

We can say that p-stable aggregation functions do not enhance input inaccuracies, since $|f(\tilde{\mathbf{x}}) - f(\mathbf{x})| \leqslant ||\tilde{\mathbf{x}} - \mathbf{x}||_p = ||\delta||_p$.

Definition 1.24 (1-Lipschitz Aggregation Functions) An aggregation function f is called 1-Lipschitz if it is p-stable with $p = 1$, i.e. for all $\mathbf{x}, \mathbf{y}$,

$$|f(\mathbf{x}) - f(\mathbf{y})| \leqslant |x_1 - y_1| + |x_2 - y_2| + \ldots + |x_n - y_n|.$$

An aggregation function that is p-stable for a given $p > 1$ will also be q-stable for any $1 \leqslant q < p$, since $||\mathbf{x}||_p \leqslant ||\mathbf{x}||_q$ for all $\mathbf{x}$. Functions such as the product, minimum, maximum, and arithmetic mean are p-stable for any p. The geometric mean is not Lipschitz-continuous, which can be observed looking at parts of the domain where one of the inputs is equal to zero.

Stability can also be considered in terms of whether adding or removing inputs (changing the cardinality) can have a big impact on the output. Yager introduced the *self-identity* property in [YR97] for extended aggregation functions.

Definition 1.25 (Self-Identity) An extended aggregation function f has the *self-identity property* if for all $n \geqslant 1$ and for all $x_1, \ldots, x_n \in \mathbb{I}^n$,

$$f(x_1, \ldots, x_n, f(x_1, \ldots, x_n)) = f(x_1, \ldots, x_n).$$

---

[5] A *norm* is a function f on a vector space with the following properties: (1) $f(\mathbf{x}) > 0$ for all nonzero $\mathbf{x}$ and $f(\mathbf{0}) = 0$, (2) $f(a\mathbf{x}) = |a| f(\mathbf{x})$, and (3) $f(\mathbf{x} + \mathbf{y}) \leqslant f(\mathbf{x}) + f(\mathbf{y})$.

In addition, extended aggregation functions with the self-identity property satisfy

$$f(x_1, \ldots, x_n, k) \geqslant f(x_1, \ldots, x_n) \quad \text{if} \quad k \geqslant f(x_1, \ldots, x_n),$$
$$f(x_1, \ldots, x_n, k) \leqslant f(x_1, \ldots, x_n) \quad \text{if} \quad k \leqslant f(x_1, \ldots, x_n).$$

It follows from the definition that such functions are necessarily idempotent and hence averaging. Examples include the arithmetic mean, minimum, and maximum. However, it is not the case that extended idempotent aggregation functions will always satisfy the self-identity property.

1.2.4 Main Families and Prototypical Examples

We now present several examples of aggregation function examples and families within each of the classes. Surface plots for the $n = 2$ case are provided along with many of the examples to help build intuition and illustrate their overall behaviour.

Min and Max

The minimum and maximum are conceptually simple but fundamentally important when it comes to the study of aggregation functions. They are the prototypical examples for conjunctive and disjunctive aggregation functions, respectively, but also demarcate between these classes and the averaging functions—the minimum is the only averaging aggregation function that is also conjunctive, while the maximum is both averaging and disjunctive. It is also important to understand their behaviour for the role they play as components in several more sophisticated aggregation functions including the Choquet and other fuzzy integrals.

In fuzzy logic, due to consistency with set-theoretical properties such as mutual distributivity [BG73], the minimum and maximum are suitable choices for generalising the intersection and union operations. As extended aggregation functions, they can be applied to any number of arguments.

$$\min(\mathbf{x}) = \min_{i=1,\ldots,n} x_i, \tag{1.5}$$

$$\max(\mathbf{x}) = \max_{i=1,\ldots,n} x_i. \tag{1.6}$$

Both functions are symmetric, associative, and Lipschitz-continuous. The min and max have neutral elements $e = 1$ and $e = 0$, respectively (on $\mathbb{I} = [0, 1]$), and

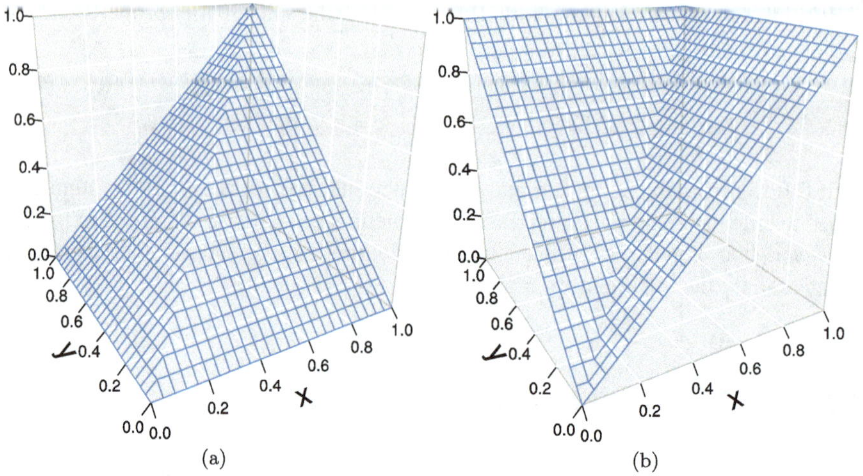

Fig. 1.4 3D surface plots showing (**a**) $\min(x, y)$ and (**b**) $\max(x, y)$

absorbent elements $a = 0$ and $a = 1$.[6] The two functions can be obtained from one another via duality using the standard negation, and in fact, any strong negation, i.e. for the standard negation we have,

$$\max(\mathbf{x}) = 1 - \min(\mathbf{1} - \mathbf{x}) = 1 - \min_{i=1,\dots,n} (1 - x_i),$$

while for any strong negation, it will hold that

$$\max(\mathbf{x}) = N(\min(N(\mathbf{x})) = N(\min_{i=1,\dots,n} (N(x_i)),$$

and vice versa (interchanging max and min in the equations).

Surface plots are provided in Fig. 1.4. We can note the rotational symmetry corresponding with the duality relationship and the idempotency property apparent from the diagonal $f(t, t) = t$.

Means

The ubiquitous *arithmetic mean* that we have mentioned throughout this chapter is commonly referred to simply as the "mean" or the "average". Even those who recognise that there are various means (such as those we included in Fig. 1.3) may

[6] For an aggregation function f, a neutral element is an element e such that $f(e, x) = f(x, e) = x$ for any number of arguments and with e in any position, while an absorbent element a satisfies $f(x, a) = f(a, x) = a$ for a in any position.

treat the concepts of means and averaging aggregation functions synonymously [Bul03]; however, there are a number of functions referred to as means that are not averaging,

We briefly introduced the following means in Example 1.1.

Definition 1.26 (Arithmetic Mean) The arithmetic mean is the function

$$AM(\mathbf{x}) = \frac{1}{n}(x_1 + x_2 + \ldots + x_n) = \frac{1}{n}\sum_{i=1}^{n} x_i.$$

Definition 1.27 (Geometric Mean) The geometric mean is the function

$$GM(\mathbf{x}) = \sqrt[n]{x_1 x_2 \ldots x_n} = \left(\prod_{i=1}^{n} x_i\right)^{1/n}.$$

Definition 1.28 (Harmonic Mean) The harmonic mean is the function

$$HM(\mathbf{x}) = n\left(\sum_{i=1}^{n}\frac{1}{x_i}\right)^{-1}.$$

The standard definitions assume symmetry; however, in some contexts, we need to allow for inputs to be weighted unequally, for example, if one decision-maker's opinion is more important or if the inputs represent values from a distribution with differing frequencies. The concept of a weighting vector helps formalise this weighting process.

Definition 1.29 (Weighting Vector) A vector $\mathbf{w} = (w_1, \ldots, w_n)$ is called a weighting vector if $w_i \in [0, 1]$ and $\sum_{i=1}^{n} w_i = 1$.

This leads to the following expression for the weighted arithmetic mean.

Definition 1.30 (Weighted Arithmetic Mean (WAM)) Given a weighting vector $\mathbf{w}$, the weighted arithmetic mean is the function

$$WAM(\mathbf{x}) = w_1 x_1 + w_2 x_2 + \ldots + w_n x_n = \sum_{i=1}^{n} w_i x_i.$$

The weighted counterparts for the geometric and harmonic means are as follows.

$$GM(\mathbf{x}) = \prod_{i=1}^{n} x_i^{w_i} \tag{1.7}$$

$$HM(\mathbf{x}) = \left(\sum_{i=1}^{n} \frac{w_i}{x_i}\right)^{-1} \tag{1.8}$$

The family of weighted power means encompass each of these as special cases.

Definition 1.31 (Weighted Power Mean (WPM)) Given a weighting vector $\mathbf{w}$ and power p, the weighted power mean is the function

$$WPM(\mathbf{x}) = \left(\sum_{i=1}^{n} w_i x_i^p\right)^{\frac{1}{p}}.$$

The value of p is a parameter that allows the power means to graduate from limiting cases of the minimum (when p approaches $-\infty$) and maximum (when p approaches ∞). When $p = 1$, we obtain the weighted arithmetic mean, while $p = -1$ gives us the harmonic mean. We can also take the limit as p approaches 0 to give us the geometric mean, and of course equal weights $w_i = 1/n$ correspond with the unweighted versions.

The radical mean and exponential mean given as examples in Fig. 1.3 are obtained from a different transformation, where instead of replacing x_i with x_i^p, we use a more general generating function $g(x_i)$ along with its inverse after taking the sum. This broader family of means is known as the quasi-arithmetic means.

Medians

After the arithmetic mean, the median is probably the most well-known averaging aggregation function used widely in everyday contexts. It is the middle number when a set of values are arranged either in ascending or descending order and is often seen as a more robust measure of centre since it is not affected by extreme values.

Definition 1.32 (Median) The median is the function

$$Med(\mathbf{x}) = \begin{cases} \frac{1}{2}(x_{(k)} + x_{(k+1)}), & \text{if } n = 2k \text{ is even} \\ x_{(k)}, & \text{if } n = 2k-1 \text{ is odd,} \end{cases}$$

where $x_{(k)}$ is the k-th largest (or smallest) component of $\mathbf{x}$.

While we usually see the median applied to numeric data, it is suitable for any kind of ordinal data, since it only requires the data to be sorted. As with the mean, there are weighted counterparts and more sophisticated averaging functions for which the median is the basis, and we will see later on that the Sugeno integral can be seen to exhibit median-like behaviour. For now, we provide only the definition of the a-median, with surface plots shown in Fig. 1.5.

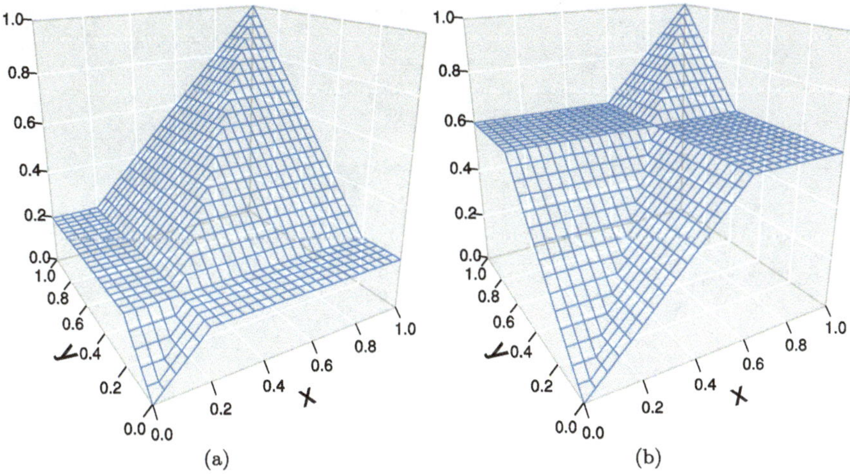

Fig. 1.5 3D surface plots showing a-medians with (**a**) $a = 0.2$ and (**b**) $a = 0.6$

Definition 1.33 (a-Median) Given a value $a \in \mathbb{I}$, the a-median is the function

$$Med_a(\mathbf{x}) = Med(x_1, \ldots, x_n, \overbrace{a, \ldots, a}^{n-1\ times}).$$

Ordered Weighted Averaging

Ordered weighted averaging (OWA) functions were introduced in the aggregation functions context by Yager in [Yag88]. Rather than applying weights to each of the inputs based on the index, the inputs are first rearranged, so that w_1 is always applied to the largest input, w_2 to the second largest, and so on.

Different choices of weighting vector allow the OWA function to graduate between the min and max functions. They are symmetric, homogeneous, and shift-invariant (i.e. they are an example of linear aggregation functions).

Definition 1.34 (OWA) Given a weighting vector $\mathbf{w}$, the OWA function is

$$OWA_{\mathbf{w}}(\mathbf{x}) = \sum_{i=1}^{n} w_i x_{(i)},$$

where $\mathbf{x}_{\searrow} = (x_{(1)}, x_{(2)}, \ldots, x_{(n)})$ in this case denotes a non increasing permutation of the inputs.[7]

[7] Note the difference to the increasing ordering used for the Choquet and Sugeno integrals.

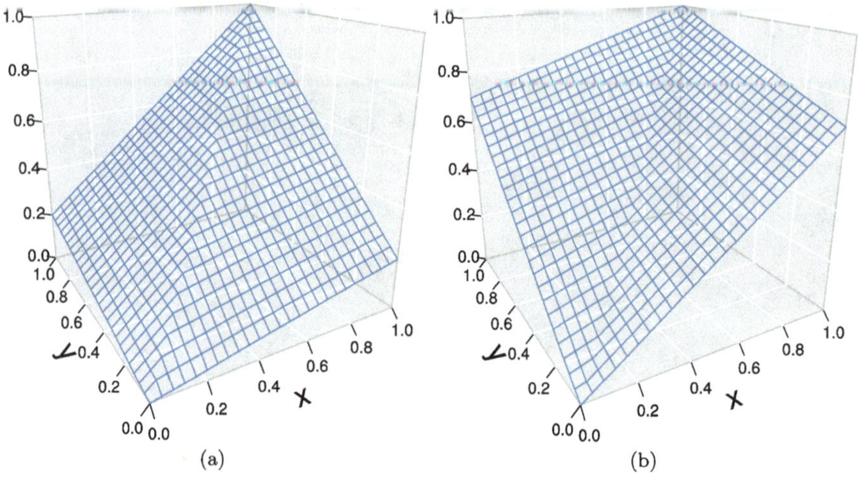

Fig. 1.6 3D surface plots showing OWA functions with respect to the weighting vectors (**a**) $\mathbf{w} = (0.2, 0.8)$ and (**b**) $\mathbf{w} = (0.7, 0.3)$

Calculation of the OWA function requires a `sort()` operation. The arithmetic mean is a special case, obtained when all weights are equal, while the median is obtained if all except the middle weight are equal to zero (or all except the middle two weights for an even number of inputs). The min and max are obtained when $w_n = 1$ and $w_1 = 1$, respectively (with all other weights equal to zero. Surface plots for some example OWA functions are shown in Fig. 1.6.

We will see later that OWA functions are also a special case of the Choquet integral when the underlying capacity is symmetric.

Conjunctive and Disjunctive Functions

In addition to the minimum and maximum, which represent the strongest and weakest conjunctive and disjunctive aggregation functions, respectively, the so-called triangular norms and conorms (t-norms and t-conorms) are also well-established examples used in applications and research.

Triangular norms include the minimum as one example and more broadly can be seen to generalise the logical `AND` operation to non-binary arguments. When introduced by Menger [Men42] for the purpose of distribution functions, t-norms were required to be symmetric, non-decreasing, and bounded by the unit interval in each argument with $f(1, 1) = 1$, and for it to hold that $f(1, a) > 0$ whenever $a > 0$; however, Schweizer and Sklar [SS83] further developed the idea and added the requirements of associativity and the neutral element $e = 1$. The associativity property allows t-norms and t-conorms to be extended to any number of arguments.

Definition 1.35 (Triangular Norm (t-Norm)) A triangular norm (t-norm for short) is a bivariate aggregation function $T : [0, 1]^2 \to [0, 1]$, which is associative and symmetric and has neutral element 1.

The neutral element of 1 results in 0 being an absorbent element since $T(1, 0) = 0$ along with monotonicity ensures that $T(x, 0) = 0$ for all x. The dual t-conorm has the neutral element 0 and hence an absorbent element $a = 1$.

Definition 1.36 (Triangular Conorm (t-Conorm)) A triangular conorm (t-conorm for short) is a bivariate aggregation function $S : [0, 1]^2 \to [0, 1]$, which is associative and symmetric and has neutral element 0.

The following example lists some well-known t-norms and t-conorms other than the min and max.

Example 1.2 Well-known t-norms and their dual t-conorms (or S-norms).

product	$T_P(\mathbf{x}) = \prod_{i=1}^{n} x_i$
dual product	$S_P(\mathbf{x}) = 1 - \prod_{i=1}^{n} (1 - x_i)$
Łukasiewicz t-norm	$T_L(\mathbf{x}) = \max\left(0, \sum_{i=1}^{n} x_i - (n-1)\right)$
Łukasiewicz t-conorm	$S_L(\mathbf{x}) = \min\left(1, \sum_{i=1}^{n} x_i\right)$

The dual product is also known as the probabilistic sum. Surface plots of the Łukasiewicz t-norm and conorm are shown in Fig. 1.7.

Archimedean t-norms have a similar construction to the quasi-arithmetic means, built from univariate generating functions that are continuous, strictly decreasing, and for which it holds that $g(1) = 0$. In terms of this generating function, they are then given by

$$g^{-1}(g(x_1) + g(x_2) + \ldots + g(x_n)). \tag{1.9}$$

Families of Archimedean t-norms are represented using an additional parameter λ, the value of which allows the functions to be stronger or weaker. A pseudo-inverse is sometimes used to extend the domain of the inverse g^{-1}.

$$g^{(-1)}(t) = \begin{cases} g^{-1}(t), & \text{if } t \in [0, a], \\ 0 & \text{otherwise.} \end{cases}$$

The dual Archimedean t-conorms are obtained using $h(t) = g(1 - t)$, resulting in generators that are also monotone increasing but which satisfy $h(0) = 0$. We present examples of Yager and Hamacher families in the following example.

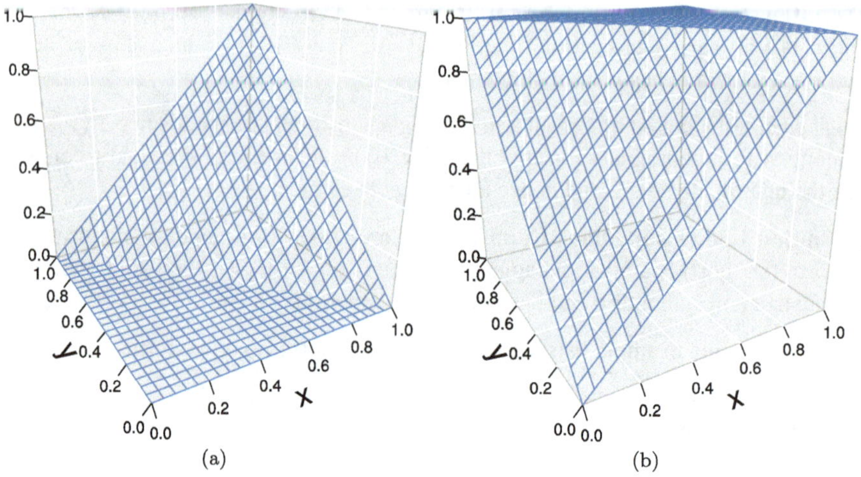

Fig. 1.7 3D surface plots showing (**a**) the Łukasiewicz t-norm $T_L(x, y)$ and (**b**) the dual $S_L(x, y)$

Example 1.3 Bivariate expressions are provided for the t-norms and t-conorms below.

Yager t-norm	$T_\lambda^Y(x, y) = \max\left\{0, 1 - \left((1-x)^\lambda + (1-y)^\lambda\right)^{\frac{1}{\lambda}}\right\}$
Yager t-conorm	$S_\lambda^Y(x, y) = \min\left\{1, \left(x^\lambda + y^\lambda\right)^{\frac{1}{\lambda}}\right\}$
Hamacher product	$T_0^H(x, y) = \begin{cases} 0, & \text{if } x = y = 0, \\ \frac{xy}{x+y-xy} & \text{otherwise.} \end{cases}$
Einstein sum	$S_2^H(x, y) = \frac{x+y}{1+xy}$

The Yager t-norms can be constructed using the generator $g_\lambda^Y(t) = (1-t)^\lambda$ with $\lambda \in]0, \infty[$. The value of $\lambda = 1$ results in the Łukasiewicz t-norm and conorm, while $\lambda \to \infty$ approaches the minimum.

The Hamacher family has the generating function $g_\lambda^H(t) = \log \frac{(1-\lambda)t+\lambda}{t}$ for $\lambda \in]0, \infty[$ and $g_\lambda^H(t) = \frac{1-t}{t}$ when $\lambda = 0$. The Hamacher product results from this latter case, while the Einstein sum is obtained from the Hamacher t-conorm when $\lambda = 2$.

Mixed Aggregation

Mixed aggregation functions are those that do not fall into any of the conjunctive, averaging, or disjunctive classes. In other words, they may exhibit different behaviours on different parts of the domain. This can be useful for some applications, with expert systems MYCIN and PROSPECTOR being classic examples [BS84, DHN76].

Uninorms and nullnorms (see [BPC07]) are particular families of mixed aggregation functions that have received much attention in research.

Example 1.4 The following functions are mixed aggregation functions.

$$3-\Pi \text{ function} \qquad f(\mathbf{x}) = \frac{\prod_{i=1}^{n} x_i}{\prod_{i=1}^{n} x_i + \prod_{i=1}^{n} (1-x_i)},$$

$$\text{Example nullnorm} \qquad f(x, y) = \begin{cases} 2xy - x - y + 1, & \min(x, y) > 0.5, \\ \min(0.5, x + y), & \max(x, y) < 0.5, \\ 0.5 & \text{otherwise.} \end{cases}$$

The $3-\Pi$ function [YR96] has the convention $\frac{0}{0} = 0$ and is discontinuous on the boundaries of $[0, 1]^n$. Its behaviour is conjunctive on $[0, \frac{1}{2}]^n$, disjunctive on $[\frac{1}{2}, 1]^n$, and averaging otherwise. It is also associative and has the neutral element $e = \frac{1}{2}$.

The example nullnorm provided has disjunctive behaviour on $[0, \frac{1}{2}]^2$ and conjunctive behaviour on $[\frac{1}{2}, 1]^2$ and is averaging otherwise.

1.3 Capacities and Fuzzy Integrals

In this book, we focus on results pertaining to the theory and application of a special family of aggregation functions referred to as fuzzy integrals, which derive their behaviour from the underlying fuzzy measure, or (more broadly) the Choquet capacities, used to define them. We recall that a fuzzy measure is just a normalised Choquet capacity.

Such functions are piecewise linear, but most notably allow for sophisticated modelling of interaction relationships between the input variables. The capacities and resulting behaviour of the fuzzy integrals defined can be analysed in terms of these relationships and measures of importance across variables and subsets, and hence we obtain functions that can strike a nice balance between flexibility, robustness, and interpretability. They represent one promising approach to explainable artificial intelligence.

The Choquet integral and Sugeno integral have received the most attention to date. The Choquet [Cho54] includes the weighted arithmetic mean and ordered weighted averaging operators (and hence the median) as special cases, making it appealing in applications that have traditionally been limited to these in aggregation calculations.

Classical measures used for integration aim to calculate lengths, areas, volumes, etc., consistently across finite and sometimes infinite sets of objects, with one of the defining properties being additivity. With this property, it should be possible to determine the measure of a set by breaking down its calculation into non-intersecting subsets that cover the set and summing these sub-calculations. Choquet capacities substitute the additivity property with monotonicity, allowing for more flexible and nonadditive relationships to hold. Such measures are also referred to as nonadditive measures (and the integrals as nonadditive integrals), since the relationship to fuzzy sets and fuzziness is not required for the theory to make sense. Overviews of various measures can be found in [Pap02, Den94, WK92, TNS14, WK09], while key results and applications relating to the use of fuzzy integrals as aggregation functions can be explored in [Gra96, Mar00a, Mar02, GKM08, GL00, Gra16].

Although there is no universal definition of what integration constitutes, it has historically been seen as relating to areas and volumes associated with functions, an idea formalised by the Lebesgue integral, which is defined with respect to an additive measure. Choquet integration, as it was defined in [Cho54], generalises Lebesgue integration by employing the broader class of Choquet capacities, referred to in different contexts as nonadditive (or "not necessarily additive") measures, monotone measures, and capacities and as fuzzy measures. We will mostly contain our results to discrete Choquet capacities, which assume a finite number of inputs and hence defined over finite sets.

While they can be defined in terms of the interaction effects between variables, at face value, a discrete Choquet capacity is an assignment of importance or value to all groups of criteria. Unfortunately, as the number of criteria increases, the number of subsets grows exponentially, as does the number of constraints that ensure monotonicity, giving rise to the main hurdle when it comes to practical implementation. Many of the results we present throughout these chapters are focused on addressing this challenge. There is a need to develop both efficient computational methods when deploying fuzzy integrals and Choquet capacities and methods allowing suitable simplifications and approximations that do not diminish their rich expressiveness.

1.3.1 Basic Definitions

Later chapters of the book will go into more detail regarding Choquet capacities, their associated properties, and how they are used in application; however, to help with readability and to provide a preliminary understanding, we recall some of the more basic definitions here.

As we have alluded to already, the definition of Choquet capacities is related to various other concepts, in particular games from cooperative game theory [Gra16] and measures.

Definition 1.37 (Game) A game is a set function[8] $\mu : 2^{\mathcal{N}} \to \mathbb{R}_+$ which is grounded $\mu(\emptyset) = 0$.

Definition 1.38 (Premeasure) A premeasure is a set function $\mu : 2^{\mathcal{N}} \to \mathbb{R}_+$, which is monotonic with respect to set inclusion (i.e. $\mu(\mathcal{A}) \leqslant \mu(\mathcal{B})$ whenever $\mathcal{A} \subset \mathcal{B}$) .

These definitions correspond with the two defining properties of Choquet capacities, the (lower) boundary condition and monotonicity, respectively.

Definition 1.39 (Choquet Capacity) Let $\mathcal{N} = \{1, 2, \dots, n\}$. Choquet capacity is a grounded monotonic set function $\mu : 2^{\mathcal{N}} \to \mathbb{R}_+$.

If we also set an upper boundary condition, we then obtain the definition that is usually used for fuzzy measures.

Definition 1.40 (Fuzzy Measure) A fuzzy measure is a grounded monotonic set function normalised to 1: $\mu(\mathcal{N}) = 1$.

Normalisation is useful for being able to more easily compare different capacities and their behaviour. When used with the Choquet integral, normalisation is equivalent to changing the input scale. Whether we view Choquet capacities as monotonic games or as grounded premeasures, their key benefit is usually seen as the facilitation of assigning weight to subsets or *coalitions* of inputs. Setting $\mu(\emptyset) = 0$ implies that no value is assigned until we have at least one element, while monotonicity means that adding new members to a coalition does not result in a reduction in value (even though it may not result in any increase).

Note 1.8 In this book, we consider finite universal sets $\mathcal{N}$ and finite set function values. As such, any Choquet capacity can be normalised to one and become a fuzzy measure, without affecting the underlying mathematics. We will generally not distinguish between normalised and non-normalised capacities and will use the term *capacity* synonymously with fuzzy measures.

In cases where the normalisation condition presents an issue for interpretation of capacity values, we will explicitly mention that the capacity is not normalised.

Let us now present some specific weighting schemes that can be used to define Choquet capacities for arbitrary n.

Example 1.5 It is often informative to consider capacities relative to the following special cases and types.

When used to define Choquet integrals, the weakest and strongest capacities correspond with the minimum and maximum, respectively. Dirac measures result in projections onto the coordinate corresponding with x_0, while additive and

[8] A set function is a function whose domain consists of all possible subsets of $\mathcal{N}$. For example, for $n = 3$, a set function is specified by $2^3 = 8$ values at $\mu(\emptyset)$, $\mu(\{1\})$, $\mu(\{2\})$, $\mu(\{3\})$, $\mu(\{1, 2\})$, $\mu(\{1, 3\})$, $\mu(\{2, 3\})$, $\mu(\{1, 2, 3\})$.

Weakest capacity	$\mu(\mathcal{A}) = \begin{cases} 1, & \text{if } \mathcal{A} = \mathcal{N}, \\ 0 & \text{otherwise.} \end{cases}$
Strongest capacity	$\mu(\mathcal{A}) = \begin{cases} 0, & \text{if } \mathcal{A} = \emptyset, \\ 1 & \text{otherwise.} \end{cases}$
Dirac measures	$\mu(\mathcal{A}) = \begin{cases} 1, & \text{if } x_0 \in \mathcal{A}, \\ 0, & \text{if } x_0 \notin \mathcal{A} \end{cases}$ for x_0 fixed in $\mathcal{N}$.
Additive capacities	$\mu(\mathcal{A}) = \sum_{i \in \mathcal{A}} \mu(\{i\})$
Symmetric capacities	$\mu(\mathcal{A}) = g(\lvert\mathcal{A}\rvert)$, for non-decreasing g.

symmetric capacities (when normalised) result in aggregation functions equivalent to the weighted arithmetic mean and ordered weighted averaging operator.

When representing capacities that are not defined according to a consistent rule as in the examples above, we often make use of the following arrangement, which is based on the Hasse diagram and subset relation:

$$\begin{array}{ccc} & \mu(\{1,2,3\}) & \\ \mu(\{1,2\}) & \mu(\{1,3\}) & \mu(\{2,3\}) \\ \mu(\{1\}) & \mu(\{2\}) & \mu(\{3\}) \\ & \mu(\emptyset) & \end{array}$$

Example 1.6 The following set function μ is a capacity on $\mathcal{N} = \{1, 2, 3\}$:

$$\begin{array}{ccc} & 1 & \\ 0.9 & 0.5 & 0.3 \\ 0.5 & 0 & 0.3 \\ & 0 & \end{array}$$

A visual representation is depicted in Fig. 1.8.

Invertible linear mappings can be used to provide alternative representations of capacities, allowing us to focus on specific relationships and properties. As we will see, these can also facilitate more efficient calculations and learning algorithms. The

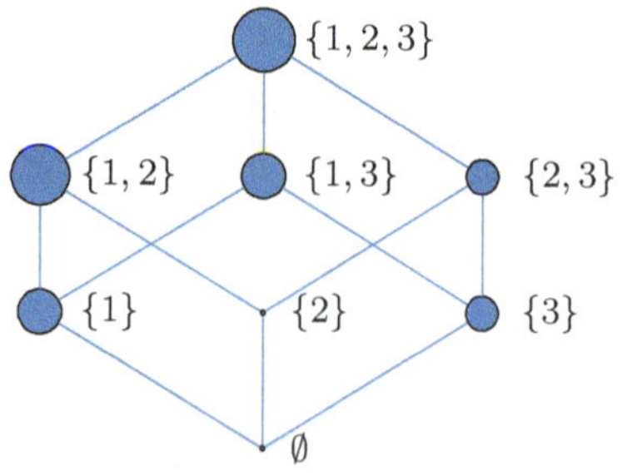

Fig. 1.8 Hasse diagram for the capacity in Example 1.6 with edges corresponding with set inclusion and where the size of each vertex is indicative of the subset weight

most well-known of these is the Möbius transform, which can be applied to any set function, but which can be defined for capacities as follows.

Definition 1.41 (Möbius Transformation) Let μ be a capacity. The Möbius transformation of μ is a set function defined for every $\mathcal{A} \subseteq \mathcal{N}$ as

$$M(\mathcal{A}) = \sum_{\mathcal{B} \subseteq \mathcal{A}} (-1)^{|\mathcal{A} \setminus \mathcal{B}|} \mu(\mathcal{B}).$$

The inverse of the Möbius transformation, which allows us to convert back into standard representation, is referred to as the Zeta transform:

$$\mu(\mathcal{A}) = \sum_{\mathcal{B} \subseteq \mathcal{A}} M(\mathcal{B}) \quad \forall \mathcal{A} \subseteq \mathcal{N}.$$

In particular, the Möbius transformation provides insight into the interaction between subsets and, as we will see in Chap. 3, is convenient for calculating interaction indices [Gra00, Gra97, Mar02, Gra16]. Since it serves as an alternative representation, we can use the following to verify the capacity monotonicity and boundary conditions:

$$\sum_{\mathcal{B} \subseteq \mathcal{A} | i \in \mathcal{B}} M(\mathcal{B}) \geqslant 0, \qquad \text{for all } \mathcal{A} \subseteq \mathcal{N} \text{ and all } i \in \mathcal{A}, \tag{1.10}$$

$$M(\emptyset) = 0 \text{ and } \sum_{\mathcal{A} \subseteq \mathcal{N}} M(\mathcal{A}) = 1.$$

1.3.2 *Fuzzy Integrals*

Fuzzy integrals, also referred to as nonadditive integrals, aggregate inputs with respect to the weight and importance information provided by the underlying capacity. A more in-depth treatment will be given in Chap. 5, while for now, we present two of the most well-known examples: the Choquet and Sugeno integrals.

Definition 1.42 (Discrete Choquet Integral) The discrete Choquet integral with respect to a capacity μ is given by

$$C_\mu(\mathbf{x}) = \sum_{i=1}^{n} x_{(i)}[\mu(\{j | x_j \geqslant x_{(i)}\}) - \mu(\{j | x_j \geqslant x_{(i+1)}\})], \tag{1.11}$$

where $\mathbf{x}_{(\cdot)} = (x_{(1)}, x_{(2)}, \ldots, x_{(n)})$ is an increasing permutation of the input $\mathbf{x}$, denoted $\mathbf{x}_{\nearrow}$ using the notation introduced in Definition 1.10 and with $x_{(n+1)} = \infty$ by convention.

A more succinct way of representing the Choquet integral is

$$C_{\mu}(\mathbf{x}) = \sum_{i=1}^{n} x_{(i)}[\mu(H_i) - \mu(H_{i+1})], \quad (1.12)$$

where $H_i = \{(i), \ldots, (n)\}$ and $H_{n+1} = \emptyset$.

By rearrangement of the terms, it can also be expressed as

$$C_{\mu}(\mathbf{x}) = \sum_{i=1}^{n} \left[x_{(i)} - x_{(i-1)}\right] \mu(H_i), \quad (1.13)$$

where $H_i = \{(i), \ldots, (n)\}$ and $x_{(0)} = 0$.

The latter two representations can be used to emphasise different decompositions of the calculation, either by measure or by the input function.

The Sugeno integral has also received a great deal of attention in fuzzy and multicriteria decision-making research [Dub+01, Mar00b].

Definition 1.43 (Discrete Sugeno Integral) The Sugeno integral with respect to a capacity μ is given by

$$S_{\mu}(\mathbf{x}) = \max_{i=1,\ldots,n} \min\{x_{(i)}, \mu(H_i)\}, \quad (1.14)$$

where $\mathbf{x}_{\nearrow} = (x_{(1)}, x_{(2)}, \ldots, x_{(n)})$ is an increasing permutation of the input $\mathbf{x}$, and $H_i = \{(i), \ldots, (n)\}$.

As with the Choquet integral, it includes the minimum, maximum, and median as special cases; however, as it is particularly well suited to aggregation of ordinal inputs, it also includes the weighted minimum and weighted maximum.

As with more classical integrals, fuzzy integrals can be interpreted as providing an overall measure of the input function, extending the "area under the curve" idea to account for interactions and dependencies between the inputs. In the discrete multicriteria decision-making or aggregation setting, the input function is the step function corresponding with the input vector $(x_1, x_2, \ldots, x_n)$.

The following section provides an example that will help illustrate these ideas and build intuition when it comes to application of Choquet capacities and integrals.

1.4 Capacities: Intuition and Example

Imagine a group of people coming to a restaurant. There is a small menu:

- Salad ... p_s
- Main ... p_m
- Dessert ... p_d

- Wine ... p_w
- Coffee ... p_c

The prices of each item are p_s, p_m, etc. Additionally, there are two tourist menus $T1$ (Salad, Main, Dessert) and $T2$ (Main and Wine), priced at p_{T1} and p_{T2}, respectively. Naturally, the rational restaurant manager sets menu prices so that $p_{T1} < p_s + p_m + p_d$ and $p_{T2} < p_m + p_w$. We pose two questions:

1. How should the prices for other meal combinations be set?
2. If the group chooses their meals, they naturally want to minimise the total price they pay. How?

For example, suppose the customers want to order items

s, s, m, m, m, d, d, w, w, c

(there can be several plates of the same). What is the bill? The prices cannot simply be added together: $2p_s + 3p_m + 2p_d + 2p_w + p_c$, because some prices are not additive. We know that we can get a cheaper price using $2p_{T1} + p_{T2} + p_w + p_c$, but we can also try $p_{T1} + 2p_{T2} + p_s + p_d + p_c$. What is the best way to calculate the total bill?

Let us introduce some terminology. Any subset of items will be called a *coalition*. The value of a coalition may be equal, smaller, or larger than the sum of items included. We call the coalition non-trivial if its value is not equal to the sum. In our case, we have two non-trivial coalitions (tourist menus) with values p_{T1} and p_{T2}. Coffee is not part of any non-trivial coalition, its price is just added to the total, and this item is called a *dummy* item (or dummy player in game theory).

We will disregard dummy items for now, as their prices are just added to the bill (no interesting phenomena) and focus on the other four plates. One way of calculating the bill is to try all possible combinations of set menus and separate items. Another way is using Choquet capacities and the Choquet integral.

1.4.1 Choquet Capacities and Integral

Recall that a Choquet capacity is a set function μ, which assigns values to all possible coalitions. In our case, the prices of groups of menu items are the coalitions, that is, $\mu(\{s\}) = p_s$, $\mu(\{s, m\}) = p_{sm}$, $\mu(\{s, m, d\}) = p_{T1}$, and so forth.

The Choquet integral of $\mathbf{x}$ is

$$C(\mathbf{x}) = \sum_{i=1}^{n} [x_{(i)} - x_{(i-1)}]\mu(\{x_{(i)}, \dots, x_{(n)}\})$$

where $x_{\nearrow} = (x_{(1)}, x_{(2)}, \dots, x_{(n)})$ is an increasing permutation and $x_{(0)} = 0$ by convention. In our case, x_i is the number of plates of type i in the order.

Let us consider a new customer order (s,s,m,m,m,d,w,w), which we represent as $\mathbf{x} = (2(s), 3(m), 1(d), 2(w))$. We can now look to calculate $C(\mathbf{x})$, which consistently regroups the items according to the largest possible subsets.

First, reorder $\mathbf{x}$ by the number of plates in the bill: $x_{\nearrow} = (1, 2, 2, 3)$, which induces an order on the plates: x_d, x_s, x_w, x_m. We have

$$\begin{aligned} C(\mathbf{x}) &= x_d\mu(\{smdw\}) + (x_s - x_d)\mu(\{smw\}) \\ &\quad +(x_w - x_s)\mu(\{mw\}) + (x_m - x_w)\mu(\{m\}) \\ &= p_{smdw} + (2-1)p_{smw} + (2-2)p_{mw} + (3-2)p_m \\ &= p_{smwd} + p_{smw} + p_m \end{aligned}$$

This illustrates how the Choquet integral decomposes the calculation into groups of items, where each group is as large as possible. We see that we can make one group of all four items (d, s, w, m), and once this group is removed, we can then make one group of (s, w, m) with a single item (m) leftover. But where did all the set menu prices go? And how can we work out p_{smdw}, and p_{smw}?

The set menus are hidden in the costs of new coalitions. We will now show how to calculate them. Consider setting prices from the manager's perspective. Recall that plates have prices p_s, p_m, p_d, p_w, p_c. In addition, there are two set tourist menus p_{T1}, p_{T2}, and naturally, $p_{mw} < p_m + p_w$ and $p_{smd} < p_s + p_m + p_d$.

A set function with this property (where a combination is cheaper than the sum) is called *submodular* , and it models discounts. If we have the opposite inequality (>), it is called *supermodular*, which models synergies. This property, in conjunction with the regrouping approach of the Choquet integral, ensures that a cheaper overall price would not be obtained by breaking up any of those groups into smaller subgroups.

The manager does not want to bother setting prices for all possible subsets, so they assume some additivity, e.g. $p_{smw} = p_{mw} + p_s$. This leads to one undefined value in the formula above and we have

$$\begin{aligned} C(\mathbf{x}) &= p_{smdw} + p_{smw} + p_m \\ &= p_{smdw} + (p_{T2} + p_s) + p_m \quad (< p_{smdw} + p_s + 2p_m + p_w). \end{aligned}$$

Now the manager could set any price for p_{smdw}. However, to be rational, it should not be larger than either $p_{smd} + p_w$ or $p_{mw} + p_s + p_d$, as customers could then opt for either of these combinations to lower their bill. One option would be to set it to the cheapest of these options:

$$p_{smdw} = \min(p_{smd} + p_w, p_{mw} + p_s + p_d).$$

But we can do even better using Möbius representation.

1.4.2 Möbius Transform

Recall the formula

$$M(\mathcal{A}) = \sum_{\mathcal{B}\subseteq\mathcal{A}} (-1)^{|\mathcal{A}|-|\mathcal{B}|}\mu(\mathcal{B}), \quad \forall \mathcal{A} \subseteq \mathcal{N},$$

along with the inverse calculation:

$$\mu(\mathcal{A}) = \sum_{\mathcal{B}\subseteq\mathcal{A}} M(\mathcal{B}), \quad \forall \mathcal{A} \subseteq \mathcal{N}.$$

Values $M(\mathcal{A})$ are also known as *Harsanyi dividends* (Nobel Prize 1994). In our case,

$$m(\{smdw\}) = -p_s - p_m - p_d - p_w + p_{sm} + p_{sd} + p_{sw} + p_{md} + p_{mw} + p_{dw}$$

$$- p_{smd} - p_{smw} - p_{sdw} - p_{mdw} + p_{smds}.$$

We will drop the curly brackets here to ease notation. We can see that for trivial coalitions, e.g. where $p_{sm} = p_s + p_m$, we have

$$M(sm) = p_s + p_m - p_{sm} = 0.$$

Similarly in the case $M(sdw)$,

$$M(sdw) = p_s + p_d + p_w - p_{sd} - p_{sw} - p_{dw} + p_{sdw}$$

$$= p_s + p_d + p_w - (p_s + p_d) - (p_s + p_w) - (p_d + p_w) + (p_s + p_d + p_w) = 0.$$

However, for tourist menus, we have $M(smd) \neq 0$, $M(mw) \neq 0$. Other Möbius values in our example are also zero, even if they include non-trivial coalitions, for example, one can expand and find $M(smw) = 0$. The fact that many $M(\mathcal{A}) = 0$ reflects lack of interaction or no added value/discount in $\mathcal{A}$.

It is therefore natural to require that $M(smdw) = 0$, and similarly for all other coalitions, we will set $M(\mathcal{A}) = 0$ as the default value when the manager does not want to set a special price.

In calculation, we require

$$M(smdw) = -p_s - p_m - p_d - p_w + p_{sm} + p_{sd} + p_{sw} + p_{md} + p_{mw} + p_{dw}$$

$$- p_{smd} - p_{smw} - p_{sdw} - p_{mdw} + p_{smds} = 0.$$

Expanding and cancelling out terms, then solving for p_{smdw}, we end up with (note the minus sign)

$$p_{smdw} = p_{smd} + p_{mw} - p_m = p_{T1} + p_{T2} - p_m.$$

1.4.3 Final Bill Calculation

Our bill is now (no minus signs at the end)

$$\begin{aligned} C(\mathbf{x}) &= p_{smdw} + p_{mw} + p_s + p_m \\ &= (p_{smd} + p_{mw} - p_m) + (p_{mw} + p_s) + p_m \\ &= p_{T1} + 2p_{T2} + p_s \end{aligned}$$

This is for the order $(2(s), 3(m), 1(d), 2(w))$. Notice that this is cheaper than either of the manager's options (breaking $p_{smdw} = p_{smd} + p_w$ and $p_{smdw} = p_{mw} + p_s + p_d$), which give the totals $p_{T1} + p_w + p_{T2} + p_s + p_m$ and $2p_{T2} + 2p_s + p_d + p_m$, respectively, because $p_{mw} < p_m + p_w$ and $p_{smd} < p_s + p_m + p_d$.

Thus, we found that by using the Choquet integral, we calculated the smallest total bill, which is still consistent with the advertised pricelist, which we take as the true value of the bill.

The formulas above lead to the best way of computing the total with just moderate arithmetic. Once we have the Möbius representation of the pricelist, the following calculation may be even simpler:

$$C(\mathbf{x}) = \sum_{\mathcal{A} \subseteq \mathcal{N}} M(\mathcal{A}) \min_{i \in \mathcal{A}} x_i$$

In our example, recalling that most Möbius values are set to zero (except for single items and the tourist menus), and hence taking only nonzero terms and substituting for $M(\mathcal{A})$, we have

$$C(2, 3, 1, 2) = 2p_s + 3p_m + p_d + 2p_w + 2M(mw) + 1M(smd),$$

which leads to $p_{smd} + 2p_{mw} + p_s$ and is the same total we obtained previously. Note that the negative Möbius values of the non-trivial coalitions (tourist menus) have lowered our bill.

An important message here is that, by using Möbius representation, one can exploit the *sparsity* of the capacity (i.e. many zero Möbius values). In such cases, there is no need to explicitly calculate all expressions (like p_{smdw}); one can simply take the pricelist (individual items and discounted menus), translate them into sparse Möbius representation, and then calculate the Choquet integral as the sum over all non-zero terms.

1.4.4 Some Conclusions

- Choquet capacities and the Choquet integral help to calculate systematically and optimally the totals when there are reasonable discounts for groups of items.
- The main tool is the *Choquet integral*, which properly accounts for all possible combinations of items.
- *Submodularity* is a key feature for rational price setting that models discounts.
- The *Möbius representation* is very useful for various manipulations with capacities.
- The manager can set the default value of 0 for Möbius values $M(\mathcal{A})$ when there is no special deal. This way the manager only needs to focus on individual prices and set menus, without worrying about all other possible subsets of plates, as they will be cancelled out in calculation.

1.4.5 Application Areas

We name just a few areas where we observe discounts and added value, redundancies, and synergies and where Choquet capacities provide an explicit model for interactions.

- Cooperative games and economics
- Operation research: modelling nonadditive objectives, portfolio selection, and nonadditive constraints
- Multiobjective optimisation problems and aggregating interacting objectives
- Social sciences and education (scoring, assessments)
- Sensor data aggregation
- Image processing and analysis
- Aggregation of ensembles in machine learning
- Statistical regression with co-dependent inputs and detection of outliers
- Aggregation in multiple criteria decision-making
- Group decision-making
- Cluster analysis and distance metrics
- Bibliometrics
- Ecological indices

Wherever we use the arithmetic means or sums to calculate indices, which implies independence of the summands, we can replace them with a more sophisticated approach to aggregation, in particular with the Choquet (and other fuzzy) integrals, which will account for synergies and redundancies.

1.4.6 Capacities: Challenges and Opportunities

While the use of Choquet capacities leads to moderate difficulty arithmethics in small problems (where the number of items is less than 6), their applications in larger problems faces serious challenges:

- Known in cooperative game theory (since the 1950s), Choquet capacities are too expensive in their general form for practical problems as they involve exponentially many parameters.
- How can we elicit the capacity values? How can we efficiently assign values to all subsets?
- Even if we are able to determine Choquet capacity values, will the Choquet integral model be too complex to calculate for practical applications?
- Can some well-established optimisation and machine learning techniques be used in one way or another when linear objectives (constraints) are replaced with fuzzy integrals?

The rest of the book addresses many of the computational challenges involving Choquet capacities and the Choquet integral and also formulates techniques that can be used in the mentioned applications.

References

[BBC16] G. Beliakov, H. Bustince, T. Calvo, *A Practical Guide to Averaging Functions* (Springer, New York, Berlin, Heidelberg, 2016)

[BJW19] G. Beliakov, S. James, J.-Z. Wu, *Discrete Fuzzy Measures: Computational Aspects* (Springer, Berlin, Heidelberg, 2019)

[BPC07] G. Beliakov, A. Pradera, T. Calvo, *Aggregation Functions: A Guide for Practitioners*, vol. 221. *Studies in Fuzziness and Soft Computing* (Springer, Berlin, 2007)

[BG73] R. Bellman, M. Giertz. On the analytic formalism of the theory of fuzzy sets. Inf. Sci. **5**, 149–156 (1973)

[BS84] B. Buchanan, E. Shortliffe, *Rule-Based Expert Systems. The MYCIN Experiments of the Stanford Heuristic Programming Project* (Addison-Wesley, Reading, MA, 1984)

[Bul03] P. Bullen, *Handbook of Means and Their Inequalities* (Kluwer, Dordrecht, 2003)

[Cal+02] T. Calvo, A. Kolesárová, M. Komorníková, R. Mesiar, Aggregation operators: properties, classes and construction methods, in *Aggregation Operators. New Trends and Applications*, eds. by T. Calvo, G. Mayor, R. Mesiar (Physica-Verlag, Heidelberg, New York, 2002), pp. 3–104

[Cho54] G. Choquet, Theory of capacities. Annales de l'institut Fourier **5**, 131–295 (1954)

[Den94] D. Denneberg, *Non-additive Measure and Integral* (Springer Science & Business Media, Dordrecht, 1994)

[Dub+01] D. Dubois, J.-L. Marichal, H. Prade, M. Roubens, R. Sabbadin, The use of the discrete Sugeno integral in decision making: a survey. Int. J. Uncertain. Fuzziness Knowledge Based Syst. **9**, 539–561 (2001)

[DP80] D. Dubois, H. Prade, *Fuzzy Sets and Systems. Theory and Applications* (Academic Press, New York, 1980)

[DP85] D. Dubois, H. Prade, A review of fuzzy set aggregation connectives. Inf. Sci. **36**, 85–121 (1985)

[DP00] D. Dubois, H. Prade, *Fundamentals of Fuzzy Sets* (Kluwer, Boston, 2000)

[DP04] D. Dubois, H. Prade. On the use of aggregation operations in information fusion processes. Fuzzy Sets Syst. **142**, 143–161 (2004)

[DHN76] R. Duda, P. Hart, N. Nilsson, Subjective Bayesian methods for rule-based inference systems. Proc. Natl. Comput. Conf. (AFIPS) **45** 1075–1082 (1976)

[Duj18] J. Dujmovic, *Soft Computing Evaluation Logic: The LSP Decision Method and Its Applications* (Wiley-IEEE Computer Society Press, Hoboken, NJ, 2018)

[Fuj14] K. Fujimoto, Cooperative game as non-additive measure, in *Non-Additive Measures*, eds. by V. Torra, Y. Narukawa, M. Sugeno (Springer, Berlin, New York, 2014), pp. 131–171.

[Gag15] M. Gagolewski, *Data Fusion. Theory, Methods and Applications* (Institute of Computer Science, Polish Academy of Sciences, 2015)

[Gra96] M. Grabisch, The applications of fuzzy integrals in multi-criteria decision making. Eur. J. Oper. Res. **89**, 445–456 (1996)

[Gra97] M. Grabisch, k-Order additive discrete fuzzy measures and their representation. Fuzzy Sets Syst. **92**, 167–189 (1997)

[Gra00] M. Grabisch, The interaction and Möbius representation of fuzzy measures on finite spaces, k-additive measures: A survey, in *Fuzzy Measures and Integrals. Theory and Applications*, eds. by M. Grabisch, T. Murofushi, M. Sugeno (Physica Verlag, Heidelberg, 2000), pp. 70–93

[Gra16] M. Grabisch, *Set Functions, Games and Capacities in Decision Making* (Springer, Berlin, New York, 2016)

[GKM08] M. Grabisch, I. Kojadinovic, P. Meyer, A review of methods for capacity identification in Choquet integral based multi-attribute utility theory. Eur. J. Oper. Res. **186**, 766–785 (2008)

[GL00] M. Grabisch, C. Labreuche, To be symmetric or asymmetric? A dilemma in decision making, in *Preferences and Decisions under Incomplete Knowledge*, eds. by J. Fodor, B. De Baets, P. Perny (Physica-Verlag, Heidelberg, 2000)

[Gra+09] M. Grabisch, J.-L. Marichal, R. Mesiar, E. Pap, *Aggregation Functions. Encyclopedia of Mathematics and Its Foundations* (Cambridge University Press, Cambridge, 2009)

[Jam16] S. James, *An Introduction to Data Analysis using Aggregation Functions in R* (Springer Nature, Cham, 2016)

[Mar00a] J.-L. Marichal, On Choquet and Sugeno integrals as aggregation functions, in *Fuzzy measures and integrals. Theory and Applications*, eds. by M. Grabisch, T. Murofushi, M. Sugeno (Physica-Verlag, Heidelberg, 2000), pp. 247–272

[Mar00b] J.-L. Marichal, On Sugeno integral as an aggregation function. Fuzzy Sets Syst. **114**, 347–365 (2000)

[Mar02] J.-L. Marichal, Aggregation of interacting criteria by means of the discrete Choquet integral, in *Aggregation Operators. New Trends and Applications*, eds. by T. Calvo, G. Mayor, R. Mesiar (Physica-Verlag, Heidelberg, New York, 2002), pp. 224–244

[MT15] J.-L. Marichal, B. Teheux, Generalizations and variants of associativity for aggregation functions, in *Proceedings of the 8th Int. Summer School on Aggregation Operators* (University of Silesia, Katowice, Poland, 2015), pp. 17–22

[MC97] G. Mayor, T. Calvo, Extended aggregation functions, in *IFSA'97*, vol. 1, Prague, 1997, pp. 281–285

[Men42] K. Menger, Statistical metrics. Proc. Natl. Acad. Sci. USA **37**, 535–537 (1942)

[Pap02] E. Pap (ed.), *Handbook of Measure Theory* (North Holland/Elsevier, Amsterdam, Boston, 2002)

[PS07] B. Peleg, P. Sudholter, *Introduction to the Theory of Cooperative Games* (Springer, Berlin, 2007)

[SS83] B. Schweizer, A. Sklar, *Probabilistic Metric Spaces* (North-Holland, New York, 1983)

[Sug74] M. Sugeno, Theory of Fuzzy Integrals and Applications. PhD Thesis, Tokyo Inst. of Technology, 1974

[TNS14] V. Torra, Y. Narukawa, M. Sugeno (eds.), *Non-Additive Measures* (Springer, Berlin, New York, 2014)

[Tor07] Y. Torra, V. Narukawa, *Modeling Decisions. Information Fusion and Aggregation Operators* (Springer, Berlin, Heidelberg, 2007)

[WK92] Z. Wang, G.J. Klir, *Fuzzy Measure Theory* (Plenum Press, New York, 1992)

[WK09] Z. Wang, G. Klir, *Generalized Measure Theory* (Springer US, New York, 2009)

[Yag88] R. Yager, On ordered weighted averaging aggregation operators in multicriteria decision making. IEEE Trans. Syst. Man Cybern. **18**, 183–190 (1988)

[YR96] R. Yager, A. Rybalov, Uninorm aggregation operators. Fuzzy Sets Syst. **80**, 111–120 (1996)

[YR97] R. Yager, A. Rybalov, Noncommutative self-identity aggregation. Fuzzy Sets Syst. **85**, 73–82 (1997)

Chapter 2
Types of Capacities

Abstract This chapter summarises the key definitions and properties of capacities, giving the foundation and background required for subsequent chapters. A number of important capacity types are defined by restrictions on the general definition, which are presented along with contextual interpretations or applications. The consequent behaviour of Choquet or nonadditive integrals when defined with respect to these capacities is also noted.

2.1 Capacity Properties and Restrictions

Choquet capacities represent a relatively wide class of set functions, with the discrete case requiring a value to be assigned to every subset. For n elements or inputs, we have 2^n such coalitions, so the computational costs can be high and interpretations become difficult when capacities are to be defined over large sets. Restricting the general definition to specific types and subclasses can alleviate both of these issues, but we also find that, either due to semantical alignment with practical specifications or because computational advantages are obtained at one or another stage of the process, certain capacity types lend themselves particularly well to particular applications. In this chapter, we present capacities that can be viewed in terms of simplifying assumptions as well as capacities that exhibit useful properties for certain applications. While we go into more detail on simplifying assumptions in later chapters, the capacities introduced here also often serve as a basis for interpreting the behaviour of more general capacities or those obtained from data-fitting.

2.2 0–1 Capacities

Some of the capacities we have already seen in Example 1.5 only allocate a measure of 0 or 1 to each subset. These are referred to as *Boolean capacities* (or Boolean measures), or simply $\{0, 1\}$-capacities.

G. Beliakov et al., *Choquet Capacities and Fuzzy Integrals*, Theory and Applications of Computability, https://doi.org/10.1007/978-3-031-97070-2_2

Definition 2.1 (Boolean Capacity) A capacity (fuzzy measure) μ is called a Boolean capacity if it holds that

$$\mu(\mathcal{A}) = 0 \text{ or } \mu(\mathcal{A}) = 1,$$

for all $\mathcal{A} \subseteq \mathcal{N}$.

For Boolean inputs, i.e. vertices on the $[0, 1]$ unit hypercube, the Choquet integral and Sugeno integral are equivalent, in both cases returning the measure of the subset corresponding with the 1-valued inputs.

2.3 Duality

Choquet and other capacity-based integrals will usually fall into the class of averaging aggregation functions and, as such, have duals that are also averaging. The dual of a Choquet integral is defined with respect to the dual capacity, and for some classes, it is easier to work with the dual and infer any behaviour accordingly rather than with the original values. We will see that this is particularly true for some simplified capacities.

Definition 2.2 (Dual (Conjugate) Capacity) Given a (normalised) capacity μ, its conjugate μ^* is defined by

$$\mu^*(\mathcal{A}) = 1 - \mu(\mathcal{A}^c), \text{ for all } \mathcal{A} \subseteq \mathcal{N},$$

where $\mathcal{A}^c = \mathcal{N} \setminus \mathcal{A}$ is the complement of $\mathcal{A}$ in $\mathcal{N}$.

The dual capacity is also sometimes referred to as the conjugate, which is used mostly in cooperative games [Gra16] where dual means something else.

Example 2.1 The capacities represented below, represented according to the arrangement in Example 1.6, are dual to one another:

$$\begin{array}{ccc} & 1 & \\ 0.7 & 0.9 & 0.3 \\ 0.5 & 0.3 & 0.2 \\ & 0 & \end{array} \qquad \begin{array}{ccc} & 1 & \\ 0.8 & 0.7 & 0.5 \\ 0.7 & 0.1 & 0.3 \\ & 0 & \end{array}$$

As you might expect, the weakest and strongest capacities presented in Example 1.5 are also dual to one another. Capacities are self-dual when (but not only) they are additive.

Definition 2.3 (Self-Dual Capacity) A normalised capacity μ is self-dual if it is equal to its dual μ^*, i.e.

$$\mu(\mathcal{A}) + \mu(\mathcal{A}^c) = 1, \text{holds for all } \mathcal{A} \subseteq \mathcal{N}.$$

2.4 Additive Measures

A defining property of classical measures is additivity, with the measure of any set being found by summing together its non-intersecting components. Examples include volume, interval length, or, for discrete sets, cardinality. Such measures (when normalised) also define probability distributions and hence additive capacities are often referred to as probability measures.

Definition 2.4 (Additive Capacity (Probability Measure)) A capacity μ is called **additive** if for any $\mathcal{A}, \mathcal{B} \subset \mathcal{N}$, $\mathcal{A} \cap \mathcal{B} = \emptyset$:

$$\mu(\mathcal{A} \cup \mathcal{B}) = \mu(\mathcal{A}) + \mu(\mathcal{B}). \tag{2.1}$$

An additive capacity is called a ***probability*** measure.

For an additive capacity, the total value for a set can be found by summing together the value of its singletons, i.e. $\mu(\mathcal{A}) = \sum\limits_{i \in \mathcal{A}} \mu(\{i\})$.

Example 2.2 The following capacity is additive:

$$\begin{array}{ccc} & 1 & \\ 0.4 & 0.7 & 0.9 \\ 0.1 & 0.3 & 0.6 \\ & 0 & \end{array}$$

A straightforward interpretation of the additivity property is that there is no interaction between elements in any of the subsets. In some cases, this may only apply to a specific criterion or input, in which case it is referred to as a dummy criterion.

Definition 2.5 (Dummy Criterion) A decision criterion $d \in \mathcal{N}$ is said to be dummy w.r.t. a capacity μ on $\mathcal{N}$ if

$$\mu(\mathcal{A} \cup \{d\}) = \mu(\mathcal{A}) + \mu(\{d\}), \forall \mathcal{A} \subseteq \mathcal{N} \backslash \{d\}.$$

This means that the contribution of d to the overall measure of the set is just its value and there are no positive or negative interaction effects. If a capacity is additive, then we could consider all elements to be dummy criteria.

2.5 Symmetric Capacities

After additive capacities, the next simplest family is that of symmetric capacities.

Definition 2.6 (Symmetric Capacity) A capacity μ is called symmetric if the value $\mu(\mathcal{A})$ depends only on the cardinality of the set $\mathcal{A}$, i.e. for any $\mathcal{A}, \mathcal{B} \subseteq \mathcal{N}$,

$$\text{if } |\mathcal{A}| = |\mathcal{B}| \text{ then } \mu(\mathcal{A}) = \mu(\mathcal{B}).$$

Alternatively, the definition can be given in terms of an underlying function referred to as a *quantifier*, Q (the name comes from the use of such functions in aggregation processes involving linguistic variables). We can say that a capacity μ is symmetric if for any $\mathcal{A} \subseteq \mathcal{N}$ it holds that

$$\mu(\mathcal{A}) = Q\left(\frac{|\mathcal{A}|}{n}\right), \tag{2.2}$$

for some monotone non-decreasing function $Q : [0, 1] \to [0, 1]$, $Q(0) = 0$ and $Q(1) = 1$.

Example 2.3 The following capacity is symmetric:

$$\begin{array}{ccc} & 1 & \\ 0.7 & 0.7 & 0.7 \\ 0.2 & 0.2 & 0.2 \\ & 0 & \end{array}$$

Balanced capacities[1] are also defined with respect to subset cardinality, with symmetric capacities constituting a special case. For these, a set of larger cardinality cannot have a lower measure than one of the smaller subsets.

Definition 2.7 (Balanced Capacity) A capacity μ is called balanced if it holds that

$$| \mathcal{A} | < | \mathcal{B} | \Longrightarrow \mu(\mathcal{A}) \leqslant \mu(\mathcal{B}), \text{ for all } \mathcal{A}, \mathcal{B} \subseteq \mathcal{N}.$$

Note 2.1 The number of monotonicity constraints for balanced capacities is reduced to $2^{n+1} + n - 2$ using extra variables.

The Choquet integral with respect to a symmetric capacity results in an OWA function. We will study such integrals in detail in Chap. 6.

---

[1] It is a different notion to that of a balanced game; see [Gra16], Ch.3.

2.6 Sugeno λ-Capacities

As with the possibility and necessity measures, Sugeno capacities (or λ-capacities) specify the relationship between the value of a set and its subsets. They were introduced by Sugeno as λ-fuzzy measures (also known as Sugeno measures) in [Sug74].

Definition 2.8 (λ-Capacity) Given a parameter $\lambda \in]-1, \infty[$, a λ-capacity is a capacity μ, which for all $\mathcal{A}, \mathcal{B} \subseteq \mathcal{N}, \mathcal{A} \cap \mathcal{B} = \emptyset$ satisfies

$$\mu(\mathcal{A} \cup \mathcal{B}) = \mu(\mathcal{A}) + \mu(\mathcal{B}) + \lambda\mu(\mathcal{A})\mu(\mathcal{B}). \tag{2.3}$$

Example 2.4 The following λ-capacity is obtained based on the singleton values $\mu(\{1\}) = 0.7, \mu(\{2\}) = 0.4, \mu(\{3\}) = 0.6$ and $\lambda = -0.8845$ (to 4 d.p.—see Sect. 7.1):

$$\begin{array}{ccc} & 1 & \\ 0.8523 & 0.9285 & 0.7877 \\ 0.7 & 0.4 & 0.6 \\ & 0 & \end{array}$$

Sugeno capacities will be covered in more depth in Sect. 7.1.

2.7 Sub- and Supermodular Capacities

The great appeal of capacities in practical and theoretical applications is often in the flexibility afforded by nonadditivity—complimentary and redundancy effects can be incorporated into our models according to the needs of the situation or based on the data we observe. In some cases, these interactions will be consistent one way or the other, leading to the important classes of submodular and supermodular capacities, as well as the weaker conditions of sub- and superadditivity.

Definition 2.9 (Submodular and Supermodular Capacity) A capacity μ is called submodular if for any $\mathcal{A}, \mathcal{B} \subseteq \mathcal{N}$

$$\mu(\mathcal{A} \cup \mathcal{B}) + \mu(\mathcal{A} \cap \mathcal{B}) \leqslant \mu(\mathcal{A}) + \mu(\mathcal{B}). \tag{2.4}$$

It is called supermodular if

$$\mu(\mathcal{A} \cup \mathcal{B}) + \mu(\mathcal{A} \cap \mathcal{B}) \geqslant \mu(\mathcal{A}) + \mu(\mathcal{B}). \tag{2.5}$$

It is called modular if

$$\mu(\mathcal{A} \cup \mathcal{B}) + \mu(\mathcal{A} \cap \mathcal{B}) = \mu(\mathcal{A}) + \mu(\mathcal{B}). \tag{2.6}$$

When $\mathcal{A} \cap \mathcal{B} = \mathcal{A}$ or $\mathcal{B}$, i.e. $\mathcal{A} \subseteq \mathcal{B}$ or $\mathcal{B} \subseteq \mathcal{A}$, the inequalities in Definition 2.9 trivially hold.

Note 2.2 For *strict* sub- and supermodularity , the strict inequalities ($<$, $>$) are required to hold for all $\mathcal{A}, \mathcal{B} \subseteq \mathcal{N}$ such that $\mathcal{A} \cap \mathcal{B} \neq \mathcal{A}, \mathcal{B}$.

In Möbius representation, a capacity is supermodular if, for all $\mathcal{A} \subseteq \mathcal{N}$ and all pairs $\{i, j\} \subseteq \mathcal{A}$, $i \neq j$, it holds that [CJ89]:

$$\sum_{\mathcal{B} \subseteq \mathcal{A} | i, j \in \mathcal{B}} M(\mathcal{B}) \geqslant 0. \tag{2.7}$$

For submodularity, the inequality is reversed.

Two weaker conditions frequently used are sub- and superadditivity. These are special cases of sub- and supermodularity that only require the inequalities to hold for disjoint subsets.

Definition 2.10 (Subadditive and Superadditive Capacity) A capacity μ is called subadditive if for any two non-intersecting subsets $\mathcal{A}, \mathcal{B} \subset \mathcal{N}$, $\mathcal{A} \cap \mathcal{B} = \emptyset$:

$$\mu(\mathcal{A} \cup \mathcal{B}) \leqslant \mu(\mathcal{A}) + \mu(\mathcal{B}). \tag{2.8}$$

It is called superadditive if

$$\mu(\mathcal{A} \cup \mathcal{B}) \geqslant \mu(\mathcal{A}) + \mu(\mathcal{B}). \tag{2.9}$$

These conditions imply that submodular capacities are subadditive and supermodular capacities are superadditive; however, as illustrated in Example 2.5, the reverse does not hold.

Example 2.5 The following capacity is superadditive but not supermodular, since $\mu(\{1, 2\}) = \mu(\{2, 3\}) = 0.6$, $\mu(\{2\}) = 0.1$ and hence $\mu(\{1, 2, 3\}) + \mu(\{2\}) \not\geqslant \mu(\{1, 2\}) + \mu(\{2, 3\})$:

$$\begin{array}{ccc} & 1 & \\ 0.6 & 0.5 & 0.6 \\ 0.3 & 0.1 & 0.2 \\ & 0 & \end{array}$$

Only additive capacities can be both sub- and superadditive, which is also true for sub- and supermodularity.

Note 2.3 The sub- and supermodularity conditions are dual to one another, with the dual of a supermodular capacity being submodular and vice versa. However, as illustrated in Example 2.6, the dual of a superadditive capacity is not necessarily subadditive and vice versa.

Example 2.6 Let μ be a $\{0, 1\}$-capacity on $\mathcal{N} = \{1, 2, 3\}$:

$$\begin{array}{ccc} & 1 & \\ 1 & 1 & 0 \\ 0 & 0 & 0 \\ & 0 & \end{array}$$

This measure is superadditive but its dual capacity μ^*, given by

$$\begin{array}{ccc} & 1 & \\ 1 & 1 & 1 \\ 1 & 0 & 0 \\ & 0 & \end{array}$$

is not subadditive, because, for instance, $\mu^*(\{2, 3\}) = 1$ and $\mu^*(\{2\}) + \mu^*(\{3\}) = 0$. It is also not superadditive, since $\mu^*(\{1, 2, 3\}) < \mu^*(\{1\}) + \mu^*(\{2, 3\})$.

The submodularity and supermodularity conditions can also be referred to describe interaction effects between specific pairs of subsets $\mathcal{A}, \mathcal{B}$, while not necessarily holding for all subsets of a more general capacity.

By comparing Eqs. (2.8) and (2.9) with Eq. (2.3), we can observe that for Sugeno capacities, $\lambda \geqslant 0$ corresponds with superadditive capacities, while $\lambda \leqslant 0$ leads to subadditivity.

The following definitions can be used to describe special cases of super- and subadditivity.

Definition 2.11 (Extremely Superadditive) A capacity on $\mathcal{N}$ is called an extremely superadditive capacity $\varepsilon_{\mathcal{S}}^{+}$ focused on subset $\mathcal{S} \subseteq \mathcal{N}$, if $\varepsilon_{\mathcal{S}}^{+}(\mathcal{A}) = 1$ if and only if $\mathcal{S} \subseteq \mathcal{A}$, $\mathcal{A} \subseteq \mathcal{N}$, and $\varepsilon_{\mathcal{S}}^{+}(\mathcal{A}) = 0$ otherwise.

Definition 2.12 (Extremely Subadditive) A capacity on $\mathcal{N}$ is called an extremely subadditive capacity $\varepsilon_{\mathcal{S}}$ focused on subset $\mathcal{S} \subseteq \mathcal{N}$, if $\varepsilon_{\mathcal{S}}^{-}(\mathcal{A}) = 1$ if and only if $\mathcal{A} \cap \mathcal{S} \neq \emptyset$, $\mathcal{A} \subseteq \mathcal{N}$, and $\varepsilon_{\mathcal{S}}^{-}(\mathcal{A}) = 0$ otherwise.

The definition for extremely superadditive capacities can be compared to that of Dirac measures provided in Example 1.5, which can be seen as a special case when $\mathcal{S}$ is a singleton. The capacity $\varepsilon_{\mathcal{S}}^{+}$ is also called a *unanimity game* in the context of game theory.

Note 2.4 With $\varepsilon_{\mathcal{S}}^{+}$ (resp. $\varepsilon_{\mathcal{S}}^{-}$), the Choquet integral collapses to the local minimum (resp. maximum) operator on the subset $\mathcal{S}$ [BBC16]. Actually, the values assigned to $\varepsilon_{\mathcal{S}}^{+}$ (resp. $\varepsilon_{\mathcal{S}}^{-}$) imply that the $|\mathcal{S}|$ criteria are fully complementary (resp. substitutive or redundant), while the other criteria not in $\mathcal{S}$ provide no useful information [MGB17].

The concepts of k-order monotone capacities and k-order alternating capacities (k-monotonicity and k-alternativity) [Cho54, CJ89] generalise the sub- and supermodularity conditions.

Definition 2.13 (k-Monotone and k-Alternating Capacity) A capacity μ is said to be k-order monotone (or k-monotone) if for all families of $k \geqslant 2$ subsets $\mathcal{A}_1, \mathcal{A}_2, \ldots, \mathcal{A}_k \in 2^{\mathcal{N}}$, it holds that

$$\mu\left(\bigcup_{i=1}^{k} \mathcal{A}_i\right) \geqslant \sum_{\mathcal{B}\subseteq\{1,\ldots,k\},\mathcal{B}\neq\emptyset} (-1)^{|\mathcal{B}|+1} \mu\left(\bigcap_{i\in\mathcal{B}} \mathcal{A}_i\right). \tag{2.10}$$

A capacity μ is said to be k-order alternating (or k-alternating) if it holds that

$$\mu\left(\bigcap_{i=1}^{k} \mathcal{A}_i\right) \leqslant \sum_{\mathcal{B}\subseteq\{1,\ldots,k\},\mathcal{B}\neq\emptyset} (-1)^{|\mathcal{B}|+1} \mu\left(\bigcup_{i\in\mathcal{B}} \mathcal{A}_i\right). \tag{2.11}$$

When $k = 2$, we obtain two-monotone (resp. two-alternating) capacities, which are equivalent to supermodular (resp. submodular) capacities. For k-monotonicity, the special case of $k = 1$ is defined as monotonicity according to the general definition (as given in the premeasure Definition 1.38).

2.8 Possibility and Necessity

Possibility and necessity measures are the basis of the theory of possibility [Zad78, WK92, DP88], providing a useful alternative to probability (additive) capacities for modelling special types of uncertainty.

Definition 2.14 (Possibility and Necessity Measures) A capacity is called a possibility, Pos, if for all $\mathcal{A}, \mathcal{B} \subseteq \mathcal{N}$, it satisfies

$$Pos(\mathcal{A} \cup \mathcal{B}) = \max\{Pos(\mathcal{A}), Pos(\mathcal{B})\}.$$

A capacity is called a necessity, Nec, if for all $\mathcal{A}, \mathcal{B} \subseteq \mathcal{N}$, it satisfies

$$Nec(\mathcal{A} \cap \mathcal{B}) = \min\{Nec(\mathcal{A}), Nec(\mathcal{B})\}.$$

Note 2.5 Possibility and necessity measures are dual to each other in the sense of Definition 2.2, that is, for all $\mathcal{A} \subseteq \mathcal{N}$,

$$Nec(\mathcal{A}) = 1 - Pos(\mathcal{A}^c).$$

A possibility measure is subadditive and a necessity measure is superadditive.

Example 2.7 The capacities below, μ and μ^*, are possibility (left) and necessity (right) measures, respectively, with μ being dual to μ^*:

$$\begin{array}{ccc} & 1 & \\ 1 & 0.3 & 1 \\ 0.3 & 1 & 0.2 \\ & 0 & \end{array} \qquad\qquad \begin{array}{ccc} & 1 & \\ 0.8 & 0 & 0.7 \\ 0 & 0.7 & 0 \\ & 0 & \end{array}$$

Possibility measures are a special case of *decomposable* capacities (Definition 7.3 in Chap. 7), since the measure of the union of any two sets can be calculated from the component values.[2] Possibility and necessity measures are also known as maxitive and minitive capacities.

Note 2.6 $\varepsilon^+_{\mathcal{S}}$ (resp. $\varepsilon^-_{\mathcal{S}}$) from Definition 2.11 (Definition 2.12) is essentially a 0–1 necessity (possibility) measure focused on subset $\mathcal{S}$ [Web88].

2.9 Belief and Plausibility

Belief and plausibility measures generalise possibility and necessity measures, respectively, constituting the basis of Dempster and Shafer evidence theory [Sha76].

Definition 2.15 (Belief Measure) A *belief measure* $Bel : 2^{\mathcal{N}} \rightarrow [0, 1]$ is a capacity that satisfies the following condition: for all $m > 1$,

$$Bel(\bigcup_{i=1}^{m} \mathcal{A}_i) \geqslant \sum_{\mathcal{B}\subseteq\{1,\ldots,m\}} (-1)^{\mathcal{B}+1} Bel(\bigcap_{i\in\mathcal{B}} \mathcal{A}_i),$$

where $\{\mathcal{A}_i\}_{i\in\{1,\ldots,m\}}$, is any finite family of subsets of $\mathcal{N}$.

It can be observed, by comparing with Definition 2.13, that this condition is equivalent to k-monotonicity holding for all $k > 1$, which is sometimes referred to as total monotonicity [Den00, Gra00]. Plausibility measures are related to belief measures by duality, i.e. $Pl : 2^{\mathcal{N}} \rightarrow [0, 1]$ is a plausibility measure if its dual set

[2] Decomposability usually requires $\mathcal{A} \cap \mathcal{B} = \emptyset$; however, this is not the case for possibility and necessity measures.

function is a belief measure.

$$Pl(\mathcal{A}) = 1 - Bel(\mathcal{A}^c),$$

for all $\mathcal{A} \subseteq \mathcal{N}$.

They also correspond with the definition for k-alternating monotonicity and can be defined independently as follows.

Definition 2.16 (Plausibility Measure) A *plausibility measure* $Pl : 2^{\mathcal{N}} \to [0, 1]$ is a capacity that satisfies the following condition: for all $m > 1$,

$$Pl(\bigcap_{i=1}^{m} \mathcal{A}_i) \leqslant \sum_{\emptyset \neq I \subset \{1,\ldots,m\}} (-1)^{|I|+1} Pl(\bigcup_{i \in I} \mathcal{A}_i),$$

where $\{\mathcal{A}_i\}_{i \in \{1,\ldots,m\}}$ is any finite family of subsets of $\mathcal{N}$.

Belief measures are superadditive and plausibility measures are subadditive, with only additive capacities satisfying conditions for both belief and plausibility. Possibility and necessity measures are special cases of plausibility and necessity measures, respectively.

The concept of a *basic probability assignment* based on the Möbius transformation (see Definition 1.41) is sometimes used to construct belief measures.

Definition 2.17 (Basic Probability Assignment) A basic probability assignment is a set function m over the powerset $2^{\mathcal{N}}$ satisfying $m(\emptyset) = 0$, $m(\mathcal{A}) \geqslant 0$ for all $\mathcal{A}$ and $\sum_{\mathcal{A} \subseteq \mathcal{N}} m(\mathcal{A}) = 1$. Focal elements of $\mathcal{N}$ are subsets $\mathcal{A}$ such that $m(\mathcal{A}) > 0$.

Given a basic probability assignment $m : 2^{\mathcal{N}} \to [0, 1]$, the belief of $\mathcal{A}$ induced by m is given by

$$Bel(\mathcal{A}) = \sum_{\mathcal{B} \subseteq \mathcal{A}} m(\mathcal{B}), \tag{2.12}$$

and the plausibility is given by

$$Pl(\mathcal{A}) = \sum_{\mathcal{B} | \mathcal{B} \cap \mathcal{A} \neq \emptyset} m(\mathcal{B}). \tag{2.13}$$

We refer the reader to the literature in this field, e.g. [WK92, GMS00, Gra16].

By comparing with the Zeta transform formula Definition 1.41, a belief measure can be seen as taking the Zeta transform of the basic probability assignment. In other words, belief measures are capacities whose Möbius transform values are all non-negative.

Example 2.8 For the following probability assignment m,

	0.1	
0.2	0.3	0.1
0.2	0	0.1
	0	

The induced belief (left) and plausibility (right) measures are given as follows:

	1				1	
0.4	0.6	0.2		0.9	1	0.8
0.2	0	0.1		0.8	0.4	0.5
	0				0	

Special cases of plausibility and belief measures can be obtained from Sugeno capacities for particular values of λ. When $-1 < \lambda \leqslant 0$, a λ-capacity is a plausibility measure, and when $\lambda \geqslant 0$, it is a belief measure.

2.10 Buoyant and Antibuoyant Capacities

Buoyant and antibuoyant capacities are subsets of submodular and supermodular capacities, respectively, and both classes are dual to one another. They are defined by means of the weighting vectors corresponding to each maximal chain in the partially ordered set of capacity values under the set inclusion relation and mimic the buoyancy property of OWA weighting vectors. For this, we introduce some extra notation to express the Choquet integral similar to the OWA functions.

Definition 2.18 (Buoyant Weighting Vector) A weighting vector $\mathbf{w}, \sum_{i=1}^{n} w_i = 1, w_i \geqslant 0$ associated with an OWA function is said to be buoyant if $w_i \geqslant w_j$ whenever $i < j$. It is said to be extensive [Yag93] if $w_i > w_j$ whenever $i < j$.

We can then draw upon the *ordered weighting vector* $\mathbf{w}^\sigma$ notation, which refers to the weights induced by the ordering $x_{(1)} \geqslant x_{(2)} \geqslant \cdots \geqslant x_{(n)}$:

$$\mathbf{w}^\sigma = (\mu(\{(1)\}), \mu(\{(1),(2)\}) - \mu(\{(1)\}), \ldots, \mu(N) - \mu(N \setminus \{(n)\})) .$$

The weights are obtained by considering a maximal chain in the lattice of capacity values, which corresponds to the ordering of $\mathbf{x}$, i.e. once the relative input ordering is fixed, the effective weight applied to each input in Eq. (1.12) forms our vector. In this notation, the Choquet integral is written as

$$C_\mu^{\mathbf{w}^\sigma}(\mathbf{x}) = \sum_{i=1}^{n} w_i^\sigma x_{(i)},$$

noting that here, $x_{(1)} \geqslant \cdots \geqslant x_{(n)}$ or alternatively, $w_i^\sigma = \mu(H_{n-i+1}) - \mu(H_{n-i+2})$. The following is the definition of buoyancy for general capacities, which was presented in [BJ21].

Definition 2.19 (Buoyant Capacity) A capacity μ is buoyant if all ordered weighting vectors $\mathbf{w}^\sigma$ associated with μ are buoyant. A capacity is antibuoyant if its conjugate is buoyant, i.e. if all ordered weighting vectors are antibuoyant.

A capacity that is buoyant is necessarily submodular and an antibuyoant capacity is necessarily supermodular. The reverse implication is false, except for symmetric capacities.

In [BJL11], it was established that, in the case of symmetric capacities, buoyancy and submodularity are equivalent, i.e. a weighting vector associated with an OWA operator is buoyant if and only if it is equivalent to the Choquet integral with respect to a symmetric and submodular capacity. However, in [BJ21], it was shown that this is not the case when buoyancy is considered for more general capacities.

One reason buoyancy is an appealing property is its relationship to the Pigou–Dalton principle. In welfare economics, the Pigou–Dalton principle [Dal20] requires that any redistribution of wealth from a richer to poorer individual should increase overall measures of welfare (or decrease measures of inequality). For an input vector $x_1 \geqslant x_2 \geqslant \cdots \geqslant x_n$, we say that a function f satisfies the Pigou–Dalton principle if for all $x_i - h \geqslant x_j + h, h > 0$, it holds that $f(x_1, \ldots, x_i, \ldots, x_j, \ldots, x_n) \leqslant f(x_1, \ldots, x_i - h, \ldots, x_j + h, \ldots, x_n)$. It is referred to as the *progressive transfers* principle, whereas the opposite inequality is referred to as regressive transfers.

Whenever a capacity is symmetric, supermodularity is also equivalent to the Pigou–Dalton principle holding for the Choquet integral, but for non-symmetric capacities, Pigou–Dalton is a stronger condition, which is equivalent to the capacity being antibuoyant [BJ21, BJ22]. Buoyant capacities are equivalent to the regressive transfers principle.

We will characterise buoyant and antibuoyant capacities in the subsequent chapters, but here we present two propositions.

Proposition 2.1 *The Choquet integral is consistent with the Pigou–Dalton principle if and only if the underlying capacity is antibuoyant.*

Proposition 2.2 *In Möbius representation, the necessary and sufficient condition for buoyancy is [BJ21]*

$$\sum_{\substack{\mathcal{B} \subseteq \mathcal{A} \cup \{i,j\} \\ \{i,j\} \subseteq \mathcal{B}}} M(\mathcal{B}) \leqslant \sum_{\substack{\mathcal{B} \subseteq \mathcal{A} \cup \{i\} \\ i \in \mathcal{B}}} M(\mathcal{B}) - \sum_{\substack{\mathcal{B} \subseteq \mathcal{A} \cup \{j\} \\ j \in \mathcal{B}}} M(\mathcal{B}), \tag{2.14}$$

for all $i, j \in \mathcal{N}$ *and all* $\mathcal{A} \subseteq \mathcal{N} \setminus \{i, j\}$.

A more convenient way to write (2.14) for computational purposes is

$$\sum_{\mathcal{B}\subseteq\mathcal{A}} M(\mathcal{B}\cup\{i,j\}) + \sum_{\mathcal{B}\subseteq\mathcal{A}} [M(\mathcal{B}\cup\{j\}) - M(\mathcal{B}\cup\{i\})] \leqslant 0.$$

We remark that the set of buoyant capacities is different from the set of plausibility measures.

Example 2.9 The following capacity is buoyant:

$$\begin{array}{ccc} & 1 & \\ 0.75 & 0.9 & 0.9 \\ 0.5 & 0.45 & 0.8 \\ & 0 & \end{array}$$

The 3! potential weighting vectors $\mathbf{w}^\sigma$ depending on the input ordering are

$$\begin{aligned} \mathbf{w}^{123} &= (0.5, 0.25, 0.25) \\ \mathbf{w}^{132} &= (0.5, 0.4, 0.1) \\ \mathbf{w}^{213} &= (0.45, 0.3, 0.25) \\ \mathbf{w}^{231} &= (0.45, 0.45, 0.1) \\ \mathbf{w}^{312} = \mathbf{w}^{321} &= (0.8, 0.1, 0.1) \end{aligned}$$

where, e.g $\mathbf{w}^{123}$ indicates that $x_1 \geqslant x_2 \geqslant x_3$.

Note that $\mu(\{3\}) > \mu(\{1,2\})$, i.e. buoyant and antibuoyant capacities are not necessarily *balanced*.

There are several relations between the behaviour of the Choquet integral and buoyancy; see [BJ21].

We conclude the chapter with a final note on convexity.

Note 2.7 The set of all capacities (for a fixed $\mathcal{N}$) is convex.[3] The sets of subadditive, superadditive, submodular, supermodular, subadditive, superadditive, additive, belief, plausibility, buoyant, and antibuoyant capacities are convex. However, the sets of possibility and necessity measures are not convex.

[3] A set $\mathcal{E}$ is convex if $\alpha x + (1-\alpha)y \in \mathcal{E}$ for all $x, y \in \mathcal{E}, \alpha \in [0, 1]$.

References

[BBC16] G. Beliakov, H. Bustince, T. Calvo, *A Practical Guide to Averaging Functions* (Springer, New York, Berlin, Heidelberg, 2016)

[BJ21] G. Beliakov, S. James, Choquet integral optimisation with constraints and the buoyancy property for fuzzy measures. Inf. Sci. **578**, 22–36 (2021)

[BJ22] G. Beliakov, S. James, Choquet integral based measures of economic welfare and species diversity. Int. J. Intell. Syst. **37**, 2849–2867 (2022)

[BJL11] G. Beliakov, S. James, G. Li, Learning Choquet-integral-based metrics for semisupervised clustering. IEEE Trans. Fuzzy Syst. **19**(3), 562–574 (2011)

[CJ89] A. Chateauneuf, J.-Y. Jaffray, Some characterizations of lower probabilities and other monotone capacities through the use of Möbius inversion. Math. Social Sci. **17**(3), 263–283 (1989)

[Cho54] G. Choquet, Theory of capacities. Annales de l'institut Fourier **5**, 131–295 (1954)

[Dal20] H. Dalton, The measurement of the inequality of incomes. Econ. J. **30**, 348–361 (1920)

[Den00] D. Denneberg, Non-additive measure and integral, basic concepts and their role for applications, in *Fuzzy Measures and Integrals. Theory and Applications*, eds. by M. Grabisch, T. Murofushi, M. Sugeno (Physica-Verlag, Heidelberg, 2000), pp. 42–69

[DP88] D. Dubois, H. Prade, *Possibility Theory* (Plenum Press, New York, 1988)

[Gra00] M. Grabisch, The interaction and Möbius representation of fuzzy measures on finite spaces, k-additive measures: A survey, in *Fuzzy Measures and Integrals. Theory and Applications*, eds. by M. Grabisch, T. Murofushi, M. Sugeno (Physica-Verlag, Heidelberg, 2000), pp. 70–93

[Gra16] M. Grabisch, *Set Functions, Games and Capacities in Decision Making* (Springer, Berlin, New York, 2016)

[GMS00] M. Grabisch, T. Murofushi, M. Sugeno (eds.), *Fuzzy Measures and Integrals. Theory and Applications* (Physica-Verlag, Heidelberg, 2000)

[MGB17] J. Murillo, S. Guillaume, P. Bulacio, k-maxitive fuzzy measures: A scalable approach to model interactions. Fuzzy Sets Syst. **324**, 33–48 (2017)

[Sha76] G. Shafer, *A Mathematical Theory of Evidence* (Princeton University Press, Princeton, 1976)

[Sug74] M. Sugeno, Theory of Fuzzy Integrals and Applications. PhD Thesis. Tokyo Inst. of Technology, 1974

[WK92] Z. Wang, G.J. Klir, *Fuzzy Measure Theory* (Plenum Press, New York, 1992)

[Web88] R.J. Weber, Probabilistic values for games, in *The Shapley Value. Essays in Honor of Lloyd S. Shapley*, eds. by A. Roth (Cambridge University Press, Cambridge, 1988), pp. 101–119

[Yag93] R.R. Yager, On the completion of qualitative possibility measures. IEEE Trans. Fuzzy Syst. **1**(3), 184–194 (1993)

[Zad78] L. Zadeh, Fuzzy sets as a basis for a theory of possibility. Fuzzy Sets Syst. **1**, 3–28 (1978)

Chapter 3
Value and Interaction Indices

Abstract Capacities assign a weight, measure of importance, or value to each subset with respect to the given reference set. This allows one to model interaction, taking into account any redundancies or complementary effects between variables, but comes at the expense of increased complexity in the number of parameters that need to be defined and interpreted. Considerable effort has gone into addressing this, resulting in some well-established theory and indices that allow the behaviour of capacities to be well understood. Many of these indices serve as alternative representations that allow for new construction methods and simplifications. This chapter presents the most well-known indices and their properties.

3.1 The Notion of Value

The departure from additivity is a distinguishing feature of Choquet capacities. The increased modelling flexibility that this affords could be sufficient justification for their deployment in decision-making or regression-type problems; however, even more appealing is that the interactions captured can be interpreted. Practitioners and policy-makers are becoming increasingly sceptical of so-called "black-box" approaches to modelling, and while the additivity property can simplify things when it comes to interpretation, it also assumes a level of independence between variables that may not be reasonable when it comes to real-world phenomena. With capacities we gain enhanced modelling power without coming at the cost of transparency.

Many of the indices proposed for understanding the behaviour of Choquet capacities focus in one way or another on how much and in which direction the capacity deviates from an additive measure. Aggregation based on the additivity assumption could be said to "double-count" criteria relating to similar concepts, e.g. "learnability" and "memorability" in evaluating a software user interface [SGBC03]. On the other hand, some criteria or features may complement one another. In medical diagnoses, for example, two symptoms together can make a strong case for a diagnosis but by themselves be considered fairly innocuous.

Capacities in their standard representation model these interactions implicitly by assigning value or importance to every coalition. If we want to ascertain the average

G. Beliakov et al., *Choquet Capacities and Fuzzy Integrals*, Theory and Applications of Computability, https://doi.org/10.1007/978-3-031-97070-2_3

importance of an individual criterion, worker, or input source, calculations such as the Shapley value are able to achieve this by averaging contributions across all subsets. Interaction indices can be seen to extend this idea, looking instead at how coalition values change when two or more criteria are combined with each subset [Gra00, Gra97]. These ideas have been instrumental in facilitating interpretation of Choquet capacities and how the Choquet integral behaves as an aggregation function. Concepts such as the nonmodularity and bipartition indices introduced more recently have sought to capture some other aspects of capacity behaviour to allow for complementary insights.

In addition to importance and interaction indices providing a solid mathematical foundation for the interpretation of capacities, they can also be used to help construct capacities, tailor-made for a given application with suitable behaviour and properties.

3.2 Derivatives of Set Functions

The contribution in value of a variable or subset of variables to a larger set naturally lends itself to the notion of discrete derivatives [Gra16, GL07].

Definition 3.1 (Discrete Derivative of a Set Function) Let μ be a set function on $\mathcal{N}$ and $\mathcal{A} \subseteq \mathcal{N}, i \in \mathcal{N}$. The derivative of μ at $\mathcal{A}$ w.r.t. i is

$$\Delta_i \mu(\mathcal{A}) = \mu(\mathcal{A} \cup \{i\}) - \mu(\mathcal{A} \setminus \{i\}).$$

The second-order derivative of μ at $\mathcal{A}$ w.r.t. $i, j \in \mathcal{N}$ is

$$\begin{aligned}\Delta_{ij}\mu(\mathcal{A}) &= \Delta_i(\Delta_j \mu(\mathcal{A}))\\ &= \mu(\mathcal{A} \cup \{i, j\}) - \mu(\mathcal{A} \cup \{i\} \setminus \{j\}) - \mu(\mathcal{A} \cup \{j\} \setminus \{i\}) + \mu(\mathcal{A} \setminus \{i, j\}).\end{aligned}$$

These calculations themselves are set functions and it need not hold that i is in $\mathcal{A}$ [1]. The i-th derivative $\Delta_i \mu(\mathcal{A})$ is interpreted as the marginal contribution of the variable i to the coalition $\mathcal{A}$.

The definition can be extended iteratively to obtain the following.

Definition 3.2 (Discrete Derivative of a Set Function (w.r.t. a Set)) Let μ be a set function on $\mathcal{N}$ and $\mathcal{A}, \mathcal{B} \subseteq \mathcal{N}$. The derivative at $\mathcal{A}$ with respect to $\mathcal{B}$ is defined as

[1] We will also use the notion of *marginal contribution* denoted by $\Delta_i \mu(\mathcal{A})$, specifying that $i \notin \mathcal{A}$. The difference will be in the factor $\frac{1}{2}$ in summation of such quantities, as the sums may involve the same quantity twice (when $i \in \mathcal{A}$ and $i \notin \mathcal{A}$). We state explicitly in the summation indices whether we require i not to belong to $\mathcal{A}$.

$$\Delta_{\mathcal{B}}\mu(\mathcal{A}) = \Delta_{\mathcal{B}\setminus\{i\}}(\Delta_{\{i\}}\mu(\mathcal{A}))$$

with the convention $\Delta_{\emptyset}\mu(\mathcal{A}) = \mu(\mathcal{A})$ and $\Delta_{\{i\}}\mu(\mathcal{A}) = \Delta_i\mu(\mathcal{A})$.

Explicitly, we have [Gra16]

$$\Delta_{\mathcal{B}}\mu(\mathcal{A}) = \sum_{\mathcal{C}\subseteq\mathcal{B}} (-1)^{|\mathcal{B}\setminus\mathcal{C}|}\mu((\mathcal{A}\setminus\mathcal{B})\cup\mathcal{C}). \tag{3.1}$$

In Möbius representation, this will be

$$\Delta_{\mathcal{B}}\mu(\mathcal{A}) = \sum_{\mathcal{C}\subseteq\mathcal{A}\setminus\mathcal{B}} M(\mathcal{B}\cup\mathcal{C}).$$

The following result can be obtained from the duality relationship for capacities (see Sect. 4.5 for a short proof). If μ^* is the dual of μ, then

$$\Delta_{\mathcal{B}}\mu^*(\mathcal{A}) = (-1)^{|\mathcal{B}|+1}\Delta_{\mathcal{B}}\mu(\mathcal{N}\setminus\mathcal{A}). \tag{3.2}$$

Many classical and contemporary indices used to quantify interaction and value can be expressed in terms of the discrete derivatives, which will be covered in more detail in the next chapter.

3.3 Shapley Value and Interaction

A fundamental question to ask when we are presented with a capacity is the importance of an input i (or equivalently a variable in a regression model or player in game theory). This is straightforward for additive models, where the contribution of an input to a subset is uniform across all coalitions. The first discrete derivative $\Delta_i\mu(\mathcal{A})$ provides us with a framework for averaging the varying contributions that we observe for nonadditive capacities.

Definition 3.3 (Value) Let μ be a capacity on $\mathcal{N}$. The value of $i \in \mathcal{N}$ w.r.t. μ is defined as

$$I_{\mathrm{p}}^{\mu}(\{i\}) = \sum_{\mathcal{A}\subseteq\mathcal{N}\setminus\{i\}} p_{\mathcal{A}}^{i}(\mathcal{N})[\mu(\mathcal{A}\cup\{i\}) - \mu(\mathcal{A})],$$

where, for any $i \in \mathcal{N}$, the family of coefficients $\{p_{\mathcal{A}}^{i}(\mathcal{N})\}_{\mathcal{A}\subseteq\mathcal{N}\setminus\{i\}}$ forms a probability distribution on $2^{\mathcal{N}\setminus\{i\}}$, i.e. the coefficients are non-negative and add to one.

The discrete derivative component $\mu(\mathcal{A}\cup\{i\}) - \mu(\mathcal{A}) = \Delta_i\mu(\mathcal{A}), i \notin \mathcal{A}$ in Definition 3.3 represents the marginal contribution of i to the subset $\mathcal{A}$ [GKM08].

The value of $I_{\mathrm{p}}^{\mu}(\{i\})$ is hence the expectation of these marginal contributions with respect to the distribution defined by the $p^{i}_{\mathcal{A}}(\mathcal{N})$ [FKM06]. The greater the value of $I_{\mathrm{p}}^{\mu}(\{i\})$, the more important the criterion i is in shaping the values of μ.

Both the Banzhaf index [Ban65] (which will be presented later on) and Shapley value [Sha53] can be expressed as special cases, differing only in the choice of coefficients $p^{i}_{\mathcal{A}}$.

Definition 3.4 (Shapley Value) Let μ be a capacity on $\mathcal{N}$. The Shapley value (or Shapley index) for every $i \in \mathcal{N}$ is

$$\varphi(i) = \sum_{\mathcal{A} \subseteq \mathcal{N} \setminus \{i\}} \frac{(n - |\mathcal{A}| - 1)!|\mathcal{A}|!}{n!} [\mu(\mathcal{A} \cup \{i\}) - \mu(\mathcal{A})]. \tag{3.3}$$

The Shapley value is the vector $\varphi(\mu) = (\varphi(1), \ldots, \varphi(n))$, with $\varphi(i) = \mu(\{i\})$ in the case of *additive* capacities.

The coefficients can also be rearranged and grouped according to subset cardinality, facilitating the following expression, which is equivalent:

$$\varphi(i) = \frac{1}{n} \sum_{t=0}^{n-1} \frac{1}{\binom{n-1}{t}} \sum_{\mathcal{A} \subseteq \mathcal{N} \setminus \{i\}, |\mathcal{A}| = t} [\mu(\mathcal{A} \cup \{i\}) - \mu(\mathcal{A})].$$

Both formulas, in turn, can be given in terms of the discrete derivatives $\Delta_i \mu(\mathcal{A})$, for example,

$$\varphi(i) = \frac{1}{n} \sum_{t=0}^{n-1} \frac{1}{\binom{n-1}{t}} \sum_{\mathcal{A} \subseteq \mathcal{N} \setminus \{i\}, |\mathcal{A}| = t} \Delta_i \mu(\mathcal{A}).$$

The following representation is also equivalent [Bel21], offering computational advantages over (3.3) due to the summation being unconditional [2]:

$$\varphi(i) = \sum_{\mathcal{A} \subseteq \mathcal{N}} \tilde{w}_{\mathcal{A}} \Delta_i \mu(\mathcal{A}), \tag{3.4}$$

where the weighting vectors $\tilde{\mathbf{w}}$ are

$$\tilde{w}_{\mathcal{A}} = \frac{1}{\binom{n+1}{1}\binom{n}{|\mathcal{A}|}}. \tag{3.5}$$

[2] Here the list of $\Delta_i \mu(\mathcal{A})$ includes both cases $i \in \mathcal{A}$ and $i \notin \mathcal{A}$.

An important property of the Shapley value is that it is normalised, i.e.

$$\sum_{i=1}^{n} \varphi(i) = \mu(\mathcal{N}) = 1. \tag{3.6}$$

$\varphi(i)$ can hence be interpreted analogously to the weights w_i of a weighted arithmetic mean, and in fact, in special cases (such as two-additive capacities) the vector $\varphi(\mu)$ coincides with the weighting vector $\mathbf{w}$ that would be obtained as a least squares approximation of the Choquet integral [Mar99].

From the notion of value of a single criterion in the presence of others, one can extend to the concept of interaction between multiple elements. At the heart of this is the progression from the marginal contributions of a single element i to other subsets to the marginal interaction of the coalition $\mathcal{A}$. An expression for the interaction between two criteria,

$$\mu(\mathcal{A} \cup \{i, j\}) - \mu(\mathcal{A} \cup \{i\}) - \mu(\mathcal{A} \cup \{j\}) + \mu(\mathcal{A}),$$

was first proposed by Owen [Owe72] and independently discovered 20 years later by Murofushi and Soneda [FKM06, MS93]. It is equivalent to the second derivative $\Delta_{\{ij\}}\mu(\mathcal{A})$ for $i, j \notin \mathcal{A}$. Grabisch [Gra97] generalised this notion of interaction between two criteria to larger coalitions by recognising the connection to the expression of value and marginal contributions, i.e.

$$\Delta_{ij}\mu(\mathcal{A}) = \Delta_j\mu(\mathcal{A} \cup \{i\}) - \Delta_j\mu(\mathcal{A}) = \Delta_i\mu(\mathcal{A} \cup \{j\}) - \Delta_i\mu(\mathcal{A}).$$

By applying the recursion formula in Definition 3.2 [FKM06, Gra97, GKM08], a consistent expression for the marginal simultaneous interaction of $\mathcal{A}$ in the presence of $\mathcal{B} \subseteq \mathcal{N}\backslash\mathcal{A}$ can be obtained as a special case of Eq. (3.1). This leads to the following general definition for simultaneous interaction in coalitions.

Definition 3.5 (Simultaneous Interaction Index) Let μ be a capacity on $\mathcal{N}$. The probabilistic simultaneous interaction index of a subset $\mathcal{A} \subseteq \mathcal{N}$ w.r.t. μ is defined as

$$I_{\mathrm{p}}^{\mu}(\mathcal{A}) = \sum_{\mathcal{B} \subseteq \mathcal{N}\backslash\mathcal{A}} p_{\mathcal{B}}^{\mathcal{A}}(\mathcal{N})\Delta_{\mathcal{A}}\mu(\mathcal{B}) = \sum_{\mathcal{B} \subseteq \mathcal{N}} q_{\mathcal{B}}(\mathcal{N})\Delta_{\mathcal{A}}\mu(\mathcal{B}),$$

where, for any $\mathcal{A} \subseteq \mathcal{N}$, the family of coefficients $\{p_{\mathcal{B}}^{\mathcal{A}}(\mathcal{N})\}_{\mathcal{B} \subseteq \mathcal{N}\backslash\mathcal{A}}$ forms a probability distribution on $2^{\mathcal{N}\backslash\mathcal{A}}$, $\{q_{\mathcal{B}}(\mathcal{N})\}_{\mathcal{B} \subseteq \mathcal{N}}$ forms a probability distribution on $2^{\mathcal{N}}$, and

$$\Delta_{\mathcal{A}}\mu(\mathcal{B}) = \sum_{\mathcal{C} \subseteq \mathcal{A}} (-1)^{|\mathcal{A}\backslash\mathcal{C}|}\mu(\mathcal{C} \cup \mathcal{B})$$

is the marginal simultaneous interaction of criteria subset $\mathcal{A}$ in the presence of $\mathcal{B} \subseteq \mathcal{N} \setminus \mathcal{A}$.

The best known interaction index is the Shapley interaction index. It is defined for pairs as follows.

Definition 3.6 (Shapley Interaction Index) Let μ be a capacity. The interaction index for every pair $i, j \in \mathcal{N}$ is

$$I_{ij} = \sum_{\mathcal{A} \subseteq \mathcal{N} \setminus \{i,j\}} \frac{(n - |\mathcal{A}| - 2)!|\mathcal{A}|!}{(n-1)!} \times$$
$$[\mu(\mathcal{A} \cup \{i, j\}) - \mu(\mathcal{A} \cup \{i\}) - \mu(\mathcal{A} \cup \{j\}) + \mu(\mathcal{A})].$$

From Definition 3.5, we can write the marginal interaction of two criteria i and j in the presence of $\mathcal{A} \subseteq \mathcal{N} \setminus \{i, j\}$ as

$$\Delta_{ij}\mu(\mathcal{A}) = \mu(\mathcal{A} \cup \{i, j\}) - \mu(\mathcal{A} \cup \{i\}) - \mu(\mathcal{A} \cup \{j\}) + \mu(\mathcal{A}) \tag{3.7}$$

This allows us to express I_{ij} in terms of the second derivatives as

$$I_{ij} = \sum_{\mathcal{A} \subseteq \mathcal{N} \setminus \{i,j\}} \frac{(n - |\mathcal{A}| - 2)!|\mathcal{A}|!}{(n-1)!} \Delta_{ij}\mu(\mathcal{A}) = \sum_{\mathcal{A} \subseteq \mathcal{N}} \tilde{w}_{\mathcal{A}} \Delta_{ij}\mu(\mathcal{A})$$

where the $\tilde{w}_{\mathcal{A}}$ coefficients are calculated using (3.5) as previously.

We will have $I_{ij} \in [-1, 1]$ for any pair $\{i, j\}$. Interaction indices satisfying $I_{ij} < 0$ indicate that the variables i and j are redundant (negative synergy), while $I_{ij} > 0$ means that they contribute to the capacity values in a complementary way (positive synergy). The following proposition relates these characterisations to the sub- and supermodularity properties.

Proposition 3.1 ([FKM06]) *For a submodular capacity μ, all interaction indices verify $I_{ij} \leqslant 0$. For a supermodular capacity, all interaction indices verify $I_{ij} \geqslant 0$.*

It should be emphasised that the reverse does not necessarily hold, i.e. $I_{ij} < 0$ for all i, j does not necessarily mean that a given capacity is submodular and similarly for $I_{i,j} > 0$ and supermodularity.

The Shapley interaction index as given in Definition 3.6 and defined by Murofushi and Soneda was extended by Grabisch to all coalitions $\mathcal{A}$ in [Gra97].

Definition 3.7 (Shapley Interaction Index for Coalitions) Let μ be a capacity. The Shapley interaction index for every set $\mathcal{A} \subseteq \mathcal{N}$ is

$$I(\mathcal{A}) = \sum_{\mathcal{B} \subseteq \mathcal{N} \setminus \mathcal{A}} \frac{(n - |\mathcal{B}| - |\mathcal{A}|)!|\mathcal{B}|!}{(n - |\mathcal{A}| + 1)!} \sum_{\mathcal{C} \subseteq \mathcal{A}} (-1)^{|\mathcal{A} \setminus \mathcal{C}|} \mu(\mathcal{B} \cup \mathcal{C}). \tag{3.8}$$

The definition incorporates both the Shapley value and the Shapley interaction index for pairs as special cases, with $I(\{i\}) = \varphi(i)$ and $I(\{i, j\}) = I_{ij}$.

It also satisfies a number of useful axioms. In particular, it is the only index satisfying the linearity, symmetry, dummy partnership, recursivity, and efficiency axioms [FKM06, GR99]; see Sect. 3.5.3.

Dummy partnership refers to the behaviour of dummy criteria (see Definition 2.5) where for a dummy criterion i, there is no interaction between i when combined with other subsets. In such cases, $\mu(\mathcal{A} \cup \{i\}) = \mu(\mathcal{A}) + \mu(\{i\})$ and $I(\mathcal{A} \cup \{i\}) = 0$ for all $\mathcal{A} \neq \emptyset$.

Note 3.1 The coefficients in Definition 3.7 can be written as

$$w_{\mathcal{A},\mathcal{B}} = \left(\binom{n - |\mathcal{A}|}{|\mathcal{B}|}(n - |\mathcal{A}| + 1)\right)^{-1},$$

and since $\sum_{b=0}^{n-a} \binom{n-a}{b}^{-1} = n - a + 1$, we get $\sum_{|\mathcal{A}|=0}^{n-|\mathcal{A}|} w_{\mathcal{A},\mathcal{B}} = 1$, which is consistent with Definition 3.5 and $w_{\mathcal{A},\mathcal{B}} = p_{\mathcal{B}}^{\mathcal{A}}(\mathcal{N})$.

The following single-sum expression for $I(\mathcal{A})$ was obtained in [GMR00]

$$I(\mathcal{A}) = \sum_{\mathcal{B} \subseteq \mathcal{N}} \frac{(-1)^{|\mathcal{A} \setminus \mathcal{B}|}}{(n - |\mathcal{A}| + 1)\binom{n-|\mathcal{A}|}{|\mathcal{B} \setminus \mathcal{A}|}} \mu(\mathcal{B}). \tag{3.9}$$

As with the Shapley index and interaction index for pairs, an alternative expression for $I(\mathcal{A})$ using unconditional summation (with $\tilde{w}_{\mathcal{B}}$ in (3.5)) is [Bel21]

$$I(\mathcal{A}) = \sum_{\mathcal{B} \subseteq \mathcal{N}} \tilde{w}_{\mathcal{B}} \Delta_{\mathcal{A}} \mu(\mathcal{B}), \tag{3.10}$$

which is consistent with the expression on the right-hand side of Definition 3.5 with $q_{\mathcal{B}}(\mathcal{N}) = \tilde{w}_{\mathcal{B}}$. The computational advantage is that the same weighting vector $\tilde{\mathbf{w}}$ can be applied irrespective of the cardinality of $\mathcal{A}$ and without the need for conditional summation.

By expressing the capacity values in Möbius representation, the Shapley value and interaction index for coalitions simplify to the following [Gra00, Gra97, Mar02a, GMR00]:

$$\varphi(i) = \sum_{\mathcal{B} | i \in \mathcal{B}} \frac{1}{|\mathcal{B}|} M(\mathcal{B}), \tag{3.11}$$

$$I(\mathcal{A}) = \sum_{\mathcal{B} | \mathcal{A} \subseteq \mathcal{B}} \frac{1}{|\mathcal{B}| - |\mathcal{A}| + 1} M(\mathcal{B}). \tag{3.12}$$

Example 3.1 A capacity μ is represented in its standard and interaction form below. For each of the values $\mu(\mathcal{A})$ shown on the left, the corresponding Shapley interaction index for coalitions $I(\mathcal{A})$ is shown on the right:

$$\begin{array}{ccc} & 1 & \\ 0.5 & 0.6 & 0.7 \\ 0.1 & 0.2 & 0.3 \\ & 0 & \end{array} \qquad\qquad \begin{array}{ccc} & -0.2 & \\ 0.1 & 0.1 & 0.1 \\ 7/30 & 1/3 & 13/30 \\ & 0 & \end{array}$$

We can observe that the singleton values $I(\{i\}) = \varphi_i$ sum to 1 as expected.

The interaction index can be inverted with the following:

$$\mu(\mathcal{A}) = \sum_{\mathcal{B}\subseteq\mathcal{N}} \beta^{|\mathcal{B}|}_{|\mathcal{A}\cap\mathcal{B}|} I(\mathcal{B}), \ \forall \mathcal{A} \subseteq \mathcal{N}$$

where $\beta_k = -\sum_{l=0}^{k-1} \binom{k}{l} \frac{\beta_l}{k-l+1}$ and $\beta^l_k = \sum_{j=0}^{k} \binom{k}{j} \beta_{l-j}$. In Möbius representation, this is [Gra97]:

$$M(\mathcal{A}) = \sum_{k=0}^{n-|\mathcal{A}|} \beta_k \sum_{\substack{\mathcal{B}\subseteq\mathcal{N}\setminus\mathcal{A} \\ |\mathcal{B}|=k}} I(\mathcal{A}\cup\mathcal{B}),$$

which can be expressed as a single-sum expression using [Gra16],

$$M(\mathcal{A}) = \sum_{\mathcal{B}|\mathcal{A}\subseteq\mathcal{B}} \beta_{|\mathcal{B}|-|\mathcal{A}|} I(\mathcal{B}).$$

3.4 Banzhaf Value and Interaction

The Banzhaf value or Banzhaf index is another well-established calculation for μ that is used to give a measure of overall importance for each of the elements in $\mathcal{N}$. It can be seen as an alternative to the Shapley value and also constitutes a special case of Definition 3.3, this time with the marginal contributions weighted equally [Ban65].

Definition 3.8 (Banzhaf Index) Let μ be a capacity. The Banzhaf index b_i for every $i \in \mathcal{N}$ is

$$b_i = \frac{1}{2^{n-1}} \sum_{\mathcal{A}\subseteq\mathcal{N}\setminus\{i\}} [\mu(\mathcal{A}\cup\{i\}) - \mu(\mathcal{A})].$$

This definition has been extended to all coalitions by Roubens in [Rou96].

Definition 3.9 (Banzhaf Interaction Index For Coalitions) Let μ be a capacity. The Banzhaf interaction index between the elements of $\mathcal{A} \subseteq \mathcal{N}$ is given by

$$J(\mathcal{A}) = \frac{1}{2^{n-|\mathcal{A}|}} \sum_{\mathcal{B} \subseteq \mathcal{N} \setminus \mathcal{A}} \sum_{\mathcal{C} \subseteq \mathcal{A}} (-1)^{|\mathcal{A} \setminus \mathcal{C}|} \mu(\mathcal{B} \cup \mathcal{C}).$$

A single-sum expression for $J(\mathcal{A})$ was given in [GMR00]:

$$J(\mathcal{A}) = \frac{1}{2^{n-|\mathcal{A}|}} \sum_{\mathcal{B} \subseteq \mathcal{N}} (-1)^{|\mathcal{A} \setminus \mathcal{B}|} \mu(\mathcal{B}).$$

The Möbius representation also allows for a simple expression here, with

$$J(\mathcal{A}) = \sum_{\mathcal{B} | \mathcal{A} \subseteq \mathcal{B}} \frac{1}{2^{|\mathcal{B}|-|\mathcal{A}|}} M(\mathcal{B}).$$

Example 3.2 (Cont'd from Example 3.1) For the same capacity, the Banzhaf indices are $b_1 = 0.25, b_2 = 0.35, b_3 = 0.4$, with $J(\mathcal{A}) = I(\mathcal{A})$ for larger subsets.

The inversion formula for the Banzhaf index is given by [Rou96, Gra16]

$$\mu(\mathcal{A}) = \sum_{\mathcal{B} \subseteq \mathcal{N}} \frac{(-1)^{|\mathcal{B} \setminus \mathcal{A}|}}{2^{|\mathcal{B}|}} J(\mathcal{B})$$

or in Möbius representation as

$$M(\mathcal{A}) = \sum_{\mathcal{B} | \mathcal{A} \subseteq \mathcal{B}} \left(-\frac{1}{2}\right)^{|\mathcal{B} \setminus \mathcal{A}|} J(\mathcal{B}).$$

The value and interaction indices of this chapter in various forms, as well as their inverses, are summarised in Table 3.1.

3.5 Nonmodularity and Nonadditivity Indices

While the value and interaction indices presented in previous sections to some extent characterise the nonadditivity of capacities, their focus is on instantaneous interaction, averaging the marginal contributions to give an idea of how a subset's elements interact and add value in the presence of others [Wu+16a, Wu+16b]. The concepts of nonadditivity and nonmodularity have been proposed to quantify the overall extent to which elements *within* a coalition interact [WB18a, WB19].

3.5.1 Nonadditivity Index

The nonadditivity index of a subset is defined as follows.

Definition 3.10 (Nonadditivity Index) Let μ be a capacity on $\mathcal{N}$. The nonadditivity index of subset $\mathcal{A} \subseteq \mathcal{N}$, $|\mathcal{A}| \geqslant 2$, with respect to μ is defined as

$$n_\mu(\mathcal{A}) = \frac{1}{2^{|\mathcal{A}|-1} - 1} \sum_{\substack{(\mathcal{B},\mathcal{A}\backslash\mathcal{B}) \\ \emptyset \neq \mathcal{B} \subset \mathcal{A}}} [\mu(\mathcal{A}) - \mu(\mathcal{B}) - \mu(\mathcal{A}\backslash\mathcal{B})].$$

Note 3.2 The notation $(\mathcal{B}, \mathcal{A}\backslash\mathcal{B})$ indicates that we sum over all distinct bipartitions, $\mathcal{B}, \mathcal{A}\backslash\mathcal{B}$, e.g. for $\mathcal{A} = \{1, 2, 3\}$, we sum over the $2^{3-1} - 1 = 3$ partitions $(\{1\}, \{2, 3\})$, $(\{2\}, \{1, 3\})$, and $(\{3\}, \{1, 2\})$. Equivalently, we could sum over all subsets $\mathcal{B} \subset \mathcal{A}$ and multiply the summation by $\frac{1}{2^{|\mathcal{A}|}-2}$.

The sign of the nonadditivity index reflects whether the interaction within a subset is superadditive (positive) or subadditive (negative). This is a deviation from the interaction index, where the indices may have opposite sign. Having the sign consistent with our intuition when it comes to super- and subadditivity is especially convenient when it comes to interpretation and analysis of the capacity values.

The following theorem establishes this more formally.

Theorem 3.1 *Let μ be a a capacity. If μ is super(sub)additive, then $n_\mu(\mathcal{A}) \geqslant (\leqslant) 0$, $\forall \mathcal{A} \subseteq \mathcal{N}$ and $|\mathcal{A}| \geqslant 2$.*

Strict super- and subadditivity result in the inequalities holding strictly, while additivity implies $n_\mu(\mathcal{A}) = 0$. As with the relationship between super- and submodularity and the interaction indices, it should be emphasised that $n_\mu(\mathcal{A}) \geqslant 0$ ($n_\mu(\mathcal{A}) \leqslant 0$) does not necessarily ensure that the capacity will be super- or subadditive.

We can also observe that the index is normalised differently to the interaction indices, with it always being the case for pairs and larger subsets that $n_\mu(\mathcal{A}) \in [-1, 1]$. The following theorem establishes this.

Theorem 3.2 *Let μ be a capacity on $\mathcal{N}$. Then $-1 \leqslant n_\mu(\mathcal{A}) \leqslant 1$, $\forall \mathcal{A} \subseteq \mathcal{N}$, $|\mathcal{A}| \geqslant 2$.*

The following expression, which is equivalent to that given in Definition 3.10 for $|\mathcal{A}| \geqslant 2$, allows the definition to be generalised to incorporate singletons and the empty set:

$$n_\mu(\mathcal{A}) = \mu(\mathcal{A}) - \frac{1}{2^{|\mathcal{A}|-1} - 1} \sum_{\mathcal{B} \subset \mathcal{A}} \mu(\mathcal{B}), \forall \mathcal{A} \subseteq \mathcal{N}. \tag{3.13}$$

We then have $n_\mu(\{i\}) = \mu(\{i\})$ and $n_\mu(\emptyset) = 0$.

The following transformation allows us to represent a capacity in terms of its nonadditivity indices:

$$\mu(\mathcal{A}) = n_\mu(\mathcal{A}) + \sum_{\mathcal{B}\subset\mathcal{A}} \sum_{i=|\mathcal{B}|}^{|\mathcal{A}|-1} \prod_{j=i}^{|\mathcal{A}|-1} \frac{|\mathcal{A}| - j}{2^j - 1} n_\mu(\mathcal{B}), \forall \mathcal{A} \subseteq \mathcal{N}. \tag{3.14}$$

In Möbius representation, we obtain the following expression [WB18a, WB19]:

$$n_\mu(\mathcal{A}) = \sum_{\mathcal{B}\subseteq\mathcal{A}} \frac{2^{|\mathcal{A}|-1} - 2^{|\mathcal{A}|-|\mathcal{B}|}}{2^{|\mathcal{A}|-1} - 1} M(\mathcal{B}), \forall \mathcal{A} \subseteq \mathcal{N}. \tag{3.15}$$

with the convention $\frac{0}{0} = 1$ when $\mathcal{A}$ is a singleton and the inversion formula

$$M(\mathcal{A}) = n_\mu(\mathcal{A}) + \sum_{\mathcal{B}\subset\mathcal{A}} (-1)^{|\mathcal{A}|-|\mathcal{B}|} \frac{2^{|\mathcal{A}|-1} - 2^{|\mathcal{A}|-|\mathcal{B}|}}{2^{|\mathcal{A}|-1} - 1} n_\mu(\mathcal{B}). \tag{3.16}$$

For dual capacities μ and μ^* on $\mathcal{N}$, the nonadditivity index satisfies $n_\mu(\mathcal{N}) = -n_{\mu^*}(\mathcal{N})$. For self-dual (i.e. additive) capacities, it follows that $n_\mu(\mathcal{N}) = 0$; however, as with the implications for sub- and superadditivity, $n_\mu(\mathcal{N}) = 0$ does not necessarily ensure that the capacity is additive.

For a dummy criterion $d \in \mathcal{N}$, we note that no interaction will occur in pairs $\{i, d\}$ for any $i \in \mathcal{N} \setminus d$ and hence it will hold that $n_\mu(\{i, d\}) = 0$; however, in larger subsets, this may not necessarily be the case since the nonadditivity index will capture interactions between the other variables.

Example 3.3 Given below are the values of a capacity with $n = 3$, expressed in standard form (left) and nonadditivity index form (right):

	1				0.2667	
0.8	0.3	0.5		0.5	−0.1	0
0.1	0.2	0.3		0.1	0.2	0.3
	0				0	

For $n = 3$, the nonadditivity index only differs to the Möbius representation for the value of the entire group, $n_\mu(\{1, 2, 3\})$. We hence also provide an example for $n = 4$.

Example 3.4 Given below are the values of a particular capacity with $n = 4$, expressed in standard form on the left and nonadditivity index form on the right:

1	−0.4071
0.9 0.9 1 1	−0.2333 − 0.2833 0.05 − 0.3
0.8 0.5 0.45 0.8 0.9 1	−0.3 − 0.1 − 0.25 − 0.1 − 0.1 0.5
0.4 0.7 0.2 0.3	0.4 0.7 0.2 0.3
0	0

Recall the extremely superadditive and extremely subadditive capacities presented in Definitions 2.11 and 2.12 (which also correspond with 0-1 necessity and possibility measures). The following theorem shows that the upper and lower boundaries of the nonadditivity index $n_\mu(\mathcal{A})$ coincide with these capacities focused on subset $\mathcal{A}$.

Theorem 3.3 ([WB18a]) *For a capacity μ and $|\mathcal{A}| \geqslant 2$, $n_\mu(\mathcal{A}) = 1 \;\Leftrightarrow\; \mu \geqslant \varepsilon^+_{\mathcal{A}}$, and $n_\mu(\mathcal{A}) = -1 \;\Leftrightarrow\; \mu \geqslant \varepsilon^-_{\mathcal{A}}$.*

Note 3.3 Corresponding with the local minimum and local maximum, respectively, the extremely superadditive and subadditive capacities represent the extreme states of interaction within the coalition $\mathcal{A}$, with $\varepsilon^+_{\mathcal{A}}$ pointing to fully complementary criteria and $\varepsilon^-_{\mathcal{A}}$ suggesting that they are redundant or substitutive. The cases focusing on $\mathcal{N}$, $\varepsilon^+_{\mathcal{N}}$, and $\varepsilon^-_{\mathcal{N}}$ represent the smallest and greatest capacities on $\mathcal{N}$, respectively.

3.5.2 *Nonmodularity Index*

Whereas nonadditivity can be related to sub- and superadditivity, nonmodularity reflects the sub- and supermodular behaviour of capacities. Nonmodularity is a stronger or more general concept, involving arbitrary subsets, while nonadditivity sums over all bipartitions and hence focuses on interaction only between disjoint subsets of criteria.

Supermodularity and submodularity can be related to the concepts of concavity and convexity, and so it is informative to express these conditions in terms of a capacity's second derivative. Proposition 3.2 follows from Corollary 2.23 in [Gra16].

Proposition 3.2 *A capacity is supermodular on $\mathcal{N}$ if and only if for any $\mathcal{A} \subseteq \mathcal{N}, i, j \notin \mathcal{A}, i \neq j$:*

$$\Delta_{ij}\mu(\mathcal{A}) = \mu(\mathcal{A} \cup \{i, j\}) - \mu(\mathcal{A} \cup \{i\}) - \mu(\mathcal{A} \cup \{j\}) + \mu(\mathcal{A}) \geqslant 0.$$

The nonmodularity index averages these second derivatives across the proper subsets of a given subset $\mathcal{A}$.

Definition 3.11 (Nonmodularity Index) Let μ be a capacity on $\mathcal{N}$. For any $\mathcal{A} \subseteq \mathcal{N}$, $|\mathcal{A}| \geqslant 2$, the nonmodularity index is defined as

$$d_\mu(\mathcal{A}) = \sum_{\substack{\mathcal{B}\subset\mathcal{A}\\ i,j\in\mathcal{A}\\ i,j\notin\mathcal{B}}} w(|\mathcal{A}|,|\mathcal{B}|)(\mu(\mathcal{B}\cup\{i,j\}) - \mu(\mathcal{B}\cup\{i\}) - \mu(\mathcal{B}\cup\{j\}) + \mu(\mathcal{B}))$$

$$= \sum_{\substack{\mathcal{B}\cup\{i,j\}\subseteq\mathcal{A}\\ i,j\notin\mathcal{B}}} w(|\mathcal{A}|,|\mathcal{B}|)\Delta_{ij}\mu(\mathcal{B}), \qquad (3.17)$$

where $w(|\mathcal{A}|,|\mathcal{B}|)$ is a weight computed as $w(|\mathcal{A}|,|\mathcal{B}|) = \left(\binom{|\mathcal{A}|}{2}\binom{|\mathcal{A}|-2}{|\mathcal{B}|}\right)^{-1}$.

The nonmodularity index takes values in the range $[-1, 1]$.

Note 3.4 There are exactly $1/w(|\mathcal{A}|, k)$ terms in the sum that correspond to subsets $\mathcal{B}$ of cardinality $|\mathcal{B}| = k$. Hence, the value $d_\mu(\mathcal{A})$ is the sum of probabilistic expectations of $\Delta_{ij}\mu(\mathcal{B})$ across all subsets $\mathcal{B}$ of $\mathcal{A}$ of cardinalities $k = 0, 1, \ldots, |\mathcal{A}| - 2$ that satisfy the condition in the summation expression (3.17). Also note that $w(|\mathcal{A}|, |\mathcal{A}| - 2 - |\mathcal{B}|) = w(|\mathcal{A}|, |\mathcal{B}|)$.

Note 3.5 The sum (3.17) involves the terms in which $|\mathcal{B}| \leqslant |\mathcal{A}| - 2$, i.e. only proper subsets of $\mathcal{A}$. This is essential, as, for example, supermodular capacities are required to have the second derivative non-negative at all $\mathcal{B} \subset \mathcal{A}$.

The following theorem formalises the relationship between the nonmodularity index and sub- and supermodular capacities.

Theorem 3.4 *Let μ be a capacity. If μ is super(sub)modular, then $d_\mu(\mathcal{A}) \geqslant (\leqslant) 0$, $\forall \mathcal{A} \subseteq \mathcal{N}$, and $|\mathcal{A}| \geqslant 2$.*

It is worth observing again that the converse does not hold, since it is possible for all nonmodularity indices to be positive and yet the capacity still not be supermodular. Capacities that are strictly super- or submodular result in $d_\mu(\mathcal{A}) > 0$ and $d_\mu(\mathcal{A}) < 0$, respectively.

As with the nonadditivity index, for dual capacities μ and μ^*, we will have $d_\mu(\mathcal{N}) = -d_{\mu^*}(\mathcal{N})$, implying that for the self-dual (additive or modular) capacity, we have $d_\mu(\mathcal{N}) = 0$, noting that here too $d_\mu(\mathcal{N}) = 0$ does not necessarily imply that the capacity is modular.

The following alternative expressions can be informative when it comes to the behaviour of the nonmodularity index and useful for implementation. We once again expand the definition to the case of singletons by augmenting the definition and setting $d_\mu(\mathcal{A}) = \mu(\mathcal{A})$ when $|\mathcal{A}| \leqslant 1$:

$$d_\mu(\mathcal{A}) = \mu(\mathcal{A}) + \mu(\emptyset) - \frac{1}{|\mathcal{A}|}\sum_{i\in\mathcal{A}}[\mu(\{i\}) + \mu(|\mathcal{A}|\setminus\{i\})]. \qquad (3.18)$$

It is worthy to observe that (3.18) does not depend on capacity values for any of the subsets with cardinality $2 \leqslant k \leqslant |\mathcal{A}| - 1$. While this might be counterintuitive, it can be seen as analogous to the Fundamental Theorem of Calculus, since the nonmodularity index is based on the second derivatives of μ and is equal to the first derivatives at the boundaries of its domain, i.e.

$$d_\mu(\mathcal{A}) = \frac{1}{|\mathcal{A}|}\left(\sum_{i\in\mathcal{A}} \Delta_i\mu(\mathcal{A}) - \sum_{i\in\mathcal{A}} \Delta_i\mu(\emptyset)\right). \tag{3.19}$$

Note 3.6 We can also express (3.17) without conditional summation as we have for other indices:

$$d_\mu(\mathcal{A}) = \frac{2}{|\mathcal{A}|}\sum_{\substack{\mathcal{B}\subseteq\mathcal{A}\\ i,j\in\mathcal{A}}} \tilde{w}_B \Delta_{ij}\mu(\mathcal{B}). \tag{3.20}$$

It will also be the case that

$$d_\mu(\mathcal{A}) = n_\mu(\mathcal{A}),\ \forall \mathcal{A}\subseteq\mathcal{N},\ |\mathcal{A}|\leqslant 3.$$

The following theorem relates the nonmodularity index to the extremely superadditive and extremely subadditive capacities, as was the case for the nonadditivity index.

Theorem 3.5 $d_\mu(\mathcal{A}) = 1 \;\Leftrightarrow\; \mu \geqslant \varepsilon^+_{\mathcal{A}}$, *and* $d_\mu(\mathcal{A}) = -1 \;\Leftrightarrow\; \mu \geqslant \varepsilon^-_{\mathcal{A}}$.

The nonmodularity index can also serve as an alternative representation for capacities. The following calculation allows $\mu(\mathcal{A})$ to be expressed in terms of the $d_\mu(\mathcal{A})$ values.

$$\mu(\mathcal{A}) = d_\mu(\mathcal{A}) + \sum_{\substack{\mathcal{B}\subset\mathcal{A}\\ |\mathcal{B}|\geqslant 2}} \binom{|\mathcal{A}|}{|\mathcal{B}|}^{-1} d_\mu(\mathcal{B}) + \sum_{\{i\}\subset\mathcal{A}} d_\mu(\{i\}), \forall\mathcal{A}\subseteq\mathcal{N}. \tag{3.21}$$

From the Möbius representation of a capacity, for any $\mathcal{A}\subseteq\mathcal{N}$, $|\mathcal{A}|\geqslant 2$, we have

$$d_\mu(\mathcal{A}) = \sum_{\substack{\mathcal{B}\subseteq\mathcal{A}\\ |\mathcal{B}|\geqslant 2}} \frac{|\mathcal{B}|}{|\mathcal{A}|} M(\mathcal{B}). \tag{3.22}$$

Then for converting from nonmodularity to Möbius representation, we also have, for any $\mathcal{A}\subseteq\mathcal{N}$, $|\mathcal{A}|\geqslant 2$,

$$M(\mathcal{A}) = \sum_{\substack{\mathcal{B}\subseteq\mathcal{A}\\ |\mathcal{B}|\geqslant 2}} \frac{(-1)^{|\mathcal{A}|-|\mathcal{B}|}|\mathcal{B}|}{|\mathcal{A}|} d_\mu(\mathcal{B}). \tag{3.23}$$

Formulas for the nonadditivity and nonmodularity indices along with their inverses are also included in Table 3.1.

3.5.3 Axiomatic Characterisation

Recall from Sect. 3.3 that the Shapley interaction index for coalitions is the only index satisfying three of the basic axioms: linearity, efficiency, and dummy partnership. In this section, we expand the definition of dummy players to dummy coalitions to characterise the Shapley interaction index and show that the non-modularity index is also an interaction index that is the only one to satisfy some combinations of basic axioms. We follow the works [FKM06, FM07, GR99, BW20].

Definition 3.12 (Dummy Criterion) The criterion r is called dummy criterion if $\mu(\mathcal{A} \cup \{r\}) = \mu(\mathcal{A}) + \mu(\{r\})$ for all $\mathcal{A} \subset \mathcal{N}$, $r \notin \mathcal{A}$.

Definition 3.13 (Independent Criteria) The criteria in the set $\mathcal{A} \subseteq \mathcal{N}$ are independent if each criterion i is dummy in $\mathcal{A}$: $\mu(\mathcal{B} \cup \{i\}) = \mu(\mathcal{B}) + \mu(\{i\})$ for all $\mathcal{B} \subset \mathcal{A}$, $i \in \mathcal{A}$ and $i \notin \mathcal{B}$.

The independent criteria in $\mathcal{A}$ need not be dummy with respect to $\mathcal{N}$. For the dummy player r, we have the derivatives $\Delta_i \mu(\mathcal{A} \cup \{r\}) = \Delta_i \mu(\mathcal{A})$ and $\Delta_r \mu(\mathcal{A}) = \mu(\{r\})$ (irrespective of whether or not $r \in A$).

Definition 3.14 (Partnership [FKM06, FM07]) A non-empty coalition $\mathcal{P}$ is called a partnership if $\mu(\mathcal{A} \cup \mathcal{S}) = \mu(\mathcal{A})$ for all $\mathcal{S} \subsetneq \mathcal{P}$ and all $\mathcal{A} \subset \mathcal{N} \setminus \mathcal{P}$.

A partnership behaves as a single hypothetical player in a reduced game.

Definition 3.15 (Restricted Game) Consider a capacity (or a game, generally) μ on $\mathcal{N}$. The restricted game $\mu^{\mathcal{A}}$ is a game on $\mathcal{A}$, such that $\mu^{\mathcal{A}}(\mathcal{S}) = \mu(\mathcal{S})$ for all $\mathcal{S} \subseteq \mathcal{A} \subset \mathcal{N}$.

The restricted game in the presence of $\mathcal{B} \subseteq \mathcal{N} \setminus \mathcal{A}$, $\mu^{\mathcal{A}}_{\cup\mathcal{B}}$ is the game on $\mathcal{A}$ such that $\mu^{\mathcal{A}}_{\cup\mathcal{B}}(\mathcal{S}) = \mu(\mathcal{S} \cup \mathcal{B}) - \mu(\mathcal{B})$ for all $\mathcal{S} \subseteq \mathcal{A} \subset \mathcal{N}$.

Definition 3.16 (Dummy Coalition) The coalition $\mathcal{R}$ is called a dummy coalition if $\mu(\mathcal{A} \cup \mathcal{R}) = \mu(\mathcal{A}) + \mu(\mathcal{R})$ for all $\mathcal{A} \subseteq \mathcal{N} \setminus \mathcal{R}$. A *dummy partnership* is simply a partnership, which is a dummy coalition.

Let $\mathcal{G}^{\mathcal{N}}$ denote the set of games on $\mathcal{N}$. We call the mapping $I : \mathcal{G}^{\mathcal{N}} \to \mathbb{R}^{2^n}$ an index. Let us specify the following axioms (we drop the subset notation for singletons):

(L) An index is a *linear* mapping $\mathcal{G}^{\mathcal{N}} \to \mathbb{R}^{2^n}$, i.e. can be written in linear form $I(\mathcal{A}) = \sum_{\mathcal{B}} c_{\mathcal{B}} \mu(\mathcal{B})$.

(S) An index is a *symmetric mapping*, i.e. $I(\mathcal{A}) = I_\pi(\pi(\mathcal{A}))$, where π is any permutation on $\mathcal{N}$ and I_π is the index w.r.t. the permuted game.

(E) An index is *efficient*, i.e. $\sum_i I(\{i\}) = \mu(\mathcal{N}) - \mu(\emptyset)$ $(= 1)$.
(D) An index I satisfies *dummy partnership*, i.e. $I(\mathcal{P}) = \mu(\mathcal{P})$ and for all $\mathcal{S} \subseteq \mathcal{N} \setminus \mathcal{P}, \mathcal{S} \neq \emptyset$: $I(\mathcal{S} \cup \mathcal{P}) = 0$, for any dummy partnership $\mathcal{P}$.
(R) An index is *recursive*, i.e. $I(\mathcal{S}) = I_{\mu^{\mathcal{N}\setminus j}_{\cup j}}(\mathcal{S} \setminus j) - I_{\mu^{\mathcal{N}\setminus j}}(\mathcal{S} \setminus j)$, where $\mu^{\mathcal{A}}_{\mathcal{B}}$ is the restricted game in the presence of $\mathcal{B}$.

Theorem 3.6 *[FKM06] The Shapley interaction index is the only index that satisfies the following axioms: linearity (L), efficiency (E), symmetry (S), recursivity (R), and dummy partnership (D).*

For the nonmodularity index, we have a different set of axioms [BW20]. We start with a few theorems.

Theorem 3.7 *For any $\mathcal{A} \subset \mathcal{N}$, $|\mathcal{A}| \geqslant 1$, and dummy $r \notin \mathcal{A}$:*

$$(|\mathcal{A}| + 1) d_\mu(\mathcal{A} \cup \{r\}) = |\mathcal{A}| d_\mu(\mathcal{A}). \tag{3.24}$$

If we think of the value $\tilde{d}_\mu(\mathcal{A}) = |\mathcal{A}| d_\mu(\mathcal{A})$ as the *scaled nonmodularity index*, then Theorem 3.7 implies that this quantity is invariant when dummy players are added to $\mathcal{A}$. So the interaction in $\mathcal{A}$ remains the same when dummy players are added. Let us also set $\tilde{d}_\mu(\{i\}) = 0$.

Corollary 3.1 *If all criteria in $\mathcal{A}$, $|\mathcal{A}| \geqslant 2$, are independent, then $d_\mu(\mathcal{A}) = 0$.*

Corollary 3.2 *When several dummy players are added to $\mathcal{A}$, the scaled index $\tilde{d}_\mu(\mathcal{A})$ does not change.*

It is also easy to check that for a dummy partnership $\mathcal{P}$ for $i \in \mathcal{A} \subseteq \mathcal{N} \setminus \mathcal{P}$

$$\Delta_i \mu(\mathcal{A} \cup \mathcal{P}) = \mu(\mathcal{A}) + \mu(P) - \mu(A \setminus \{i\}) - \mu(\mathcal{P}) = \Delta_i \mu(\mathcal{A}),$$

whereas when $i \in \mathcal{P}$ we have

$$\Delta_i \mu(\mathcal{A} \cup \mathcal{P}) = \mu(\mathcal{A}) + \mu(\mathcal{P}) - \mu(\mathcal{A}) = \mu(\mathcal{P}).$$

Furthermore, it follows that $d_\mu(\mathcal{P}) = \mu(\mathcal{P})$ for a partnership $\mathcal{P}$.

Theorem 3.8 *For a dummy partnership $\mathcal{P}$, $|\mathcal{P}| \geqslant 2$ and any $\mathcal{A} \subseteq \mathcal{N} \setminus \mathcal{P}$, $|\mathcal{A}| \geqslant 1$:*

$$(|\mathcal{A}| + |\mathcal{P}|) d_\mu(\mathcal{A} \cup \mathcal{P}) = |\mathcal{A}| d_\mu(\mathcal{A}) + |\mathcal{P}| d_\mu(\mathcal{P}). \tag{3.25}$$

Combining the fact that any coalition $\mathcal{P}$ is a dummy partnership in a corresponding unanimity game $\varepsilon^+_{\mathcal{P}}$ with Theorem 3.5, we get

Corollary 3.3 *For any unanimity game $\mu = \varepsilon^+_{\mathcal{P}}$, we have $d_\mu(\mathcal{A}) = \frac{|\mathcal{P}|}{|\mathcal{A}|}$ for any $\mathcal{A} \supseteq \mathcal{P}$ and 0 otherwise.*

With $\tilde{d}_\mu(\{i\}) = 0$, we obtain a consistent formula for dummy partnerships, including dummy singletons:

$$\tilde{d}_\mu(\mathcal{A} \cup \mathcal{P}) = \tilde{d}_\mu(\mathcal{A}) + \tilde{d}_\mu(\mathcal{P}).$$

The above result was further extended to dummy coalitions that are not necessarily partnerships.

Definition 3.17 (Noninteracting Coalition) The coalition $\mathcal{R}$ is called non-interacting with another coalition $\mathcal{A}$ if all its subsets satisfy $\mu(\mathcal{A} \cup \mathcal{S}) = \mu(\mathcal{A}) + \mu(\mathcal{S})$ for all $\mathcal{S} \subseteq \mathcal{R}$.

Definition 3.18 (Independent Coalitions) Two coalitions $\mathcal{A}$ and $\mathcal{B}$, $\mathcal{A} \cap \mathcal{B} = \emptyset$ are mutually independent if for any $\mathcal{S} \subseteq \mathcal{A}$, we have $\mu(\mathcal{S} \cup \mathcal{B}) = \mu(\mathcal{S}) + \mu(\mathcal{B})$ and for any $\mathcal{T} \subseteq \mathcal{B}$, we have $\mu(\mathcal{A} \cup \mathcal{T}) = \mu(\mathcal{A}) + \mu(\mathcal{T})$.

Theorem 3.9 *For a coalition $\mathcal{R}$ non-interacting with $\mathcal{A} \subseteq \mathcal{N} \setminus \mathcal{R}$, $|\mathcal{A}| \geqslant 1$,*

$$(|\mathcal{A}| + |\mathcal{R}|)d_\mu(\mathcal{A} \cup \mathcal{R}) = |\mathcal{A}|d_\mu(\mathcal{A}) + |\mathcal{R}|d_\mu(\mathcal{R}). \tag{3.26}$$

Theorem 3.10 *For two mutually independent coalitions $\mathcal{A}$, $\mathcal{B}$,*

$$(|\mathcal{A}| + |\mathcal{A}|)d_\mu(\mathcal{A} \cup \mathcal{B}) = |\mathcal{A}|d_\mu(\mathcal{A}) + |\mathcal{A}|d_\mu(\mathcal{B}). \tag{3.27}$$

The main result of this section is that the (scaled) nonmodularity index is the only index that satisfies the properties of dummy partnership and maximality at unanimity games along with the linearity, which establishes its axiomatic characterisation similar to the Shapley interaction index.

Theorem 3.11 *The scaled nonmodularity index of a capacity μ, $\tilde{d}_\mu(\mathcal{A})$ for $|\mathcal{A}| \geqslant 2$ is the only index d that satisfies the following three axioms:*

(L) *Linearity: d is a linear function of μ.*

(DPA) *Dummy partnership additivity: for a dummy partnership $\mathcal{P}$, any $\mathcal{A} \subseteq \mathcal{N} \setminus \mathcal{P}$, $|\mathcal{A}| \geqslant 1$, $d(\mathcal{A} \cup \mathcal{P}) = d(\mathcal{A}) + d(\mathcal{P})$ with the assumption $d(\{i\}) = 0$.*

(M) *Maximality: $d(\mathcal{A}) = |\mathcal{A}|$ for any unanimity game $\mu = \varepsilon^+_{\mathcal{A}}$, $|\mathcal{A}| \geqslant 2$.*

The maximality axiom has a natural interpretation: in a unanimity game, the interaction of the players is the largest, as the worth of any player or any subset of players is zero, and the maximum synergy is achieved when all players form a coalition. Similarly, dummy partnership additivity has a natural interpretation: the addition of an independent partnership $\mathcal{D}$ to coalition $\mathcal{A}$ adds both the worth of $\mathcal{D}$ and interaction in $\mathcal{D}$ to that in $\mathcal{A}$.

Some consequences of the axioms also seem quite natural: the interaction within a coalition remains unchanged when dummy players are added; interactions of two mutually independent coalitions are just added, etc.

3.5.4 Some Relations Between Interaction Indices

The fact that interaction indices are essentially weighted sums of discrete derivatives over the subsets of $\mathcal{N}$ leads to some relations between them. Below we list some of the formulas established in [Bel21]:

- The nonmodularity index d_μ is related to pairwise Shapley interaction indices I_{ij} as follows:

$$d_\mu(\mathcal{N}) = \frac{2}{n} \sum_{i,j\in\mathcal{N}} I_{ij}, \tag{3.28}$$

  compare to (3.20).
- For any set function μ on $\mathcal{N}$ and fixed $k \leqslant n$, we have

$$\sum_{\substack{\mathcal{B}\subseteq\mathcal{N}\\|\mathcal{B}|=k}} I(\mathcal{B}) = \sum_{\substack{\mathcal{B}\subseteq\mathcal{N}\\|\mathcal{B}|=k}} \sum_{\mathcal{A}\subseteq\mathcal{N}} \tilde{w}_{\mathcal{A}} \Delta_{\mathcal{B}}(\mathcal{A}) = \frac{1}{k} \sum_{|\mathcal{B}|=k-1} \Delta_{\mathcal{B}}(\mathcal{N}) - \Delta_{\mathcal{B}}(\emptyset) \tag{3.29}$$

  where the coefficients $\tilde{w}_{\mathcal{A}}$ are as in (3.5).
- A special case of pairs [3] is

$$\sum_{i,j\in\mathcal{N}} I_{ij} = \sum_{\mathcal{A}\subseteq\mathcal{N}} \sum_{i,j\in\mathcal{N}} \tilde{w}_{\mathcal{A}} \Delta_{ij}\mu(\mathcal{A}) = \sum_{i\in\mathcal{N}} \Delta_i\mu(\mathcal{N}) - \Delta_i\mu(\emptyset). \tag{3.30}$$

- For the derivatives of a set function μ, we have

$$\sum_{\substack{\mathcal{B}\subseteq\mathcal{N}\\|\mathcal{B}|=k}} \sum_{\mathcal{A}\subseteq\mathcal{N}\setminus\mathcal{B}} w^k_{\mathcal{A}} \Delta_{\mathcal{B}}(\mathcal{A}) = \frac{n-k+1}{k\binom{n}{k}} \sum_{|\mathcal{B}|=k-1} \Delta_{\mathcal{B}}(\mathcal{N}) - \Delta_{\mathcal{B}}(\emptyset) \tag{3.31}$$

  where the coefficients $w^k_{\mathcal{A}} = \frac{1}{\binom{n}{k}\binom{n-k}{|\mathcal{A}|}}$.

[3] Here both indices in the sum run over $\mathcal{N}$, so each I_{ij} is included twice. The extra factor $\frac{1}{k}$ in (3.29) is due to the difference in summation $\mathcal{B} \subseteq \mathcal{N}$ versus $i_1, i_2, \ldots, i_k \in N$.

It happens that the mentioned formulas relating the summations of the derivatives of set functions on the left and the difference of lower-order derivatives on the right are consequences of the classical divergence theorem of calculus [Bel21].

Consider the Choquet integral C_μ as a function of the input variables $\mathbf{x} = (x_1, \ldots, x_n)$ and its partial derivatives $\Delta_i C_\mu(\mathbf{x})$ written here in Möbius representation

$$\Delta_i C_\mu(\mathbf{x}) = \sum_{\mathcal{A}|i \in \mathcal{A}} M(\mathcal{A}) \min_{j \in \mathcal{A} \setminus \{i\}} x_j.$$

It is known that the Shapley values (and Shapley interaction indices in general) can be written as the definite integrals of the derivatives of the function C_μ:

$$\varphi(i) = \int_{[0,1]^n} \Delta_i C_\mu(\mathbf{x})\, \mathrm{d}\mathbf{x}. \tag{3.32}$$

Let us now consider the vector field $\mathbf{C}_\mu = (C_\mu, \ldots, C_\mu)$, and let S be the surface of the unit cube, with $\mathbf{n}$ being the normal vectors pointing outwards. The divergence theorem of calculus connects the integral of a vector field over the volume with the integral of its divergence over the bounding surface. The analogue of the divergence theorem for the discrete derivatives of set functions was established in [Bel21], namely,

$$\int_{[0,1]^n} \mathrm{div}(\mathbf{C}_\mu)\, \mathrm{d}\mathbf{x} = \int_{[0,1]^n} \sum_i \frac{\partial C}{\partial x_i}\, \mathrm{d}\mathbf{x} = \int_S \mathbf{C}_\mu \cdot \mathbf{n}\, \mathrm{d}S. \tag{3.33}$$

By integrating over the boundary of the domain in the right-hand side, we obtain [Bel21]

$$\sum_i \varphi(i) = \sum_i \int_{[0,1]^n} \Delta_i C_\mu(\mathbf{x})\, \mathrm{d}\mathbf{x} = \int_S \mathbf{C}_\mu \cdot \mathbf{n}\, \mathrm{d}S = 1 \quad (= \mu(\mathcal{N}) - \mu(\emptyset)).$$

In fact, the set function μ need not be normalised, grounded, or even monotone; the value of the surface integral is always $\mu(\mathcal{N}) - \mu(\emptyset)$. Thus, we obtain the equation (3.6) as a *consequence* of the divergence theorem. Furthermore, we can now replace the capacity μ with another set function, for example, the derivative, or sum of derivatives of μ, and obtain analogous results expressed in (3.29)–(3.31).

3.6 Bipartition Interaction Indices

The definitions for additivity and sub- and superadditivity are based on the relationships between disjoint subsets, i.e. we look to whether $\mu(\mathcal{A}) + \mu(\mathcal{B})$ is smaller, greater than, or equal to $\mu(\mathcal{A} \cup \mathcal{B})$ when $\mathcal{A} \cap \mathcal{B} = \emptyset$. This leads to the concept of characterising interaction in terms of set partitions.

A set of non-empty subsets $\{\mathcal{A}_1, \mathcal{A}_2, \ldots, \mathcal{A}_k\}$, $k \geqslant 2$, is called a partition of the set $\mathcal{A}$ if $\mathcal{A}_i \neq \emptyset$, $\mathcal{A}_i \cap \mathcal{A}_j = \emptyset$ and $\bigcup_{i=1}^{k} \mathcal{A}_i = \mathcal{A}$, where $i, j = 1, 2, \ldots, k$ and $i \neq j$. A *proper* partition of $\mathcal{A}$ involves only proper subsets of $\mathcal{A}$, and a *bipartition* is any proper partition that includes two subsets only. We denote the set of all bipartitions of $\mathcal{A} \subseteq \mathcal{N}$ by $\pi(\mathcal{A})$, i.e.

$$\pi(\mathcal{A}) = \{\{\mathcal{C}, \mathcal{A}\backslash\mathcal{C}\} | \mathcal{C} \subset \mathcal{A}, \mathcal{C} \neq \emptyset\}.$$

The sets $\pi(\emptyset)$ and $\pi(\{i\})$, $i \in \mathcal{N}$ are always empty. The number of bipartitions for a set $\mathcal{A}$, i.e. the number of elements in $\pi(\mathcal{A})$, is $2^{|\mathcal{A}|-1} - 1$.

Definition 3.19 (Marginal Interaction w.r.t. a Bipartition) Let μ be a capacity on $\mathcal{N}$. The marginal interaction of $\mathcal{A} \subseteq \mathcal{N}$ w.r.t. its bipartition $\{\mathcal{C}, \mathcal{A}\backslash\mathcal{C}\} \in \pi(\mathcal{A})$ in the presence of $\mathcal{B} \subseteq \mathcal{N}\backslash\mathcal{A}$ is defined as

$$\begin{aligned}\hat{\Delta}_{\mathcal{A},\mathcal{C}}\mu(\mathcal{B}) &= [\mu(\mathcal{B} \cup \mathcal{A}) - \mu(\mathcal{B})] - \\ &\quad \Big([\mu(\mathcal{B} \cup \mathcal{C}) - \mu(\mathcal{B})] + [\mu(\mathcal{B} \cup (\mathcal{A}\backslash\mathcal{C})) - \mu(\mathcal{B})]\Big).\end{aligned} \tag{3.34}$$

We can see that $\forall \mathcal{A} \subseteq \mathcal{N}, \forall \{\mathcal{C}, \mathcal{A}\backslash\mathcal{C}\} \in \pi(\mathcal{A})$,

$$\hat{\Delta}_{\mathcal{A},\mathcal{C}}\mu(\mathcal{B}) = \mu(\mathcal{B} \cup \mathcal{A}) + \mu(\mathcal{B}) - [\mu(\mathcal{B} \cup \mathcal{C}) + \mu(\mathcal{B} \cup (\mathcal{A}\backslash\mathcal{C}))].$$

In order to collect the marginal bipartition interactions w.r.t. all bipartitions into an integrated marginal interaction, we can suppose each bipartition has the same weight and get the concept of marginal bipartition interaction of a subset.

Definition 3.20 (Marginal Bipartition Interaction) Let μ be a capacity on $\mathcal{N}$. The marginal bipartition interaction of $\mathcal{A} \subseteq \mathcal{N}$ in the presence of $\mathcal{B} \subseteq \mathcal{N}\backslash\mathcal{A}$ w.r.t. μ is defined as

$$\begin{aligned}\hat{\Delta}_{\mathcal{A}}\mu(\mathcal{B}) &= [\mu(\mathcal{B} \cup \mathcal{A}) - \mu(\mathcal{B})] \\ &\quad - \frac{1}{2^{|\mathcal{A}|-1} - 1} \sum_{\{\mathcal{C},\mathcal{A}\backslash\mathcal{C}\} \in \pi(\mathcal{A})} ([\mu(\mathcal{B} \cup \mathcal{C}) - \mu(\mathcal{B})] + [\mu(\mathcal{B} \cup (\mathcal{A}\backslash\mathcal{C})) - \mu(\mathcal{B})]),\end{aligned} \tag{3.35}$$

where $\pi(\mathcal{A})$ is the set of all bipartitions of the subset $\mathcal{A}$.

This can be simplified to

$$\hat{\Delta}_{\mathcal{A}}\mu(\mathcal{B}) = \mu(\mathcal{B} \cup \mathcal{A}) + \mu(\mathcal{B}) - \frac{1}{2^{|\mathcal{A}|-1} - 1} \sum_{\emptyset \neq \mathcal{C} \subset \mathcal{A}} \mu(\mathcal{B} \cup \mathcal{C}), \ \forall \mathcal{A} \subseteq \mathcal{N}, |\mathcal{A}| \geqslant 2.$$

Analogous to Definition 3.5, we can define the probabilistic bipartition interaction index [WB18b].

Definition 3.21 (Probabilistic Bipartition Interaction Index) Let μ be a capacity on $\mathcal{N}$. The probabilistic bipartition interaction index of a subset $\mathcal{A} \subseteq \mathcal{N}$ w.r.t. μ is defined as

$$\hat{I}^{\mu}_{\text{pr}}(\mathcal{A}) = \sum_{\mathcal{B} \subseteq \mathcal{N} \setminus \mathcal{A}} p^{A}_{B}(N) \hat{\Delta}_{\mathcal{A}} \mu(\mathcal{B}),$$

where, for any $\mathcal{A} \subseteq \mathcal{N}$, the family of coefficients $\{p^{\mathcal{A}}_{\mathcal{B}}(\mathcal{N})\}_{\mathcal{B} \subseteq \mathcal{N} \setminus \mathcal{A}}$ forms a probability distribution on $2^{\mathcal{N} \setminus \mathcal{A}}$.

The following definition summarises the three special cases.

Definition 3.22 (Bipartition Interaction Special Cases) Let μ be a capacity on $\mathcal{N}$; the Shapley bipartition interaction index of a subset $\mathcal{A} \subseteq \mathcal{N}$ is defined as

$$\hat{I}^{\mu}_{\text{Sh}}(\mathcal{A}) = \sum_{\mathcal{B} \subseteq \mathcal{N} \setminus \mathcal{A}} \frac{1}{|\mathcal{N}| - |\mathcal{A}| + 1} \binom{|\mathcal{N}| - |\mathcal{A}|}{|\mathcal{B}|}^{-1} \hat{\Delta}_{\mathcal{A}} \mu(\mathcal{B}).$$

The Banzhaf bipartition interaction index of a subset $\mathcal{A} \subseteq \mathcal{N}$ is defined as

$$\hat{I}^{\mu}_{\text{Ba}}(\mathcal{A}) = \sum_{\mathcal{B} \subseteq \mathcal{N} \setminus \mathcal{A}} \frac{1}{2^{(|\mathcal{N}| - |\mathcal{A}|)}} \hat{\Delta}_{\mathcal{A}} \mu(\mathcal{B}).$$

The internal bipartition interaction index of a subset $\mathcal{A} \subseteq \mathcal{N}$ is defined as

$$\hat{I}^{\mu}_{\text{in}}(\mathcal{A}) = \hat{\Delta}_{\mathcal{A}} \mu(\emptyset).$$

The marginal bipartition interaction includes the cases of singletons and pairs in its definition, so these special cases of probabilistic bipartition interaction indices can also be seen as extensions of the respective concepts as applied to singletons and pairs.

The following theorem establishes that dummy criteria are independent from all other single decision criteria and do not interact.

Theorem 3.12 *[WB18b] Let μ be a capacity on $\mathcal{N}$ and $d \in \mathcal{N}$ be a dummy criterion. Then $\hat{I}^{\mu}_{\text{pr}}(\{d\}) = \mu(\{d\})$, $\hat{I}^{\mu}_{\text{pr}}(\{i, d\}) = 0$, $\forall i \in N \setminus \{d\}$.*

For larger subsets, however, the value of $\hat{I}^{\mu}_{\text{pr}}(\{\mathcal{A} \cup d\})$ is not usually zero. This is a deviation from the dummy property when considering probabilistic simultaneous interaction.

The following theorem establishes the range for the probabilistic bipartition indices.

Theorem 3.13 ([WB18b]) *Let μ be a capacity on $\mathcal{N}$; then*

$$\hat{I}^{\mu}_{\text{pr}}(\emptyset) = 0,$$

$$\hat{I}^{\mu}_{\text{pr}}(\{i\}) \in [0, 1], \forall i \in \mathcal{N},$$

and

$$\hat{I}^{\mu}_{\text{pr}}(\mathcal{A}) \in [-1, 1], \forall \mathcal{A} \subseteq \mathcal{N}, |\mathcal{A}| \geqslant 2.$$

Corollary 3.4 *Let μ be a capacity on $\mathcal{N}$; then for any $\mathcal{A} \subseteq \mathcal{N}$, $\hat{I}^{\mu}_{\text{Sh}}(\mathcal{A})$, $\hat{I}^{\mu}_{\text{Ba}}(\mathcal{A})$, and $\hat{I}^{\mu}_{\text{in}}(\mathcal{A}) \in [-1, 1]$.*

This is convenient for being able to compare interaction indices of different forms as well as subsets with varying cardinality. It is also worth noting that the probabilistic bipartition interaction index of the empty set is zero, deviating again from the probabilistic simultaneous interaction index.

As with the nonadditivity and nonmodularity indices, we can establish relationships between the extremely superadditive and extremely subadditive capacities and the index $\hat{I}^{\mu}_{\text{pr}}$ reaching the minimum and maximum of its range.

Theorem 3.14 *Let $\varepsilon^{+}_{\mathcal{S}}$ and $\varepsilon^{-}_{\mathcal{S}}$ be the extremely superadditive and subadditive capacities focused on subset $\mathcal{S} \subseteq \mathcal{N}$; then $\hat{I}^{\varepsilon^{+}_{\mathcal{S}}}_{\text{pr}}(\mathcal{S}) = 1$, and $\hat{I}^{\varepsilon^{-}_{\mathcal{S}}}_{\text{pr}}(\mathcal{S}) = -1$ when $|\mathcal{S}| \geqslant 2$. Furthermore, if the associated probability distribution coefficients are all nonzero, i.e. if $p^{\mathcal{S}}_{\mathcal{B}}(\mathcal{N}) > 0$, $\forall \mathcal{B} \subseteq \mathcal{N} \setminus \mathcal{S}$, then $\hat{I}^{\mu}_{\text{pr}}(\mathcal{S}) = 1 \;\Leftrightarrow \mu = \varepsilon^{+}_{\mathcal{S}}$, and for $\mathcal{S} \subseteq \mathcal{N}$ with $|\mathcal{S}| \geqslant 2$, $\hat{I}^{\mu}_{\text{pr}}(\mathcal{S}) = -1 \;\Leftrightarrow \mu = \varepsilon^{-}_{\mathcal{S}}$.*

Theorem 3.15 *Let μ be a capacity on $\mathcal{N}$, $\varepsilon^{+}_{\mathcal{S}}$ and $\varepsilon^{-}_{\mathcal{S}}$ be the extremely superadditive and subadditive capacities focused on subset $\mathcal{S} \subseteq \mathcal{N}$; then $\hat{I}^{\mu}_{\text{Sh}}(\mathcal{S})$, $\hat{I}^{\mu}_{\text{Ba}}(\mathcal{S}) = 1 \Leftrightarrow \mu = \varepsilon^{+}_{\mathcal{S}}$, and for $\mathcal{S} \subseteq \mathcal{N}$ with $|\mathcal{S}| \geqslant 2$, $\hat{I}^{\mu}_{\text{Sh}}(\mathcal{S})$, $\hat{I}^{\mu}_{\text{Ba}}(\mathcal{S}) = -1 \Leftrightarrow \mu = \varepsilon^{-}_{\mathcal{S}}$.*

3.7 Cardinality Index

The interaction and value of previous sections focus on marginal contributions of inputs and their coalitions. From an interpretation perspective, the Shapley value provides a vector reflecting the relative importance of each input and so can be viewed similar to the vector of weights of a weighted mean (e.g. as in Definition 1.30). For the ordered weighted averaging operator (Definition 1.34), the weights are indicative of the importance of the relative size of an input rather than the source of the input itself. This led to the proposal of a cardinality index in [Yag02], resulting in a vector that characterises a capacity in terms of the average effective weight assigned to inputs when they are the largest, second largest, and so on.

Definition 3.23 (Cardinality Index) For a given capacity μ, the cardinality index ϑ_i for each i in $\{1, 2, \ldots, n\}$ is given by

$$\vartheta_i = \left(\frac{1}{\binom{n}{i}} \sum_{|\mathcal{A}|=i} \mu(\mathcal{A}) \right) - \left(\frac{1}{\binom{n}{i-1}} \sum_{|\mathcal{A}|=i-1} \mu(\mathcal{A}) \right).$$

The two components average the weight of coalitions of size $|\mathcal{A}| = i$ and $|\mathcal{A}| = i - 1$.

Example 3.5 For the following capacity defined on $\mathcal{N} = \{1, 2, 3\}$,

$$\begin{array}{ccc} & 1 & \\ 0.6 & 0.3 & 0.9 \\ 0.1 & 0 & 0.5 \\ & 0 & \end{array}$$

The cardinality indices would be $\vartheta = (0.2, 0.6 - 0.2, 1 - 0.6) = (0.2, 0.4, 0.4)$.

The cardinality index can alternatively be expressed as

$$\vartheta_i = \frac{1}{i\binom{n}{i}} \sum_{|\mathcal{A}|=i} \sum_{j \in \mathcal{A}} (\mu(\mathcal{A}) - \mu(\mathcal{A} \setminus \{j\})),$$

which makes it clear that this is just the average of first-order derivatives for sets of cardinality i. It can be expressed in terms of the marginal contributions as

$$\vartheta_i = \frac{1}{i\binom{n}{i}} \sum_{|\mathcal{A}|=i} \sum_{j \in \mathcal{A}} \Delta_j \mu(\mathcal{A}),$$

and in Möbius representation, this becomes (see Table 3.1)

$$\vartheta_i = \frac{1}{i\binom{n}{i}} \sum_{|\mathcal{A}|=i} \sum_{j \in \mathcal{A}} \sum_{\mathcal{B} \subseteq \mathcal{A} | j \in \mathcal{B}} M(\mathcal{B}).$$

Table 3.1 Different index calculations in standard and Möbius representation with inverses

Index	Standard/Möbius form	Inverse
Möbius $M(\mathcal{A})$	$M(\mathcal{A}) = \sum_{\mathcal{B}\subseteq\mathcal{A}} (-1)^{\|\mathcal{A}\setminus\mathcal{B}\|}\mu(\mathcal{B})$	$\mu(\mathcal{A}) = \sum_{\mathcal{B}\subseteq\mathcal{A}} M(\mathcal{B})$
Co-Möbius (see Sect. 4.1) $\check{M}(\mathcal{A})$	$\check{M}(\mathcal{A}) = \sum_{\mathcal{B}\subseteq\mathcal{A}} (-1)^{\|\mathcal{B}\|}\mu(\mathcal{N}\setminus\mathcal{B})$	$\mu(\mathcal{A}) = \sum_{\mathcal{B}\subseteq\mathcal{N}\setminus\mathcal{A}} (-1)^{\|\mathcal{B}\|}\check{M}(\mathcal{B})$
	$\check{M}(\mathcal{A}) = \sum_{\mathcal{B}\supseteq\mathcal{A}} M(\mathcal{B})$	$M(\mathcal{A}) = \sum_{\mathcal{B}\supseteq\mathcal{A}} (-1)^{\|\mathcal{B}\setminus\mathcal{A}\|}\check{M}(\mathcal{B})$
Marginal contribution (of a set) $\Delta_{\mathcal{B}}\mu(\mathcal{A})$	$\Delta_{\mathcal{B}}\mu(\mathcal{A}) = \sum_{\mathcal{C}\subseteq\mathcal{B}} (-1)^{\|\mathcal{B}\setminus\mathcal{C}\|}\mu((\mathcal{A}\setminus\mathcal{B})\cup\mathcal{C})$	Eq. (4.24)
	$\Delta_{\mathcal{B}}\mu(\mathcal{A}) = \sum_{\mathcal{C}\subseteq\mathcal{A}\setminus\mathcal{B}} M(\mathcal{B}\cup\mathcal{C})$	$M(\mathcal{A}) = \sum_{\mathcal{B}\subseteq\mathcal{A}\setminus\{i\}} (-1)^{\|\mathcal{A}\setminus\mathcal{B}\|-1}\Delta_i\mu(\mathcal{B}), \forall i\in\mathcal{A}$
Shapley value $\varphi(i)$	$\varphi(i) = \sum_{\mathcal{A}\subseteq\mathcal{N}} \frac{1}{\binom{n+1}{1}\binom{n}{\|\mathcal{A}\|}}[\mu(\mathcal{A}\cup\{i\}) - \mu(\mathcal{A}\setminus\{i\})] =$ $\sum_{\mathcal{A}\subseteq\mathcal{N}\setminus\{i\}} \frac{1}{\binom{n-1}{\|\mathcal{A}\|}n}[\mu(\mathcal{A}\cup\{i\}) - \mu(\mathcal{A})].$	
Shapley interaction for sets $I(\mathcal{A})$	$I(\mathcal{A}) = \sum_{\mathcal{B}\subseteq\mathcal{N}\setminus\mathcal{A}} \frac{1}{\binom{n-\|\mathcal{A}\|}{\|\mathcal{B}\|}(n-\|\mathcal{A}\|+1)} \sum_{\mathcal{C}\subseteq\mathcal{A}} (-1)^{\|\mathcal{A}\setminus\mathcal{C}\|}\mu(\mathcal{B}\cup$ $\mathcal{C}) = \sum_{\mathcal{B}\subseteq\mathcal{N}} \frac{1}{\binom{n+1}{1}\binom{n}{\|\mathcal{B}\|}} \sum_{\mathcal{C}\subseteq\mathcal{A}} (-1)^{\|\mathcal{A}\setminus\mathcal{C}\|}\mu((\mathcal{B}\setminus A)\cup\mathcal{C}).$	
	$I(\mathcal{A}) = \sum_{\mathcal{B}\subseteq\mathcal{N}} \frac{(-1)^{\|\mathcal{A}\setminus\mathcal{B}\|}}{(n-\|\mathcal{A}\|+1)\binom{n-\|\mathcal{A}\|}{\|\mathcal{B}\setminus\mathcal{A}\|}}\mu(\mathcal{B})$	$\mu(\mathcal{A}) = \sum_{\mathcal{B}\subseteq\mathcal{N}} \beta^{\|\mathcal{B}\|}_{\|\mathcal{A}\cap\mathcal{B}\|} I(\mathcal{B})$, with $\beta^l_k = \sum_{j=0}^{k}\binom{k}{j}\beta_{l-j}$ and β_k as below
	$I(\mathcal{A}) = \sum_{\mathcal{B}\|\mathcal{A}\subseteq\mathcal{B}} \frac{1}{\|\mathcal{B}\|-\|\mathcal{A}\|+1} M(\mathcal{B})$	$M(\mathcal{A}) = \sum_{\mathcal{B}\|\mathcal{A}\subseteq\mathcal{B}} \beta_{\|\mathcal{B}\|-\|\mathcal{A}\|} I(\mathcal{B})$, with $\beta_k = -\sum_{l=0}^{k-1}\binom{k}{l}\frac{\beta_l}{k-l+1}$

Banzhaf Interaction $J(\mathcal{A})$	$J(\mathcal{A}) = \frac{1}{2^{n-\vert\mathcal{A}\vert}} \sum_{\mathcal{B}\subseteq\mathcal{N}} (-1)^{\vert\mathcal{A}\setminus\mathcal{B}\vert}\mu(\mathcal{B})$	$\mu(\mathcal{A}) = \sum_{\mathcal{B}\subseteq\mathcal{N}} \frac{(-1)^{\vert\mathcal{B}\setminus\mathcal{A}\vert}}{2^{\vert\mathcal{B}\vert}} J(\mathcal{B})$
	$J(\mathcal{A}) = \sum_{\mathcal{B}\vert\mathcal{A}\subseteq\mathcal{B}} \frac{1}{2^{\vert\mathcal{B}\vert-\vert\mathcal{A}\vert}} M(\mathcal{B})$	$M(\mathcal{A}) = \sum_{\mathcal{B}\vert\mathcal{A}\subseteq\mathcal{B}} \left(-\frac{1}{2}\right)^{\vert\mathcal{B}\setminus\mathcal{A}\vert} J(\mathcal{B})$
Nonadditivity $n_\mu(\mathcal{A})$	$n_\mu(\mathcal{A}) = \frac{1}{2^{\vert\mathcal{A}\vert-1}-1} \sum_{\substack{(\mathcal{B},\mathcal{A}\setminus\mathcal{B})\\ \emptyset\neq\mathcal{B}\subset\mathcal{A}}} [\mu(\mathcal{A}) - \mu(\mathcal{B}) - \mu(\mathcal{A}\setminus\mathcal{B})]$	$\mu(\mathcal{A}) = n_\mu(\mathcal{A}) + \sum_{\mathcal{B}\subset\mathcal{A}} \sum_{i=\vert\mathcal{B}\vert}^{\vert\mathcal{A}\vert-1} \prod_{j=i}^{\vert\mathcal{A}\vert-1} \frac{\vert\mathcal{A}\vert-j}{2^j-1} n_\mu(\mathcal{B})$
	$n_\mu(\mathcal{A}) = \sum_{\mathcal{B}\subseteq\mathcal{A}} \frac{2^{\vert\mathcal{A}\vert-1}-2^{\vert\mathcal{A}\vert-\vert\mathcal{B}\vert}}{2^{\vert\mathcal{A}\vert-1}-1} M(\mathcal{B})$, (with $\frac{0}{0} = 1$)	$M(\mathcal{A}) = n_\mu(\mathcal{A}) + \sum_{\mathcal{B}\subset\mathcal{A}} (-1)^{\vert\mathcal{A}\vert-\vert\mathcal{B}\vert} \frac{2^{\vert\mathcal{A}\vert-1}-2^{\vert\mathcal{A}\vert-\vert\mathcal{B}\vert}}{2^{\vert\mathcal{A}\vert-1}-1} n_\mu(\mathcal{B})$
Nonmodularity $d_\mu(\mathcal{A})$	$d_\mu(\mathcal{A}) = \frac{2}{\vert\mathcal{A}\vert} \sum_{\mathcal{B}\subseteq A, i,j\in A} \frac{1}{\binom{n+1}{1}\binom{n}{\vert\mathcal{B}\vert}} (\mu(\mathcal{B}\cup\{i,j\}) - \mu(\mathcal{B}\cup\{i\}) -$ $\mu(\mathcal{B}\cup\{j\}) + \mu(\mathcal{B})) = \frac{1}{\vert\mathcal{A}\vert}\left(\sum_{i\in\mathcal{A}} \Delta_i\mu(\mathcal{A}) - \sum_{i\in\mathcal{A}} \Delta_i\mu(\emptyset)\right)$	
	$d_\mu(\mathcal{A}) = \mu(\mathcal{A}) - \frac{1}{\vert\mathcal{A}\vert} \sum_{i\in\mathcal{A}} [\mu(\{i\}) + \mu(\vert\mathcal{A}\vert\setminus\{i\})]$, $(\vert\mathcal{A}\vert \geqslant 2)$	$\mu(\mathcal{A}) = d_\mu(\mathcal{A}) + \sum_{\mathcal{B}\subset\mathcal{A}, \vert\mathcal{B}\vert\geqslant 2} \binom{\vert\mathcal{A}\vert}{\vert\mathcal{B}\vert}^{-1} d_\mu(\mathcal{B}) + \sum_{\{i\}\subset\mathcal{A}} d_\mu(\{i\})$
	$d_\mu(\mathcal{A}) = \sum_{\mathcal{B}\subseteq\mathcal{A}, \vert\mathcal{B}\vert\geqslant 2} \frac{\vert\mathcal{B}\vert}{\vert\mathcal{A}\vert} M(\mathcal{B})$, $(\vert\mathcal{A}\vert \geqslant 2)$	$M(\mathcal{A}) = \sum_{\mathcal{B}\subseteq\mathcal{A}, \vert\mathcal{B}\vert\geqslant 2} \frac{(-1)^{\vert\mathcal{A}\vert-\vert\mathcal{B}\vert}\vert\mathcal{B}\vert}{\vert\mathcal{A}\vert} d_\mu(\mathcal{B})$
Cardinality index ϑ_i	$\vartheta_i = \frac{1}{i\binom{n}{i}} \sum_{\vert\mathcal{A}\vert=i} \sum_{j\in\mathcal{A}} (\mu(\mathcal{A}) - \mu(\mathcal{A}\setminus\{j\}))$	
	$\frac{1}{i\binom{n}{i}} \sum_{\vert\mathcal{A}\vert=i} \sum_{j\in\mathcal{A}} \sum_{\mathcal{B}\subseteq\mathcal{A}\vert j\in\mathcal{B}} M(\mathcal{B})$	

3.8 Entropy

Another way for understanding some of the overall behaviour of a capacity is in terms of its entropy. For a standard weight distribution, entropy can be seen as the extent to which importance is shared by the inputs or variables, reaching a maximum when weights are equal and a minimum when all value is assigned to a single input. As well as facilitating interpretation of a capacity's overall behaviour, entropy can also be used as an objective when learning capacity values from data.

The following definition extends Shannon's entropy to capacities in the context of Choquet integration [Mar02b, KMR05].

Definition 3.24 (Shannon Entropy) The Shannon entropy of a capacity μ is

$$H(\mu) = \sum_{i \in \mathcal{N}} \sum_{\mathcal{A} \subseteq \mathcal{N} \setminus \{i\}} \frac{(n - |\mathcal{A}| - 1)!|\mathcal{A}|!}{n!} h(\mu(\mathcal{A} \cup \{i\}) - \mu(\mathcal{A})), \qquad (3.36)$$

with $h(t) = -t \ln t$, if $t > 0$ and $h(0) = 0$.

Observe that the argument of h is the i-th descrete derivative $\Delta_i \mu(\mathcal{A})$. For capacities that are additive or symmetric, the calculation collapses to the definition of weights dispersion used for the weighted arithmetic mean and OWA functions. The maximum value, $\ln n$, is reached only when the capacity is both additive and symmetric and the minimum of 0 is obtained only when μ is a Boolean capacity (which means that the output of the Choquet integral will coincide with one of its inputs).

An alternative to Shannon entropy is Renyi entropy, which also includes Shannon entropy as a special case.

Definition 3.25 (Renyi Entropy) For a given $\alpha > 0, \alpha \neq 1$, the Renyi entropy of a capacity μ is

$$H_\alpha(\mu) = \frac{1}{1-\alpha} \ln \left(\sum_{i=1}^{n} \sum_{\mathcal{A} \subseteq \mathcal{N} \setminus \{i\}} \frac{(n - |\mathcal{A}| - 1)!|\mathcal{A}|!}{n!} (\mu(\mathcal{A} \cup \{i\}) - \mu(\mathcal{A}))^\alpha \right),$$

The Shannon entropy calculation is obtained as the limiting case $\alpha \to 1$. The case $\alpha \to \infty$ is known as the min-entropy H_∞.

3.9 Core

Some important theoretical results relating to capacities can be established in terms of the core, which is a concept that has been developed both for capacities and games. The core can be conceptualised as the set of all additive (or probability) measures that dominate a capacity μ on $\mathcal{N}$. It can be defined in this way [MG12];

however, the set is usually expressed in terms of vectors in $\mathbb{R}^n$ corresponding with the singleton values rather than the measures themselves, i.e.

$$Core(\mu) = \{\nu \in \mathbb{R}^n | \nu(\mathcal{A}) \geqslant \mu(\mathcal{A})\},$$

where $\nu(\mathcal{A}) = \sum_{i \in \mathcal{A}} \nu_i$, $\nu(\mathcal{N}) = 1$, and $\nu(\emptyset) = 0$.

Each vector ν hence generates an additive capacity, with Core(μ) defining a convex closed bounded polyhedron in $\mathbb{R}^n$ (which for some capacities is empty). The convexity allows it to be expressed as the set of extreme points in the polytope.

The formal definition involves the notion of pre-imputation and applies to both capacities and games (see [Gra16, MG12]).

Definition 3.26 (Pre-imputation of a Game) Let μ be a game. We say that a vector $x = (x_1, \cdots, x_n) \in \mathbb{R}^n$ is a pre-imputation for μ if it satisfies

$$\sum_{i=1}^{n} x_i \geqslant \mu(\mathcal{N}).$$

Note 3.7 It is convenient to use the notation $x(\mathcal{A}) := \sum_{i \in \mathcal{A}} x_i$, for all $\mathcal{A} \subseteq \mathcal{N}$, with the convention $x(\emptyset) = 0$. Thus, x corresponds with an additive game.

Definition 3.27 (Core of a Game) Let μ be a game. The core of μ, denoted by $Core(\mu)$, is defined by

$$Core(\mu) := \{x \in R^n | x(\mathcal{A}) \geqslant \mu(\mathcal{A}), \forall \mathcal{A} \subseteq \mathcal{N}, x(\mathcal{N}) = \mu(\mathcal{N})\}.$$

The core for strictly submodular capacities is empty, while the core for supermodular capacities is always non-empty. In the latter case, the core can be obtained as the convex hull of vectors of the marginal contributions (defined in Sect. 4.8). In such cases, the Shapley value also belongs to the core. The necessary and sufficient condition of non-emptiness of the core is established by the Bondareva–Shapley theorem (see [Gra16], Ch.3, for details).

References

[Ban65] J. Banzhaf, Weight voting doesn't work: A mathematical analysis. Rutgers Law Rev. **19**, 317–343 (1965)

[Bel21] G. Beliakov, On the derivatives of set functions in matrix representation. Inf. Sci. **546**, 469–480 (2021)

[BW20] G. Beliakov, J.-Z. Wu, The axiomatic characterisations of non-modularity index. Int. J. General Syst. **49**, 675–688 (2020)

[FKM06] K. Fujimoto, I. Kojadinovic, J.-L. Marichal, Axiomatic characterizations of probabilistic and cardinal-probabilistic interaction indices. Games Econ. Behav. **55**(1), 72–99 (2006)

[FM07] K. Fujimoto, T. Murofushi, Some relations among values, interactions, and decomposability of non-additive measures. Int. J. Uncertain. Fuzziness Knowledge Based Syst. **15**, 175–191 (2007)

[Gra97] M. Grabisch, k-Order additive discrete fuzzy measures and their representation. Fuzzy Sets Syst. **92**, 167–189 (1997)

[Gra00] M. Grabisch, The interaction and Möbius representation of fuzzy measures on finite spaces, k-additive measures: A survey, in *Fuzzy Measures and Integrals. Theory and Applications*, eds. by M. Grabisch, T. Murofushi, M. Sugeno (Physica-Verlag, Heidelberg, 2000), pp. 70–93

[Gra16] M. Grabisch, *Set Functions, Games and Capacities in Decision Making* (Springer, Berlin, New York, 2016)

[GKM08] M. Grabisch, I. Kojadinovic, P. Meyer, A review of methods for capacity identification in Choquet integral based multi-attribute utility theory: Applications of the Kappalab R package. Eur. J. Oper. Res. **186**(2), 766–785 (2008)

[GL07] M. Grabisch, C. Labreuche, Derivative of functions over lattices as a basis for the notion of interaction between attributes. Ann. Math. Artif. Intell. **49**, 1–4 (2007)

[GMR00] M. Grabisch, J.-L. Marichal, M. Roubens, Equivalent representations of set functions. Math. Oper. Res. **25**, 157–178 (2000)

[GR99] M. Grabisch, M. Roubens, An axiomatic approach to the concept of interaction among players in cooperative games. Int. J. Game Theory **28**(4), 547–565 (1999)

[KMR05] I. Kojadinovic, J.-L. Marichal, M. Roubens, An axiomatic approach to the definition of the entropy of a discrete Choquet capacity. Inf. Sci. **172**, 131–153 (2005)

[Mar99] J.-L. Marichal, Aggregation operators for multicriteria decision aid. PhD Thesis. Universite de Liege, 1999

[Mar02a] J.-L. Marichal, Aggregation of interacting criteria by means of the discrete Choquet integral, in *Aggregation Operators. New Trends and Applications*, eds. by T. Calvo, G. Mayor, R. Mesiar (Physica-Verlag, Heidelberg, New York, 2002), pp. 224–244

[Mar02b] J.-L. Marichal, Entropy of discrete Choquet capacities. Eur. J. Oper. Res. **137**, 612–624 (2002)

[MG12] P. Miranda, M. Grabisch, An algorithm for finding the vertices of the k-additive monotone core. Discrete Appl. Math. **160**(4-5), 628–639 (2012)

[MS93] T. Murofushi, S. Soneda, Techniques for reading fuzzy measures (iii): Interaction index, in *9th Fuzzy Systems Symposium*, Sapporo, Japan (1993), pp. 693–696

[Owe72] G. Owen, Multilinear extensions of games. Manag. Sci. **18**(5-part-2), 64–79(1972)

[Rou96] M. Roubens, Interaction between criteria and definition of weights in MCDA problems, in *44th Meeting of the European Working Group Multicriteria Aid for Decisions*, Brussels, Belgium (1996)

[Sha53] L.S. Shapley, A value for n-person games. Contribut. Theory Games **2**(28), 307–317 (1953)

[SGBC03] M. Sicilia, E. García Barriocanal, T. Calvo, An inquiry-based method for Choquet integral-based aggregation of interface usability parameters. Kybernetika **39**, 601–614 (2003)

[WB18a] J.-Z. Wu, G. Beliakov, Nonadditivity index and capacity identification method in the context of multicriteria decision making. Inf. Sci. **467**, 398–406 (2018)

[WB18b] J.-Z. Wu, G. Beliakov, Probabilistic bipartition interaction index of multiple decision criteria associated with the nonadditivity of fuzzy measures. Int. J. Intell. Syst. **34**(2), 247–270 (2018)

[WB19] J.-Z. Wu, G. Beliakov, Nonmodularity index for capacity identifying with multiple criteria preference information. Inf. Sci. **492**, 164–180 (2019)

[Wu+16a] J.-Z. Wu, L. Yu, G. Li, J. Jin, B. Du, The sum interaction indices of some particular families of monotone measures. J. Intell. Fuzzy Syst. **31**(3), 1447–1457 (2016)

[Wu+16b] J.-Z. Wu, L. Yu, G. Li, J. Jin, B. Du, Using the monotone measure sum to enrich the measurement of the interaction of multiple decision criteria. J. Intell. Fuzzy Syst. **30**(5), 2529–2539 (2016)

[Yag02] R. Yager, On the cardinality index and attitudinal character of fuzzy measures. Int. J. Gen. Syst. **31**(3), 303–329 (2002)

Chapter 4
Representations

Abstract Here, we present various transformations and capacity representations, including the Möbius and interaction indices of the previous chapter. While we have seen that these alternative forms can help facilitate interpretation and provide insight into different aspects of a capacity's composition, the focus of this chapter is on the simplifications and improvements in efficiency that can be obtained from a practical perspective. Vector representation and linear transformation matrices are introduced, which allow for efficient computer implementation, as well as expressions for different indices in marginal contribution representation and an overview on the polytope of normalised capacities.

4.1 Standard and Möbius Representations

In Chap. 3, we introduced several interaction and value indices from the perspective of interpreting the nonadditive behaviour and properties of Choquet capacities. We also saw that we can express these indices in different ways, with it being possible to perform calculations in terms of the capacity values in standard representation (μ), Möbius representation (M), Shapley interaction representation (I), and so on (see Table 3.1). While each of the indices can provide different theoretical insights concerning a given capacity, it happens that these alternative representations can also facilitate simplified calculations when working with capacities in various applications. They hence can help us in addressing two of the key challenges in interpretability and tractability.

Let us first revisit the standard and Möbius representation. Recall that a capacity μ is defined on every subset $\mathcal{A} \subseteq \mathcal{N}$ (also referred to as the powerset $2^{\mathcal{N}}$) with $\mathcal{A} \subseteq \mathcal{B} \rightarrow \mu(\mathcal{A}) \leqslant \mu(\mathcal{B})$. The Hasse diagram and graphical representations based on it (see Fig. 4.1) can be useful for presenting small examples and showing some of the relationships that exist between subsets.

The Möbius transformation serves as an alternative representation for μ, simplifying many of the calculations we saw for value and interaction indices in Chap. 3 (see also [Gra97, Gra00, Mar02]).

G. Beliakov et al., *Choquet Capacities and Fuzzy Integrals*, Theory and Applications of Computability, https://doi.org/10.1007/978-3-031-97070-2_4

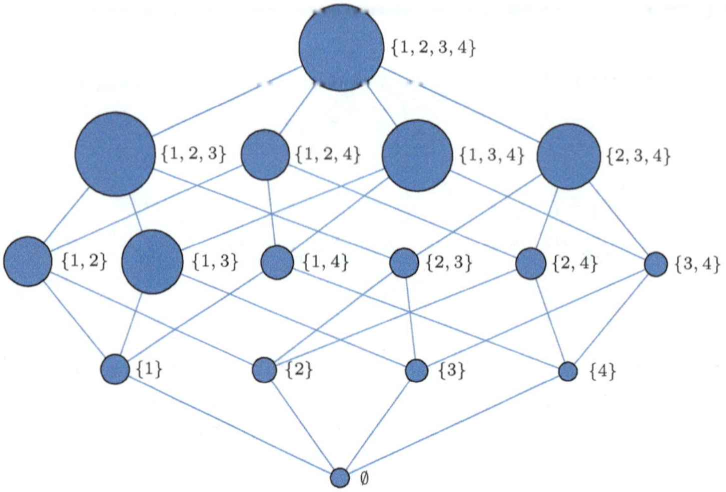

Fig. 4.1 Hasse diagram of a set function. The size of the circles indicates the magnitude of $\mu(\mathcal{A})$, and the monotonicity can be verified from the edges in the figure indicating subsets

Definition 4.1 (Möbius Transformation) Let μ be a capacity. The Möbius transformation of μ is a set function defined for every $\mathcal{A} \subseteq \mathcal{N}$ as

$$M(\mathcal{A}) = \sum_{\mathcal{B} \subseteq \mathcal{A}} (-1)^{|\mathcal{A} \setminus \mathcal{B}|} \mu(\mathcal{B}).$$

The invertibility of the Möbius transformation is what allows it to serve as an alternative representation, with there being a one-to-one mapping between μ and M. The so-called Zeta transform allows us to recover μ,

$$\mu(\mathcal{A}) = \sum_{\mathcal{B} \subseteq \mathcal{A}} M(\mathcal{B}) \quad \forall \mathcal{A} \subseteq \mathcal{N}.$$

By using these linear transformations, we can work in whichever representation is more convenient. The monotonicity and boundary conditions $\mu(\emptyset) = 0$, $\mu(\mathcal{N}) = 1$ are expressed in Möbius representation, respectively, as

$$\sum_{\mathcal{B} \subseteq A | i \in \mathcal{B}} M(\mathcal{B}) \geqslant 0, \qquad \text{for all } \mathcal{A} \subseteq \mathcal{N} \text{ and all } i \in \mathcal{A}, \tag{4.1}$$

$$M(\emptyset) = 0 \text{ and } \sum_{\mathcal{A} \subseteq \mathcal{N}} M(\mathcal{A}) = 1.$$

Example 4.1 A capacity on $\mathcal{N} = \{1, 2, 3\}$ is shown in its standard representation μ on the left and Möbius representation M on the right:

$$\begin{array}{ccc} & 1 & \\ 0.9 & 0.5 & 0.3 \\ 0.5 & 0 & 0.3 \\ & 0 & \end{array} \qquad\qquad \begin{array}{ccc} & 0.1 & \\ 0.4 & -0.3 & 0 \\ 0.5 & 0 & 0.3 \\ & 0 & \end{array}$$

We can verify that Möbius values on the right sum to 1, consistent with the boundary condition in (4.1). Observe also that the singleton values coincide in both representations.

Table 4.1 summarises the monotonicity and boundary conditions in Möbius as well as other representations that will be presented throughout this chapter.

The duality relationship can also be a useful transformation, with some special families being easier to express in terms of the dual capacity:

$$\mu^*(\mathcal{A}) = 1 - \mu(\mathcal{N} \setminus \mathcal{A}).$$

In Möbius representation, this is [Gra16]

$$M^*(\mathcal{A}) = (-1)^{|\mathcal{A}|+1} \sum_{\mathcal{B} \supseteq \mathcal{A}} M(\mathcal{B}). \tag{4.2}$$

We saw in Sect. 2.9 that subset values for belief measures can be more naturally expressed in Möbius representation, since it corresponds with the definition of a basic probability assignment. Some other examples where the Möbius representation is useful include extremely superadditive capacities, $\mu = \varepsilon^+_{\mathcal{S}}$ (or unanimity games) whose Möbius values for each subset $\mathcal{A}$ are given by

$$M(\mathcal{A}) = \begin{cases} 1, & \text{if } \mathcal{A} = \mathcal{S} \\ 0, & \text{otherwise,} \end{cases}$$

and λ-capacities, where for all $\emptyset \neq \mathcal{A} \subseteq \mathcal{N}$, $\lambda \neq 0$ we have

$$M(\mathcal{A}) = \lambda^{|\mathcal{A}|-1} \prod_{i \in \mathcal{A}} \mu(\{i\}). \tag{4.3}$$

Given the correspondence between $M(\{i\})$ and $\mu(\{i\})$ for all singletons, this facilitates a direct calculation from the singleton capacity values, whereas the definition in standard form implies the need for subset values to be determined recursively.

The co-Möbius transformation is a related concept that applies the same calculation but to the complements of each subset.

Definition 4.2 (Co-Möbius Transformation) Let μ be a capacity. The co-Möbius transformation of μ is a set function defined for every $\mathcal{A} \subseteq \mathcal{N}$ as

$$\check{M}(\mathcal{A}) = \sum_{\mathcal{B}\subseteq\mathcal{A}} (-1)^{|\mathcal{B}|}\mu(\mathcal{N}\setminus\mathcal{B}).$$

Its inverse is

$$\mu(\mathcal{A}) = \sum_{\mathcal{B}\subseteq\mathcal{N}\setminus\mathcal{A}} (-1)^{|\mathcal{B}|}\check{M}(\mathcal{B}).$$

The relations between Möbius and co-Möbius representations are

$$\check{M}(\mathcal{A}) = \sum_{\mathcal{B}\supseteq\mathcal{A}} M(\mathcal{B}) \text{ and } M(\mathcal{A}) = \sum_{\mathcal{B}\supseteq\mathcal{A}} (-1)^{|\mathcal{B}\setminus\mathcal{A}|}\check{M}(\mathcal{B})$$

$$\check{M}(\mathcal{A}) = (-1)^{|\mathcal{A}|+1} M^*(\mathcal{A})$$

In terms of discrete derivatives,

$$M(\mathcal{A}) = \Delta_{\mathcal{A}}\mu(\emptyset) \text{ and } \check{M}(\mathcal{A}) = \Delta_{\mathcal{A}}\mu(\mathcal{N}).$$

4.2 Possibilistic Möbius Transform

Definition 4.3 (Possibilistic Möbius Transform) For a capacity μ on $\mathcal{N}$, the possibilistic Möbius transform is a mapping $M_p : 2^{\mathcal{N}} \to [0, 1]$ defined by

$$M_p(\mathcal{A}) = \begin{cases} \mu(\mathcal{A}) & \text{if } \mu(\mathcal{A}) > \max_{\mathcal{B}\subset\mathcal{A}} \mu(\mathcal{B}), \\ 0 & \text{otherwise.} \end{cases} \tag{4.4}$$

Such capacities will allocate 0 to subsets satisfying $\mu(\mathcal{A}) = \max_{\mathcal{B}\subset\mathcal{A}} \mu(\mathcal{B})$, i.e. as is the case for all subsets in the case of possibility measures (Definition 2.14).

Example 4.2 The following capacity is given in standard representation (left) and possibilistic Möbius representation (right). Note that it is only $\mu(\{1, 3\})$ and $\mu(\{2, 3\})$ that are set to 0:

	1				1	
0.8	0.6	0.7		0.8	0	0
0.3	0.7	0.6		0.3	0.7	0.6
	0				0	

The possibilistic Zeta transform of M_p is the mapping $Z_{M_p} : 2^{\mathcal{N}} \to [0, 1]$ defined by

$$Z_{M_p}(\mathcal{A}) = \max_{\mathcal{B} \subseteq \mathcal{A}} M_p(\mathcal{B}). \tag{4.5}$$

Definition 4.4 (Necessity Möbius Transform) The necessity Möbius transform of a capacity μ on $\mathcal{N}$ is a mapping $M_n : 2^{\mathcal{N}} \to [0, 1]$ defined by

$$M_n(\mathcal{A}) = \begin{cases} \mu(\mathcal{A}) & \text{if } \mu(\mathcal{A}) < \min_{\mathcal{B} \supset \mathcal{A}} \mu(\mathcal{B}), \\ 1 & \text{otherwise.} \end{cases} \tag{4.6}$$

The necessity Zeta transform of M_n is the mapping $Z_{M_n} : 2^{\mathcal{N}} \to [0, 1]$ defined by

$$Z_{M_n}(\mathcal{A}) = \min_{\mathcal{B} \subseteq \mathcal{A}} M_n(\mathcal{B}). \tag{4.7}$$

4.3 Vector Representation

When performing calculations using software or implementing algorithms in different coding platforms, it is usually helpful to store the capacity values of μ in an array $\mathbf{v}$ of size 2^n.

This requires a consistent indexing system so that the capacity values in different representations can be recovered for interpretation or in performing different calculations. In the following, we detail some of the more common choices, each of which provides a convenient one-to-one mapping between the subsets $\mathcal{A} \subseteq \mathcal{N}$ and the set of integers $\mathcal{I} = \{0, \ldots, 2^n - 1\}$.

4.3.1 Binary Ordering

With this indexing system, we use the binary representation of an integer to determine the presence or absence of elements in the subset. For each index j of our integer set $\mathcal{I}$, we consider the binary representation, e.g. $j = 5 = 101$, with the first position from the right corresponding with $\{1\}$, the second with $\{2\}$, and so on. The digits of the binary number can be used to construct a characteristic vector $\mathbf{c} \in \{0, 1\}^n : c_{n-i+1} = 1$ if $i \in \mathcal{A}$ and 0 otherwise, with each 1 indicating that the element is present in the subset and each 0 indicating that it is not. In this case, the 1s correspond with $\{1\}$ and $\{3\}$ and so our set is $\mathcal{A} = \{1, 3\}$.

Binary ordering hence indexes the subsets of $\mathcal{N}$ in the order:

$$\emptyset, \{1\}, \{2\}, \{1,2\}, \{3\}, \{1,3\}, \{2,3\}, \{1,2,3\}, \{4\}, \ldots, \{1,2,\ldots,n\}.$$

The values of the capacity μ are mapped to the elements of the vector $\mathbf{v}$ as follows:

v_0 $= v_{(0000)}$	v_1 $= v_{(0001)}$	v_2 $= v_{(0010)}$	v_3 $= v_{(0011)}$	v_4 $= v_{(0100)}$	v_5 $= v_{(0101)}$	...
$\mu(\emptyset)$	$\mu(\{1\})$	$\mu(\{2\})$	$\mu(\{1,2\})$	$\mu(\{3\})$	$\mu(\{1,3\})$	...

An advantage of this system is that the index assigned to a subset is independent of the cardinality of $\mathcal{N}$, i.e. the capacity value $\mu(\{2,3\})$ will always be mapped to v_6 regardless of whether $\mathbf{v}$ represents a capacity on a set of three, five, or ten elements. It is hence well suited to progressively building capacities for larger and larger n, with each new variable doubling the length of $\mathbf{v}$ to accommodate the new subsets but all of the original assignments staying fixed. This is very efficient for implementation, especially for large n, and works well on shared and distributed memory architectures.

4.3.2 *Cardinality Ordering*

Another intuitive way of indexing subsets is by grouping subsets according to cardinality and then applying a lexicographic order to subsets of the same size. We refer to this as cardinality ordering:

$$\emptyset, \underbrace{\{1\}, \{2\}, \ldots, \{n\}}_{n \text{ singletons}}, \underbrace{\{1,2\}, \{1,3\}, \ldots, \{1,n\}, \{2,3\}, \ldots, \{n-1,n\}}_{\binom{n}{2} \text{ pairs}}, \{1,2,3\}, \ldots$$

Clearly, the index of a subset in this system will depend on the value of n. For $n = 3$, the value of $\{1,3\}$ will be mapped to v_5, but for $n = 4$, the same subset will be mapped to v_6. However, this ordering is particularly convenient when working with k-order capacities (see Chap. 7) where some values are fixed for a given cardinality in one or another representation. For example, with *two*-additive capacities, we have $M(\mathcal{A}) = 0$ for all $|\mathcal{A}| \geqslant 3$, and so in learning and computation, we can work with a vector storing only the Möbius values for singletons and pairs, reducing the required size of $\mathbf{v}$ significantly.

Note 4.1 It is worth mentioning that while one can convert between indices in binary and cardinality ordering using explicit transformation formulas, in our experience and to optimise for speed, it is best to use a computational algorithm to precompute an array of integers that maps one ordering to the other, or to optimise for storage it can be computed on the fly. An illustration is provided in the FMTool package mentioned later in Chap. 8.

4.3.3 Sparse Representation

This ordering is particularly amenable to capacities in Möbius representation in which many values $M(\mathcal{A}) = 0$, but not necessarily based on subset cardinalities. We can then choose to avoid explicitly storing the zero (or other fixed) values and only operate with non-zeros. In sparse vector representation, we store the array of pairs $(\mathcal{A}, M(\mathcal{A}))$, where the subsets $\mathcal{A}$ are usually represented according to the binary form detailed above, e.g. $j = 6 = 110$ for $\mathcal{A} = \{2, 3\}$. The length of such an array is hence the number of non-zero values $M(\mathcal{A})$.

It can also make sense to order the pairs $(\mathcal{A}, M(\mathcal{A}))$ in cardinality ordering, so that we first list the non-zero singletons, then non-zero pairs, and so on, because often operations are performed with subsets of the same cardinality (e.g. in (3.11) or (3.22) the coefficients in the sums depend on subset cardinality).

In the following sections, we will see that once a capacity is in vector representation, many of the computations required in working with them can be achieved through matrix-vector multiplication and other linear algebra operations. This provides a tremendous advantage, especially given the availability of a number of efficient linear algebra software libraries.

4.4 Transformations

Conversion formulas between the different capacity representations allow us to work with whichever is more convenient for calculation or interpretation. The following definition from [Gra96, GMR00] specifies a general condition for a capacity transformation to be considered a representation.

Definition 4.5 A set function $v : 2^{\mathcal{N}} \to \mathbf{R}$ is a representation of μ if there exists an invertible transform $\mathbf{T}$ such that $v = \mathbf{T}(\mu)$ and $\mu = \mathbf{T}^{-1}(v)$.

We can further refine this to focus on linear transformations.

Definition 4.6 A set function $v : 2^{\mathcal{N}} \to \mathbf{R}$ is a representation of μ if there exists an invertible linear transform $\mathbf{T}$ such that $v = \mathbf{T}(\mu)$ and $\mu = \mathbf{T}^{-1}(v)$.

From this definition, it follows that such a transformation $\mathbf{T}$ can be defined in terms of an invertible matrix $2^n \times 2^n$ such that

$$\mathbf{u} = \mathbf{T}\mathbf{v} \text{ and } \mathbf{v} = \mathbf{T}^{-1}\mathbf{u}.$$

Theorem 4.1 *A set of values $\{v(\mathcal{A})\}_{\mathcal{A} \subseteq \mathcal{N}}$ encoded into an array $\mathbf{u}$ is an equivalent representation of a capacity on $\mathcal{N}$ if and only if*

$$\mathbf{T}^{-1}_{pos(\emptyset)}\mathbf{u} = 0, \qquad \mathbf{T}^{-1}_{pos(\mathcal{N})}\mathbf{u} = 1 \qquad \textit{(boundary conditions)}$$

$$\mathbf{T}^{-1}_{pos(\{i\}\cup\mathcal{A})}\mathbf{u} - \mathbf{T}^{-1}_{pos(\mathcal{A})}\mathbf{u} \geqslant 0, \quad \forall i \in \mathcal{N}, \mathcal{A} \subseteq \mathcal{N} \setminus \{i\}, \qquad \textit{(monotonicity)}$$

where $\mathbf{T}^{-1}_j$ *is the* j*th row vector in the inverse matrix* $\mathbf{T}^{-1}$ *and* $pos(\mathcal{A})$ *is the position of subset* $\mathcal{A}$ *in the corresponding ordering of all subsets of* $\mathcal{N}$*.*

Both the dual capacity and Möbius transform, two of the more traditional capacity representations, can be expressed in this framework. Let $argpos : \{0, \ldots, 2^n\} \to \mathcal{P}(\mathcal{N})$ define a function that takes an input i, which denotes the position in the chosen ordering, and returns the corresponding subset $\mathcal{A}$. Its inverse is the pos function. For example, in binary ordering $argpos(0) = \emptyset$, $argpos(5) = \{1, 3\}$ and so on.

The following provides the transformation matrices for some of the capacity representations we have presented here and in Chap. 3:

- The duality transform matrix is $\mathbf{V} = [v_{ij}]_{2^n \times 2^n}$, $i, j = 1, ..., n$,

$$v_{ij} = \begin{cases} 1 & \text{if } i = j = pos(\emptyset) \\ 1 & \text{if } j = pos(\mathcal{N}) \text{ and } argpos(i)) \neq \emptyset \\ -1 & \text{if } j = pos(\mathcal{N} \setminus argpos(i)) \text{ and } argpos(i)) \neq \emptyset \\ 0 & \text{otherwise.} \end{cases} \tag{4.8}$$

- The transform matrix of the Möbius representation is $\mathbf{M} = [m_{ij}]_{2^n \times 2^n}$, $i, j = 0, ..., n$,

$$m_{ij} = \begin{cases} (-1)^{(|argpos(i)|-|argpos(j)|)} & \text{if } argpos(j) \subseteq argpos(i) \\ 0 & \text{otherwise.} \end{cases} \tag{4.9}$$

- The transform matrix of the co-Möbius representation is $\mathbf{C} = [c_{ij}]_{2^n \times 2^n}$, $i, j = 0, ..., n$,

$$c_{ij} = \begin{cases} (-1)^{|argpos(i) \setminus argpos(j)|} & \text{if } argpos(j) \cup argpos(i) = \mathcal{N} \\ 0 & \text{otherwise.} \end{cases} \tag{4.10}$$

- The transform matrix of the Shapley simultaneous interaction index is $\mathbf{S} = [s_{ij}]_{2^n \times 2^n}$, $i, j = 0, ..., n$,

$$s_{ij} = \frac{(-1)^{|argpos(i) \setminus argpos(j)|}}{n - |argpos(i)| + 1} \binom{n - |argpos(i)|}{|argpos(j) \setminus argpos(i)|}^{-1} \tag{4.11}$$

- The transform matrix of the Banzhaf simultaneous interaction index is $\mathbf{B} = [b_{ij}]_{2^n \times 2^n}$, $i, j = 0, ..., n$,

$$b_{ij} = \left(\frac{1}{2}\right)^{n-argpos(i)} (-1)^{|argpos(i)\setminus argpos(j)|} \tag{4.12}$$

- The transform matrix of the nonadditivity index is $\mathbf{N} = [n_{ij}]_{2^n \times 2^n}$, $i, j = 0, ..., n$,

$$n_{ij} = \begin{cases} 1 & \text{if } i = j \\ -(2^{|argpos(i)|-1})^{-1} & \text{if } argpos(j) \subset argpos(i) \\ 0 & \text{otherwise.} \end{cases} \tag{4.13}$$

- The transform matrix of the nonmodularity index is $\mathbf{D} = [d_{ij}]_{2^n \times 2^n}$, $i, j = 0, ..., n$,

$$d_{ij} = \begin{cases} 1 & \text{if } i = j \\ -1 & \text{if } argpos(j) \subset argpos(i),\ |argpos(i)| \leqslant 2 \\ -|argpos(i)|^{-1} & \text{if } argpos(j) \subset argpos(i),\ |argpos(i)| \geqslant 3, \\ & |argpos(j)| = 1 \text{ or } |argpos(i)| - 1 \\ 0 & \text{otherwise.} \end{cases} \tag{4.14}$$

4.5 Derivatives in Matrix Form

We have seen that the derivatives of set functions (Sect. 3.2) are helpful in expressing some of the transformations between different capacity representations. Recalling this definition in matrix-vector form, for the derivative $\Delta_1 \mu$, we have

$$\Delta_1 \mu = \mathbf{D}_1 \mathbf{v},$$

where $\mathbf{v}$ is the vector representation of a set function in binary ordering and

$$\mathbf{D}_1 = \begin{bmatrix} -1 & 1 & 0 & 0 & 0 & \dots & 0 \\ -1 & 1 & 0 & 0 & 0 & \dots & 0 \\ 0 & 0 & -1 & 1 & 0 & \dots & 0 \\ 0 & 0 & -1 & 1 & 0 & \dots & 0 \\ 0 & 0 & 0 & 0 & -1 & 1 \dots & 0 \\ \vdots & \vdots & & \dots & & & \vdots \\ 0 & 0 & \dots & & & -1 & 1 \\ 0 & 0 & \dots & & & -1 & 1 \end{bmatrix}$$

which is a block diagonal matrix. The repetition of rows is due to $\Delta_i \mu(\mathcal{A})$ being the same regardless of whether or not $i \in \mathcal{A}$.

The differentiation matrices for other variables $\mathbf{D}_i$ can be determined through permutations of the rows and columns of $\mathbf{D}_1$, i.e. by setting $\mathbf{D}_i = \mathbf{P}_i \mathbf{D}_1 \mathbf{P}_i$ where $\mathbf{P}_i$ is a permutation matrix. The required permutation and hence the entries of $\mathbf{P}_i$ can be obtained by exchanging the role of the first and i-th variable.

Note that the matrix $\mathbf{D}_1$ satisfies $\mathbf{D}_1\mathbf{D}_1 = \mathbf{0}$, and hence this will also be true for all $\mathbf{D}_i$, which is consistent with Definition 3.1. The second derivative can be obtained by applying the transformation iteratively, so that we have

$$\Delta_{ij}\mu = \mathbf{D}_j \mathbf{D}_i \mathbf{v} = \mathbf{D}_i \mathbf{D}_j \mathbf{v},$$

and

$$\Delta_{\mathcal{B}}\mu = \prod_{i \in \mathcal{B}} \mathbf{D}_i \mathbf{v}.$$

We shall now provide the proof for (3.2), which gives the relationship between the derivatives of a capacity and the derivatives of its dual. We will use the duality transformation matrix $\mathbf{V}$ and the anti-diagonal matrix $\mathbf{A}$, $a_{ij} = 1$ if $i = n - j + 1$ and 0 otherwise. We have

$$\mathbf{D}_i \mathbf{V} \mathbf{v} = \mathbf{A} \mathbf{D}_i \mathbf{v}$$

which can be proven directly. Next let us calculate the second derivative

$$\mathbf{D}_1 \mathbf{D}_i \mathbf{V} \mathbf{v} = \mathbf{D}_1 \mathbf{A} \mathbf{D}_i \mathbf{v},$$

(and $\mathbf{D}_1$ can be replaced by $\mathbf{D}_j$ with the help of the relevant permutation matrix). One can check directly that $\mathbf{D}_1 \mathbf{A} = -\mathbf{A}\mathbf{D}_1$ and therefore

$$(\mathbf{D}_1 \mathbf{D}_i)\mathbf{V}\mathbf{v} = -\mathbf{A}(\mathbf{D}_1 \mathbf{D}_i)\mathbf{v}.$$

Hence, $\Delta_{ij}\mu^*(\mathcal{A}) = -\Delta_{ij}\mu(\mathcal{N} \setminus \mathcal{A})$. Mathematical induction results in (3.2).

Let us now express several relations between the derivatives of set functions in matrix form following [Bel21]. We recall the definition of vector $\tilde{\mathbf{w}}$ (3.5) and other relations from Sect. 3.3:

$$\tilde{w}_{[pos(\mathcal{A})]} = \tilde{w}_{\mathcal{A}} = \frac{1}{\binom{n+1}{1}\binom{n}{|\mathcal{A}|}}.$$

and also $\mathbf{u}^t = (-1, 0, \ldots, 0, 1)$.

Then we state those relations as follows:

- The Shapley values are

$$\varphi(i) = \tilde{\mathbf{w}}^t \mathbf{D}_i \mathbf{v}.$$

- The sum of Shapley values

$$\tilde{\mathbf{w}}^t (\sum_i \mathbf{D}_i)\mathbf{v} = \mathbf{u}^t \mathbf{v}. \tag{4.15}$$

- Recursive application leads to

$$\tilde{\mathbf{w}}^t (\sum_i \mathbf{D}_i)(\sum_i \mathbf{D}_i \mathbf{v}) = \mathbf{u}^t \sum_i \mathbf{D}_i \mathbf{v}. \tag{4.16}$$

- From the properties of matrices $\mathbf{D}_i$, we get

$$\tilde{\mathbf{w}}^t (\sum_i \mathbf{D}_i)^2 \mathbf{v} = \mathbf{u}^t \sum_i \mathbf{D}_i \mathbf{v}. \tag{4.17}$$

  Hence,

$$\tilde{\mathbf{w}}^t (\sum_{i,j} \mathbf{D}_i \mathbf{D}_j)\mathbf{v} = \tilde{\mathbf{w}}^t (\sum_{i,j} \mathbf{D}_{ij})\mathbf{v} = \mathbf{u}^t \sum_i \mathbf{D}_i \mathbf{v}. \tag{4.18}$$

$$\tilde{\mathbf{w}}^t \sum_{\substack{\mathcal{B}\subseteq\mathcal{N} \\ |\mathcal{B}|=k}} \prod_{i\in\mathcal{B}} \mathbf{D}_i \mathbf{v} = \mathbf{u}^t \sum_{|\mathcal{B}|=k-1} \prod_{i\in\mathcal{B}} \mathbf{D}_i \mathbf{v}.$$

- The integral of the Choquet integral C_μ over the unit cube can be written as the scalar product

$$\tilde{\mathbf{w}}^t \mathbf{v} = \int_{[0,1]^n} C_\mu(\mathbf{x}) d\mathbf{x}. \tag{4.19}$$

4.6 Interaction Representation

While linear mappings such as those provided in previous sections allow us to verify the satisfaction of capacity properties in any representation, clearly it is useful if one can check for monotonicity and boundary conditions in the given form (Table 4.1). For the Shapley and Banzhaf interaction indices (see Definitions 3.7 and 3.9 and Table 3.1), we have the following theorems.

Theorem 4.2 ([GMS00]) *The coefficient set* $\{I(\mathcal{A})\}_{\mathcal{A}\subseteq\mathcal{N}}$ *corresponds to the Shapley simultaneous interaction index of a capacity on* $\mathcal{N}$ *if and only if*

Table 4.1 Conditions of monotonicity and convexity in different representations

Representation	Monotonicity	Convexity
Capacity $\mu(A)$	$\mu(\mathcal{A} \cup \{i\}) - \mu(\mathcal{A}) \geqslant 0,\ \forall i \in \mathcal{N}, \mathcal{A} \subseteq \mathcal{N} \setminus \{i\}$	$\mu(\mathcal{A} \cup \mathcal{B}) + \mu(\mathcal{A} \cap \mathcal{B}) \geqslant \mu(\mathcal{A}) + \mu(\mathcal{B})$
Möbius $M(\mathcal{A})$	$\sum_{\mathcal{B} \subseteq A \mid i \in \mathcal{B}} M(\mathcal{B}) \geqslant 0, \quad \forall \mathcal{A} \subseteq \mathcal{N},\ \forall i \in \mathcal{A}$	$\sum_{\mathcal{B} \subseteq A \mid i,j \in \mathcal{B}} M(\mathcal{B}) \geqslant 0, \quad \forall \mathcal{A} \subseteq \mathcal{N},\ \forall i, j \in \mathcal{A}$
Co-Möbius $\check{M}(\mathcal{A})$	$\sum_{\mathcal{B} \subseteq A \mid i \in \mathcal{B}} (-1)^{\lvert A \rvert + 1} \check{M}(\mathcal{N} \setminus \mathcal{B}) \geqslant 0, \quad \forall \mathcal{A} \subseteq \mathcal{N},\ \forall i \in \mathcal{A}$	$\sum_{\mathcal{B} \subseteq A \mid i,j \in \mathcal{B}} (-1)^{\lvert B \rvert + 1} \check{M}(\mathcal{B}) \leqslant 0, \quad \forall \mathcal{A} \subseteq \mathcal{N},\ \forall i, j \in \mathcal{A}$
Marginal Contribution $\Delta_i \mu(\mathcal{A})$	$\Delta_i \mu(\mathcal{A}) \geqslant 0,\ \forall i \in \mathcal{N}, \mathcal{A} \subseteq \mathcal{N} \setminus \{i\}$	$\Delta_i \mu(\mathcal{A}) \leqslant \Delta_i \mu(\mathcal{A} \cup \{j\}), \quad \forall\, i, j \in \mathcal{N}, \mathcal{A} \subseteq \mathcal{N} \setminus \{i, j\}$
Shapley Interaction[a] $I(\mathcal{A})$	$\sum_{\mathcal{B} \subseteq \mathcal{N} \setminus i} \beta^{\lvert \mathcal{B} \rvert}_{\lvert \mathcal{B} \cap \mathcal{A} \rvert} I(\mathcal{B} \cup \{i\}) \geqslant 0, \forall i \in \mathcal{N}, \mathcal{A} \subseteq \mathcal{N} \setminus \{i\}$	
Banzhaf Interaction $J(\mathcal{A})$	$\sum_{\mathcal{B} \subseteq \mathcal{N} \setminus i} (1/2)^{\lvert \mathcal{A} \rvert} (-1)^{\lvert \mathcal{B} \cap \mathcal{A} \rvert} J(\mathcal{B} \cup \{i\}) \geqslant 0,$ $\forall i \in \mathcal{N}, \mathcal{A} \subseteq \mathcal{N} \setminus \{i\}$	
	Antibuoyancy	
Capacity	$\mu(\mathcal{A} \cup \{i, j\}) - \mu(\mathcal{A} \cup \{j\}) \geqslant$ $\mu(\mathcal{A} \cup \{j\}) - \mu(\mathcal{A}),\ \forall \mathcal{A} \subseteq \mathcal{N} \setminus \{i, j\}\ \forall i, j \notin \mathcal{A}, i \neq j$	
Möbius	$\sum_{\mathcal{B} \subseteq \mathcal{A}} M(\mathcal{B} \cup \{i, j\}) + \sum_{\mathcal{B} \subseteq \mathcal{A}} [M(\mathcal{B} \cup \{j\}) - M(\mathcal{B} \cup \{i\})] \geqslant 0$	
Marginal contribution	$\Delta_i(\mathcal{A} \cup \{j\}) \geqslant \Delta_j(\mathcal{A})\ \forall \mathcal{A} \subseteq \mathcal{N} \setminus \{i, j\}$ and pairs $i, j \notin \mathcal{A}$	

[a] $\beta_k = -\sum_{l=0}^{k-1} \binom{k}{l} \frac{\beta_l}{k-l+1}$ and $\beta_k^l = \sum_{j=0}^{k} \binom{k}{j} \beta_{l-j}$

$$\sum_{\mathcal{A}\subseteq\mathcal{N}} \beta_{|\mathcal{A}|} I(\mathcal{A}) = 0, \quad \sum_{i\in\mathcal{N}} I(\{i\}) = 1$$

$$\sum_{\mathcal{B}\subseteq\mathcal{N}\setminus i} \beta^{|\mathcal{B}|}_{|\mathcal{B}\cap\mathcal{A}|} I(\mathcal{B}\cup\{i\}) \geqslant 0, \quad \forall i \in \mathcal{N}, \mathcal{A} \subseteq \mathcal{N}\setminus\{i\}$$

where $\beta_k = -\sum_{l=0}^{k-1} \frac{\beta_l}{k-l+1}\binom{k}{l}$, $\beta^l_k = \sum_{j=0}^{k} \binom{k}{j}\beta_{l-j}$.

Theorem 4.3 (**[GMS00]**) *The coefficient set* $\{J(\mathcal{A})\}_{\mathcal{A}\subseteq\mathcal{N}}$ *corresponds to the Banzhaf simultaneous interaction index of a capacity on* $\mathcal{N}$ *if and only if*

$$\sum_{\mathcal{A}\subseteq\mathcal{N}} (-1/2)^{|\mathcal{A}|} J(\mathcal{A}) = 0, \quad \sum_{\mathcal{A}\subseteq\mathcal{N}} (1/2)^{|\mathcal{A}|} J(\mathcal{A}) = 1$$

$$\sum_{\mathcal{B}\subset\mathcal{N}\setminus i} (1/2)^{|\mathcal{A}|}(-1)^{|\mathcal{B}\cap\mathcal{A}|} J(\mathcal{B}\cup\{i\}) \geqslant 0, \quad \forall i \in \mathcal{N}, \mathcal{A} \subseteq \mathcal{N}\setminus\{i\}$$

4.7 Nonmodularity and Nonadditivity Representations

The theorems establishing the capacity conditions for nonmodularity and nonadditivity are provided below [WB18, WB19]. In each case, the conditions can be expressed in alternate ways.

Theorem 4.4 *The coefficient set* $\{n_\mu(\mathcal{A})\}_{\mathcal{A}\subseteq\mathcal{N}}$ *corresponds to the nonadditivity index of a capacity on* $\mathcal{N}$ *if and only if*

$$n_\mu(\emptyset) = 0, \quad n_\mu(\mathcal{N}) + \sum_{\mathcal{A}\subset\mathcal{N}} \sum_{i=|\mathcal{A}|}^{n-1} \prod_{j=i}^{n-1} \frac{n-j}{2^j-1} n_\mu(\mathcal{A}) = 1$$

$$n_\mu(\mathcal{A}\cup\{i\}) + \sum_{\mathcal{B}\subset\mathcal{A}\cup\{i\}} \sum_{i=|\mathcal{B}|}^{|\mathcal{A}|} \prod_{j=i}^{|\mathcal{A}|} \frac{|\mathcal{A}|+1-j}{2^j-1} n_\mu(\mathcal{B}) - n_\mu(\mathcal{A})$$

$$-\sum_{\mathcal{B}\subset\mathcal{A}} \sum_{i=|\mathcal{B}|}^{|\mathcal{A}|-1} \prod_{j=i}^{|\mathcal{A}|-1} \frac{|\mathcal{A}|-j}{2^j-1} n_\mu(\mathcal{B}) \geqslant 0, \quad \forall i \in \mathcal{N}, \mathcal{A} \subseteq \mathcal{N}\setminus\{i\}$$

Theorem 4.5 *The coefficient set* $\{n_\mu(\mathcal{A})\}_{\mathcal{A}\subseteq\mathcal{N}}$ *corresponds to the nonadditivity index of a capacity on* $\mathcal{N}$ *if and only if*

$$n_\mu(\emptyset) = 0, \quad \sum_{\mathcal{A}\subset\mathcal{N}} (-1)^{n-|\mathcal{A}|} \frac{2^{n-1} - 2^{n-|\mathcal{A}|}}{2^{n-1} - 1} n_\mu(\mathcal{A}) = 1$$

$$\sum_{\mathcal{B}\subseteq\mathcal{A}\cup\{i\}} (-1)^{|\mathcal{A}|+1-|\mathcal{B}|} \frac{2^{|\mathcal{A}|} - 2^{|\mathcal{A}|+1-|\mathcal{B}|}}{2^{|\mathcal{A}|} - 1} n_\mu(\mathcal{B})$$

$$- \sum_{\mathcal{B}\subseteq\mathcal{A}} (-1)^{|\mathcal{A}|-|\mathcal{B}|} \frac{2^{|\mathcal{A}|-1} - 2^{|\mathcal{A}|-|\mathcal{B}|}}{2^{|\mathcal{A}|-1} - 1} n_\mu(\mathcal{B}) \geqslant 0, \quad \forall i \in \mathcal{N}, \mathcal{A} \subseteq \mathcal{N} \setminus \{i\}$$

Theorem 4.6 *The coefficient set* $\{d_\mu(\mathcal{A})\}_{\mathcal{A}\subseteq\mathcal{N}}$ *corresponds to the nonmodularity index of a capacity on* $\mathcal{N}$ *if and only if*

$$d_\mu(\emptyset) = 0, \quad \sum_{\substack{\mathcal{B}\subseteq\mathcal{N}\\|\mathcal{B}|\geqslant 2}} \binom{|\mathcal{N}|}{|\mathcal{B}|}^{-1} d_\mu(\mathcal{B}) + \sum_{i\in\mathcal{N}} d_\mu(\{i\}) = 1$$

$$\sum_{\substack{\mathcal{B}\subseteq\mathcal{A}\\|\mathcal{B}|\geqslant 1}} \binom{|\mathcal{A}|+1}{|\mathcal{B}|+1}^{-1} d_\mu(\mathcal{B}\cup\{i\})$$

$$- \sum_{\substack{\mathcal{B}\subseteq\mathcal{A}\\|\mathcal{B}|\geqslant 2}} \frac{|\mathcal{B}|}{|\mathcal{A}|+1} \binom{|\mathcal{A}|}{|\mathcal{B}|}^{-1} d_\mu(\mathcal{B}) + d_\mu(\{i\}) \geqslant 0, \quad \forall i \in \mathcal{N}, \mathcal{A} \subseteq \mathcal{N} \setminus \{i\}$$

Theorem 4.7 *The coefficient set* $\{d_\mu(\mathcal{A})\}_{\mathcal{A}\subseteq\mathcal{N}}$ *corresponds to the nonmodularity index of a capacity on* $\mathcal{N}$ *if and only if*

$$d_\mu(\emptyset) = 0, \quad \sum_{\substack{\mathcal{A}\subseteq\mathcal{N}\\|\mathcal{A}|\geqslant 2}} \sum_{k=|\mathcal{A}|}^{n} \frac{(-1)^{n-k}(n-k)|\mathcal{A}|}{k} d_\mu(\mathcal{A}) + \sum_{i\in\mathcal{N}} d_\mu(\{i\}) = 1$$

$$\sum_{\substack{\mathcal{B}\subseteq\mathcal{A}\\|\mathcal{B}|\geqslant 1}} \sum_{k=|\mathcal{B}|+1}^{|\mathcal{A}|+1} \frac{(-1)^{|\mathcal{A}|+1-k}(|\mathcal{A}|+1-k)(|\mathcal{B}|+1)}{k} d_\mu(\mathcal{B}\cup\{i\}) + d_\mu(\{i\}) \geqslant 0,$$

$$\forall i \in \mathcal{N}, \mathcal{A} \subseteq \mathcal{N} \setminus \{i\}$$

4.8 Marginal Contribution Representation

In this section, we show yet another application of matrix representation of derivatives, which is in representing capacities in terms of a marginal representation

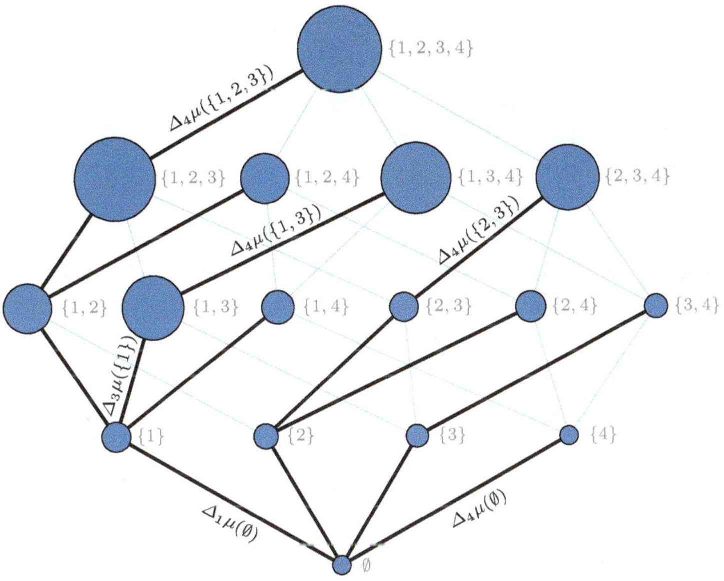

Fig. 4.2 A capacity for $n = 4$. The values of this set function are reflected in the sizes of the circles. The new basis variables are marked with thick lines (only some are labelled)

basis [BD20]. Consider the following subset of capacity derivatives that satisfy additional conditions:

$$MC = \{\Delta_i\mu(\mathcal{A}) : \forall e \in \mathcal{A} \subsetneq \mathcal{N},\ e < i\}. \tag{4.20}$$

That is, the elements in $\mathcal{A}$ are ordered lexicographically, and i needs to be larger than the largest element in $\mathcal{A}$. For example, the above-mentioned variables $\Delta_1\mu(\emptyset)$, $\Delta_3\mu(\{1\})$, $\Delta_6\mu(\{1,3\})$, $\Delta_7\mu(\{1,3,6\})$ are elements of the subset specified in (4.20), but $\Delta_1\mu(\{3\})$, $\Delta_6\mu(\{1,3,7\})$ are not. See Fig. 4.2 for a graphical illustration.

Theorem 4.8 ([BD20]) *The subset of marginal contributions* (4.20) *forms a basis in the space of capacities.*

It is convenient to represent the marginal contributions in a vector following this ordering. We place $\Delta_i\mu(\mathcal{A})$ at the position occupied in $\mathbf{v}$ by $\mu(\mathcal{A} \cup \{i\})$, with the first few elements of the vector being

$$\mathbf{d} = (\mu(\emptyset), \Delta_1\mu(\emptyset), \Delta_2\mu(\emptyset), \Delta_2\mu(\{1\}), \ldots).$$

Note that only the values from (4.20) are encoded, along with the redundant $\mu(\emptyset) = 0$.

Consider now the following block matrix:

$$\mathbf{M} = \begin{bmatrix} I_1 & & & \cdots & \\ -I_1 & I_1 & & \cdots & \\ -I_2 & & I_2 & \cdots & \\ -I_3 & & & I_3 & \\ \vdots & & \cdots & & \vdots \\ -I_n & & \cdots & & I_n \end{bmatrix} \tag{4.21}$$

where I_k are identity matrices of size k. The product $\mathbf{d} = \mathbf{Mv}$ results precisely in the vector of marginal contributions $\mathbf{d}$ in the above-described ordering. The monotonicity conditions in this representation involve a number of linear constraints that we establish in the sequel.

4.8.1 The Set of All Marginal Contributions

In this section, we look at some relations involving the whole set of marginal contributions, not only the basis mentioned earlier.

Monotonicity conditions for capacities can be expressed in terms of marginal contributions as (also see Table 4.1)

$$\Delta_i \mu(\mathcal{B}) = \mu(\mathcal{B} \cup \{i\}) - \mu(\mathcal{B}) \geqslant 0, \ \forall i \in \mathcal{N}, \mathcal{B} \subseteq \mathcal{N} \setminus \{i\}. \tag{4.22}$$

This can be simply interpreted as requiring the marginal contribution of any criterion i to a subset $\mathcal{B}$ being non-negative:

$$\Delta_i \mu(\mathcal{B}) \geqslant 0, \forall i \in \mathcal{N}, \mathcal{B} \subseteq \mathcal{N} \setminus \{i\}.$$

We have the following results from [WB20].

Theorem 4.9 *Let $\mathcal{N} = \{1, 2, ..., n\}$, and let π denote a permutation of $(1, 2, \ldots, n)$. Then μ is a capacity on $\mathcal{N}$ if and only if $\Delta_i \mu(B) \geqslant 0$ for all $i \in \mathcal{N}$ and $B \subseteq N \setminus \{i\}$ and*

$$\sum_{i=1}^{n} \Delta_{\pi(i)} \mu(\mathcal{N}_{\pi(i-1)}) = 1 \text{ for all } \pi, \tag{4.23}$$

where $\mathcal{N}_{\pi(i)} = \{\pi(1), \ldots, \pi(i)\}$, $\mathcal{N}_{\pi(0)} = \emptyset$.

Thus, whereas the number of (unfixed) parameters in standard representation is $2^n - 2$ for a normalised capacity on $\mathcal{N}$, in marginal contribution representation, this becomes $n \times 2^{(n-1)}$, which is the same as the number of monotonicity constraints in standard form. From Theorem 4.9, we can also see that in marginal contribution representation, the number of constraints is $n!$—one for each permutation.

From Theorem 4.9, we have the following corollary.

Corollary 4.1 *Let* μ *be a capacity on* $\mathcal{N}$, $\mathcal{A} = \{a_1, ..., a_{|\mathcal{A}|}\} \subseteq \mathcal{N}$, π *be a permutation of* $(1, 2, \ldots, |\mathcal{A}|)$. *Then*

$$\mu(\mathcal{A}) = \sum_{i=1}^{|\mathcal{A}|} \Delta_{a_{\pi(i)}} \mu(\mathcal{A}_{\pi(i-1)}), \ \forall \pi, \tag{4.24}$$

where $\mathcal{A}_{\pi(i)} = \{a_{\pi(1)}, ..., a_{\pi(i)}\}$, $\mathcal{A}_{\pi(0)} = \emptyset$.

We also have the following, based on the the permutation symmetry of Eq. (4.24).

Corollary 4.2

$$\mu(\mathcal{A}) = \sum_{i=1}^{|\mathcal{A}|} \Delta_{a_{\pi(i)}} \mu(\overline{\mathcal{A}}_{\pi(i)}), \ \forall \pi, \tag{4.25}$$

where $\overline{\mathcal{A}}_{\pi(i)} = \mathcal{A} \backslash \mathcal{A}_{\pi(i)}$.

Combining Theorem 4.9 and Corollary 4.2 then leads to Corollary 4.3.

Corollary 4.3 μ *is a capacity on* $\mathcal{N}$ *if and only if* $\Delta_i \mu(\mathcal{B}) \geqslant 0$ *for all* $i \in \mathcal{N}$, $\mathcal{B} \subseteq \mathcal{N} \setminus \{i\}$, *and*

$$\sum_{i=1}^{n} \Delta_{\pi(i)} \mu(\overline{\mathcal{N}}_{\pi(i)}) = 1 \textit{ for all } \pi, \tag{4.26}$$

where $\overline{\mathcal{N}}_{\pi(i)} = \mathcal{N} \backslash \mathcal{N}_{\pi(i)}$.

The following theorems establish some important relationships and conditions in different capacity representations.

Theorem 4.10 *Let* $\mathcal{A} = \{a_1, ..., a_{|\mathcal{A}|}\} \subset \mathcal{N}$. *For all permutations* π *of* $(1, 2, \ldots, |\mathcal{A}|)$,

$$\mu^*(\mathcal{A}) = \sum_{i=1}^{|\mathcal{A}|} \Delta_{a_{\pi(i)}} \mu(\mathcal{N} \backslash \mathcal{A}_{\pi(i)}) = \sum_{i=1}^{|\mathcal{A}|} \Delta_{a_{\pi(i)}} \mu(\mathcal{N} \backslash \overline{\mathcal{A}}_{\pi(i-1)}). \tag{4.27}$$

Theorem 4.11

$$M(\mathcal{A}) = \sum_{\mathcal{B} \subseteq \mathcal{A} \backslash \{i\}} (-1)^{|\mathcal{A} \backslash \mathcal{B}|-1} \Delta_i \mu(\mathcal{B}), \forall i \in \mathcal{A}. \tag{4.28}$$

Based on the fact that $M(\mathcal{A}) = \Delta_{\mathcal{A}} \mu(\emptyset)$, we also have

Corollary 4.4

$$\Delta_{\mathcal{A}}\mu(\mathcal{B}) = \sum_{\mathcal{C}\subseteq\mathcal{A}\setminus\{i\}} (-1)^{|\mathcal{A}\setminus\mathcal{C}|-1}\Delta_i\mu(\mathcal{C}\cup\mathcal{B}), \forall i \in \mathcal{A}.$$

Theorem 4.12

$$I(\mathcal{A}) = \frac{1}{|\mathcal{N}| - |\mathcal{A}| + 1}\sum_{\mathcal{B}\subseteq\mathcal{N}\setminus\mathcal{A}}\sum_{\mathcal{C}\subseteq\mathcal{A}\setminus\{i\}} \frac{(-1)^{|\mathcal{A}\setminus\mathcal{C}|}}{\binom{|\mathcal{N}|-|\mathcal{A}|}{|\mathcal{B}|}}\Delta_i\mu(\mathcal{C}\cup\mathcal{B}), \forall i \in \mathcal{A}. \tag{4.29}$$

$$J(\mathcal{A}) = \frac{1}{2^{(|\mathcal{N}|-|\mathcal{A}|)}}\sum_{\mathcal{B}\subseteq\mathcal{N}\setminus\mathcal{A}}\sum_{\mathcal{C}\subseteq\mathcal{A}\setminus\{i\}} (-1)^{|\mathcal{A}\setminus\mathcal{C}|}\Delta_i\mu(\mathcal{C}\cup\mathcal{B}), \forall i \in \mathcal{A}. \tag{4.30}$$

where $I(\mathcal{A})$ and $J(\mathcal{A})$ are the Shapley and Banzhaf interaction indices for coalitions.

Along with the expression provided for the Shapley value in Sect. 3.3, Theorem 4.12 also suggests that we can calculate the Shapley and Banzhaf values in terms of the marginal contributions as follows:

$$I(\{i\}) = \varphi(i) = \sum_{\mathcal{A}\subseteq\mathcal{N}\setminus\{i\}} \frac{|\mathcal{A}|!(|\mathcal{N}| - |\mathcal{A}| - 1)!}{|\mathcal{N}|!}\Delta_i\mu(\mathcal{A}), \tag{4.31}$$

$$J(\{i\}) = b_i = \sum_{\mathcal{A}\subseteq\mathcal{N}\setminus\{i\}} \frac{1}{2^{(|\mathcal{N}|-1)}}\Delta_i\mu(\mathcal{A}). \tag{4.32}$$

We now consider the weighted sum or average of marginal contributions. For a given capacity μ, the weighted sum of all marginal contributions is given by

$$WS_{MCh} = \sum_{i\in\mathcal{N}}\sum_{\mathcal{A}\subseteq\mathcal{N}\setminus\{i\}} w_{i,\mathcal{A}}\Delta_i\mu(\mathcal{A}), \tag{4.33}$$

where $w_{i,\mathcal{A}} \geqslant 0$, $\sum_{i\in\mathcal{N}}\sum_{\mathcal{A}\subseteq\mathcal{N}\setminus\{i\}} w_{i,\mathcal{A}} = 1$. From Eq. (4.22), we have $0 \leqslant WS_{MCh} \leqslant 1$.

Theorem 4.13 *If $w_{i,\mathcal{A}} = \frac{|\mathcal{A}|!(|\mathcal{N}|-|\mathcal{A}|-1)!}{|\mathcal{N}|!}$, then $WS_{MCh} = 1$.*

From Theorem 4.9, we have already observed that the marginal contributions on each maximal chain sum to 1, with each permutation on $\mathcal{N}$ corresponding with a maximal chain. Since we have $n!$ maximal chains for a capacity on $\mathcal{N}$, the average of all maximal chains will still be 1, i.e.

$$Ave_{MCh} = \frac{1}{n!}\sum_{\pi}\sum_{i=1}^{n} \Delta_{\pi(i)}\mu(\mathcal{N}_{\pi(i-1)}) = 1 = \sum_{i=1}^{n} I(\{i\}). \tag{4.34}$$

Since each $\Delta_i\mu(\mathcal{A})$ will be included in multiple maximal chains, these marginal contributions will appear multiple times (more precisely, $|\mathcal{A}|!(|\mathcal{N}| - |\mathcal{A}| - 1)!$ times) in the above sum. Recall also that there are $n2^{(n-1)}$ marginal contributions for any given capacity on $\mathcal{N}$. From this, we can calculate the average of all marginal contributions with equal weights as

$$Ave_{MCo} = \frac{1}{n2^{(n-1)}}\sum_{i\in\mathcal{N}}\sum_{\mathcal{A}\subseteq\mathcal{N}\setminus\{i\}} \Delta_i\mu(\mathcal{A}) = \sum_{i=1}^{n} J(\{i\}). \tag{4.35}$$

The following theorems provide conditions for the nonadditive, nonmodularity, and bipartition indices from Sects. 3.5–3.6.

Theorem 4.14 *The marginal contribution representation of the nonadditivity index is*

$$\begin{aligned} n_\mu(\mathcal{A}) &= \frac{1}{2^{|\mathcal{A}|}-2}\sum_{\mathcal{B}\subset\mathcal{A}}\sum_{k=1}^{|\mathcal{B}|}(\Delta_{a_{\pi(k)}}\mu(\mathcal{A}_{\pi(|\mathcal{A}|-|\mathcal{B}|+k-1)}) - \Delta_{a_{\pi(k)}}\mu(\mathcal{B}_{\pi(k-1)}))\\ &= \frac{1}{2^{|\mathcal{A}|}-2}\sum_{\mathcal{B}\subset\mathcal{A}}\sum_{k=1}^{|\mathcal{B}|}(\Delta_{a_{\pi(k)}}\mu(\overline{\mathcal{A}}_{\pi(k)}) - \Delta_{a_{\pi(k)}}\mu(\overline{\mathcal{B}}_{\pi(k)})). \end{aligned} \tag{4.36}$$

Theorem 4.15 *The marginal contribution representation of the nonmodularity index is*

$$d_\mu(\mathcal{A}) = \frac{1}{|\mathcal{A}|}\left(\sum_{i\in\mathcal{A}}\Delta_i\mu(\mathcal{A}) - \sum_{i\in\mathcal{A}}\Delta_i\mu(\emptyset)\right). \tag{4.37}$$

Theorem 4.16 *The marginal contribution representation of the marginal bipartition interaction is*

$$\begin{aligned} \hat{\Delta}_{\mathcal{A}}\mu(\mathcal{B}) &= \frac{1}{2^{|\mathcal{A}|}-2}\sum_{\mathcal{C}\subset\mathcal{A}}\sum_{k=1}^{|\mathcal{C}|}(\Delta_{a_{\pi(k)}}\mu(\mathcal{A}_{\pi(|\mathcal{A}|-|\mathcal{C}|+k-1)}\cup\mathcal{B}) - \Delta_{a_{\pi(k)}}\mu(\mathcal{C}_{\pi(k-1)}\cup\mathcal{B})\\ &= \frac{1}{2^{|\mathcal{A}|}-2}\sum_{\mathcal{C}\subset\mathcal{A}}\sum_{k=1}^{|\mathcal{C}|}(\Delta_{a_{\pi(k)}}\mu(\overline{\mathcal{A}}_{\pi(k)}\cup\mathcal{B}) - \Delta_{a_{\pi(k)}}\mu(\overline{\mathcal{C}}_{\pi(k)})\cup\mathcal{B}). \end{aligned} \tag{4.38}$$

4.8.2 Using a Basis in Marginal Contribution Representation

We previously introduced the idea of a subset of marginal contributions forming a basis in the space of capacities (see Theorem 4.8 and Eq. (4.20)). Besides the non-negativity of the elements in that basis, the monotonicity conditions can be given as follows. We will use matrix notation, and matrices **M**, $\mathbf{D}_i$, and the basis variables encoded in the vector **d**.

Then non-negativity of the derivatives is written as

$$\begin{bmatrix} \mathbf{D}_1 \\ \mathbf{D}_2 \\ \vdots \\ \mathbf{D}_n \end{bmatrix} \mathbf{M}^{-1}\mathbf{d} \geqslant \mathbf{0}.$$

These constraints can also be easily derived from

$$\mu(\mathcal{A}) = \sum_{i \in \mathcal{A}} \Delta_i \left(\bigcup_{j=1}^{i-1} \{j\} \right),$$

noting that the sum contains the elements of MC only and using (4.22).

We can also look as the set of supermodular capacities, in which case monotonicity conditions become redundant. The supermodularity condition becomes a partial ordering on the components of $\mathbf{d} = \mathbf{Mv}$ (see Fig. 4.3), which are also conveniently organised in blocks, namely, for all $\Delta_i\mu(\mathcal{A}) \in MC$:

$$\Delta_i\mu(\mathcal{A}) \leqslant \Delta_i\mu(\mathcal{A} \cup \{j\}), \quad \forall\, i, j \in \mathcal{N}, \mathcal{A} \subseteq \mathcal{N} \setminus \{i, j\}. \tag{4.39}$$

Theorem 4.17 *If the marginal contributions MC in* (4.20) *are non-negative, the supermodularity constraints imply monotonicity.*

Proof By Corollary 2.23 in [Gra16], supermodularity is equivalent to the requirement that $\Delta_i\mu(\mathcal{A}) \leqslant \Delta_i\mu(\mathcal{B})$ for all $\mathcal{A} \subset \mathcal{N}$ and $\mathcal{B} = \mathcal{A} \cup \{j\}$, $j \notin \mathcal{A}$, $i \in \mathcal{N} \setminus \mathcal{B}$. Then the marginal contributions (for a fixed i) form a non-decreasing sequence with subset cardinality. Since all $\Delta_i\mu(\emptyset) \geqslant 0$, we obtain the result.

Consequently, in the new capacity representation in terms of $\Delta_i\mu(\mathcal{A})$, which satisfy (4.20), specifying conditions that characterise supermodular capacities requires only the supermodularity constraints $\Delta_i\mu(\mathcal{A}) \leqslant \Delta_i\mu(\mathcal{A} \cup \{j\})$, but no monotonicity constraints, which is a significant improvement over [BJL11]. Note that the constraints (4.39) need to hold not only for $\Delta_i\mu(\mathcal{A}) \in MC$ but for all discrete derivatives. For example, in terms of the variables from MC, we need $\Delta_3\mu(\{1\}) + \Delta_4\mu(\emptyset) + \Delta_4\mu(\{1,3\}) - \Delta_3\mu(\emptyset) - \Delta_4\mu(\{1\}) - \Delta_4\mu(\{3\}) \geqslant 0$, which is not automatically satisfied.

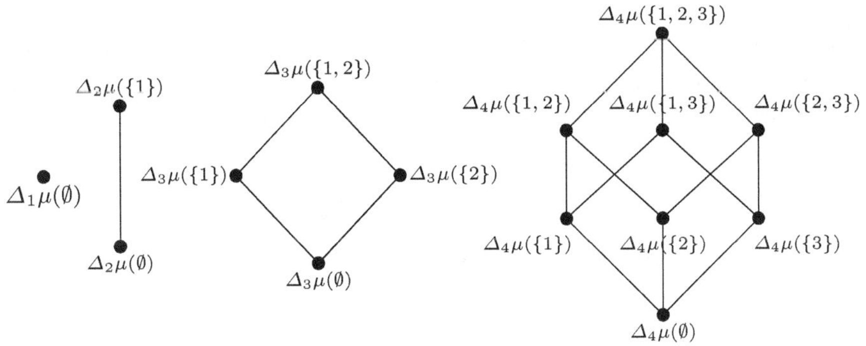

Fig. 4.3 The partial order imposed on the marginal contributions by supermodularity

In summary, here we have 2^n variables and $O(n^2 2^n)$ supermodularity constraints and no monotonicity. The case of submodular capacities is equivalent by using duality.

In terms of the discrete derivatives, an antibuoyant capacity (Sect. 2.10) is characterised by [BJ21]

$$\Delta_i(\mathcal{A} \cup \{i, j\}) \geqslant \Delta_j(\mathcal{A} \cup \{j\}) \text{ for all } \mathcal{A} \subseteq \mathcal{N} \setminus \{i, j\} \text{ and pairs } i, j \notin \mathcal{A}. \quad (4.40)$$

That is, the discrete derivatives are increasing with set cardinality.

4.9 Polytope of Normalised Capacities

It is sometimes convenient to represent capacities as points in $d = 2^n - 1$ or $d = 2^n - 2$-dimensional space $\mathbb{R}^d$ whose coordinates are the values $\mu(\mathcal{A})$ (or sometimes $M(\mathcal{A})$). The set of capacities is a polyhedron determined by the monotonicity conditions $\mu(\mathcal{A}) \leqslant \mu(\mathcal{A} \cup \{i\})$. The normalisation condition $\mu(\mathcal{N}) = 1$ leads to boundedness; hence, the set of normalised capacities is a polytope P.

The vector space of games $\mathbb{G}$ (not necessarily monotone, but grounded set functions) is spanned by unanimity games (it is a standard basis in Möbius representation) and contains P, so every capacity can be written as a linear combination of unanimity games. However, not every vertex of P is a unanimity game (hence some capacities cannot be written as convex linear combinations of unanimity games).

P is a special type of polytope known as an order polytope [Sta86, CSD19, GLS23]. In our case,

$$\mathcal{O}(2^{\mathcal{N}}) = \{f : 2^{\mathcal{N}} \to [0, 1] \mid f(x) \leqslant f(y) \text{ if } x \preceq y\}.$$

It is defined by the $n2^{n-1}$ non-redundant linear constraints.

Stanley demonstrates in [Sta86] that an order polytope can be triangulated into simplices, with each simplex having an equal volume. The number of elements of the simplicial partition of this polytope is given by the OEIS[1] sequence A046873, which starts from $n = 1$ and reads

$$1, 2, 48, 1680384, 14807804035657359360,$$

$$141377911697227887117195970316200795630205476957716480$$

and so forth (the actual subsequent values are currently unknown). Consequently, these simplices cannot be explicitly enumerated for $n > 4$.

A polytope can also be represented by its vertices. The order polytope P has vertices that are all 0–1 capacities (different from 0), and their number is given by the OEIS sequence A000372 (Dedekind numbers):[2] $2, 3, 6, 20, 168, 7581, 7828354,$ 2414682040998, and continues indefinitely. Due to the exceptionally large number of vertices, it is not practical to represent this polytope through its vertices, but solely as a set of half-spaces, at least for $n \geqslant 5$.

The Chebyshev centre of the polytope of capacities is the additive symmetric capacity $\mu(\mathcal{A}) = \frac{|\mathcal{A}|}{n}$.

Within the polytope of capacities lie their mentioned subsets that are also polytopes, often with a special structure, which will be mentioned where relevant further on. For now, we note some of these subsets:

- The set of belief measures is a simplex (in Möbius representation, it is the unit simplex (a set of non-negative values that add to one), whose vertices are unanimity games). So is the set of plausibility measures.
- The set of additive measures is the unit simplex of dimension n and is the intersection of the sets of belief and plausibility measures.
- The set of symmetric capacities is also the unit simplex of dimension n.
- The set of supermodular capacities is no longer an order polytope.
- The set of balanced capacities is an order polytope.
- The set of k-additive capacities (see Chap. 7) is a rather complex polytope studied in [CM06], whose vertices can be different from 0–1 capacities.
- The set of two-additive capacities is the convex hull of 0–1 two-additive capacities. These are precisely two-additive unanimity games, dual unanimity games, and additive 0–1 measures, and there are n^2 such vertices.
- p-symmetric and k-interactive capacities form order polytopes of the respective dimensions.

[1] The Online Encyclopedia of Integer Sequences, http://oeis.org.

[2] This sequence starts with $n = 0$ and lists the values of all 0–1 monotone set functions. For capacities, we exclude constant functions $\mu = 0$ and $\mu = 1$, so the actual number of 0–1 capacities is two less than the number in that sequence.

References

[Bel21] G. Beliakov, On the derivatives of set functions in matrix representation. Inf. Sci. **546**, 469–480 (2021)

[BD20] G. Beliakov, D. Divakov, On representation of fuzzy measures for learning Choquet and Sugeno integrals. Knowledge Based Syst. **189**, 105134 (2020)

[BJ21] G. Beliakov, S. James, Choquet integral optimisation with constraints and the buoyancy property for fuzzy measures. Inf. Sci. **578**, 22–36 (2021)

[BJL11] G. Beliakov, S. James, G. Li, Learning Choquet integral-based metrics in semi-supervised classification. IEEE Trans. Fuzzy Syst. **19**, 562–574 (2011)

[CSD19] E.F. Combarro, J. Hurtado de Saracho, I. Díaz, Minimals Plus: An improved algorithm for the random generation of linear extensions of partially ordered sets. Inf. Sci. **501**, 50–67 (2019)

[CM06] E.F. Combarro, P. Miranda, Identification of fuzzy measures from sample data with genetic algorithms. Comput. Oper. Res. **33**(10), 3046–3066 (2006)

[Gra97] M. Grabisch, k-Order additive discrete fuzzy measures and their representation. Fuzzy Sets Syst. **92**, 167–189 (1997)

[Gra00] M. Grabisch, The interaction and Möbius representation of fuzzy measures on finite spaces, k-additive measures: A survey, in *Fuzzy Measures and Integrals. Theory and Applications*, eds. by M. Grabisch, T. Murofushi, M. Sugeno (Physica-Verlag, Heidelberg, 2000), pp. 70–93

[Gra16] M. Grabisch, *Set Functions, Games and Capacities in Decision Making* (Springer, Berlin, New York, 2016)

[GMR00] M. Grabisch, J.-L. Marichal, M. Roubens, Equivalent representations of set functions. Math. Oper. Res. **25**, 157–178 (2000)

[GMS00] M. Grabisch, T. Murofushi, M. Sugeno (eds.), *Fuzzy Measures and Integrals. Theory and Applications* (Physica-Verlag, Heidelberg, 2000)

[Gra96] M. Grabisch, The representation of importance and interaction of features by fuzzy measures. Pattern Recognit. Lett. **17**(6), 567–575 (1996)

[GLS23] M. Grabisch, C. Labreuche, P. Sun, An approximation algorithm for random generation of capacities. Order, 1–26 (2023)

[Mar02] J.-L. Marichal, Aggregation of interacting criteria by means of the discrete Choquet integral. Aggregation Operators, in *New Trends and Applications*, eds. by T. Calvo, G. Mayor, R. Mesiar (Physica-Verlag, Heidelberg, New York, 2002), pp. 224–244

[Sta86] R.P. Stanley, Two poset polytopes. Discrete Comput. Geometry **1**, 9–23 (1986)

[WB18] J.-Z. Wu, G. Beliakov, Nonadditivity index and capacity identification method in the context of multicriteria decision making. Inf. Sci. **467**, 398–406 (2018)

[WB19] J.-Z. Wu, G. Beliakov, Nonmodularity index for capacity identifying with multiple criteria preference information. Inf. Sci. **492**, 164–180 (2019)

[WB20] J.-Z. Wu, G. Beliakov, Marginal contribution representation of capacity based multicriteria decision making. Int. J. Intell. Syst. **35**(3), 373–400 (2020)

Chapter 5
Fuzzy Integrals

Abstract Integration defined with respect to a capacity allows for aggregation to be performed in a way that accounts for interaction between variables. The Choquet and Sugeno integrals are two such examples of what are often referred to as "fuzzy integrals", with their behaviour as aggregation functions understood from the properties of the underlying capacity and related value and interaction indices. This chapter gives an overview of these functions, including some alternative expressions and practical steps for calculation.

5.1 The Many Fuzzy Integrals

The term "fuzzy integral" is often used broadly to refer to the various integration-like functions that use capacities, fuzzy measures, nonadditive measures, and other related concepts in their definition. Sugeno's fuzzy integral was introduced in [Sug74], defined with respect to monotone, but not necessarily additive, measures he called "fuzzy measures". These were later identified (e.g. in [Web84, MS89]) as having an equivalent form to Choquet's capacities [Cho54], which at a similar time had also seen revived attention from Schmeidler in the field of economics [Sch86]. He is often credited as being the first to use the name "Choquet integral" [WK09],[1] while Sugeno's function is often referred to simply as the Sugeno integral.

The Choquet and Sugeno integrals can be seen as capturing two distinct approaches to nonadditive integration, with the Choquet integral generalising Lebesgue integration by relaxing the additivity property and the Sugeno integral representing the ordinal approach [Dub96]. Both functions have since been studied extensively, with numerous practical applications as well as generalisations and other theoretical results. While the more general theory of fuzzy and nonadditive integration is rich, in this chapter, we will contain our considerations to the Choquet and Sugeno integral, focusing primarily on their discrete and finite cases.

[1] Wang and Klir also note that the Choquet integral is equivalent to a function introduced by Vitali in 1925 (with the paper translated into English in 1997 [Vit97]).

G. Beliakov et al., *Choquet Capacities and Fuzzy Integrals*, Theory and Applications of Computability, https://doi.org/10.1007/978-3-031-97070-2_5

Further reading on the topic can be found in [GMS00, Gra16, WK09]. Different mathematical representations of the Choquet and Sugeno integral are summarised in Table 5.1 at the end of the chapter.

5.2 Choquet Integral

In the discrete case, the Choquet integral allows us to aggregate a set of finite inputs with weights and behaviour determined by a capacity. The various properties and indices presented in Chaps. 2–4 can be used to help construct capacities so that certain behaviours or relationships are modelled, or when the capacity values and learned from data, such indices can be used to help interpret the importance and interactions between variables. The appeal of being able to replace simpler functions such as the weighted arithmetic mean with the Choquet integral, or designing new modelling processes around it, has led to a number of successful applications in multicriteria evaluation [NT07, GKM08], data modelling [Yan+08, RSMG09, BJ11], attitude analysis [CCT02, LO05, LO06], and classification [NT05, KCP05].

Let us first recall the definition of the Choquet integral as was presented in Chap. 1 (Definition 1.42).

Definition 5.1 (Discrete Choquet Integral) The discrete Choquet integral with respect to a capacity μ is given by

$$C_\mu(\mathbf{x}) = \sum_{i=1}^{n} x_{(i)}[\mu(\{j|x_j \geqslant x_{(i)}\}) - \mu(\{j|x_j \geqslant x_{(i+1)}\})], \tag{5.1}$$

where $\mathbf{x}_\nearrow = (x_{(1)}, x_{(2)}, \ldots, x_{(n)})$ is a non-decreasing permutation of the input $\mathbf{x}$, and $x_{(n+1)} = \infty$ by convention.

Examples of the Choquet integral when $n = 2$ are depicted as surface plots in Fig. 5.1.

5.2.1 *Alternative Expressions*

By introducing the notation H_i to denote the subsets associated with each $x_{(i)}$, referred to as the decumulative distribution in [Gra16], Eq. (5.1) can be rewritten as

$$C_\mu(\mathbf{x}) = \sum_{i=1}^{n} x_{(i)}[\mu(H_i) - \mu(H_{i+1})], \tag{5.2}$$

where $H_i = \{(i), \ldots, (n)\}$ is the subset of indices of the $n-i+1$ largest components of $\mathbf{x}$ and $H_{n+1} = \emptyset$.

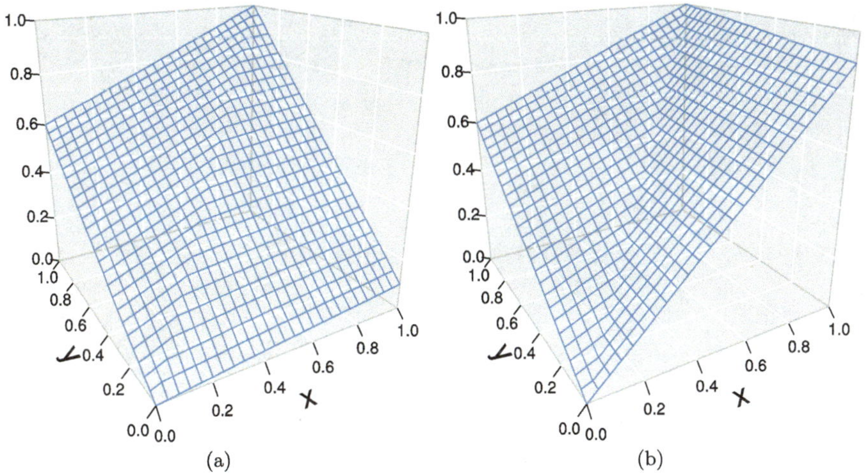

Fig. 5.1 3D plots of Choquet integrals for (**a**) $\mu(\{1\}) = 0.1, \mu(\{2\}) = 0.6$ and (**b**) $\mu(\{1\}) = 0.9, \mu(\{2\}) = 0.6$. The value of $\mu(\{1, 2\})$ is 1 in both cases

Using marginal contribution representation, the $[\mu(H_i) - \mu(H_{i+1})]$ terms can be expressed succinctly as $\Delta_{(i)}\mu(H_i)$, so we can have

$$C_\mu(\mathbf{x}) = \sum_{i=1}^{n} x_{(i)} \cdot \Delta_{(i)}\mu(H_i). \tag{5.3}$$

The summation in (5.2) can also be rearranged to obtain

$$C_\mu(\mathbf{x}) = \sum_{i=1}^{n} \left[x_{(i)} - x_{(i-1)}\right] \mu(H_i). \tag{5.4}$$

where $x_{(0)} = 0$ by convention, and $H_i = \{(i), \ldots, (n)\}$.

Figures 5.2 and 5.3 help illustrate the differences in calculation approach between Eqs. (5.2) and (5.4). Both can be seen in terms of dissecting the (nonadditive) area defined by the inputs, with (5.2) and (5.4), respectively, leading to the summation of vertical and horizontal "slices".

These representations make it easy to see the connection to traditional integration, where the value (height) of each contribution x_i would be multiplied by its density (width). The key difference is that these "widths" are no longer additive.

Note 5.1 In the case of ties, both calculations for $C_\mu(\mathbf{x})$ will be equivalent, although they may call on different subsets. For example, suppose $x_1 = x_3$ and the ordering $x_1 \leqslant x_3 < x_2$ is invoked for $\mathbf{x}_\nearrow$. Equation (5.2) gives

$$C_\mu(\mathbf{x}) = x_1[\mu(\{1, 2, 3\}) - \mu(\{1, 2, 3\})] + x_3[\mu(\{1, 2, 3\}) - \mu(\{2\})] + x_2\mu(\{2\}),$$

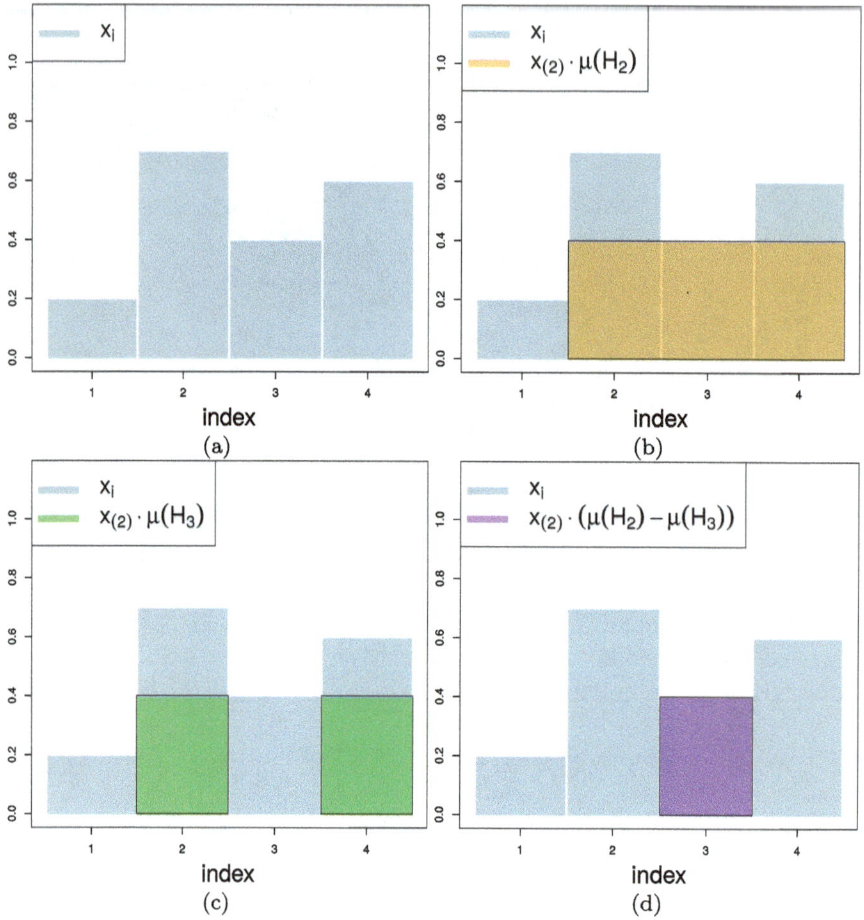

Fig. 5.2 For an input vector $\mathbf{x} = (0.2, 0.7, 0.4, 0.6)$ (depicted in (**a**)), the contribution associated with x_3 (which will be $x_{(2)}$ in the calculation) is calculated by starting with the rectangle $x_3 \cdot \mu(H_2)$ (shown in (**b**)) and subtracting the rectangle with its "width" measured according to set H_3, in this case consisting of the second and fourth elements (shown in (**c**)). The contribution is shown as (**d**); however, note that these rectangle "widths" are not proportional to the visual representation

while Eq. (5.4) would be calculated as

$$C_\mu(\mathbf{x}) = x_1[\mu(\{1, 2, 3\}) - \mu(\{2, 3\})] + x_3[\mu(\{2, 3\}) - \mu(\{2\})] + x_2\mu(\{2\}).$$

In both cases, the repeated value of x_1, x_3 is multiplied by $[\mu(\{1, 2, 3\}) - \mu(\{2\})]$ overall.

While Eqs. (5.1)–(5.4) only use the capacity values induced by the input ordering in their calculation, the following approach allows us to express the Choquet integral

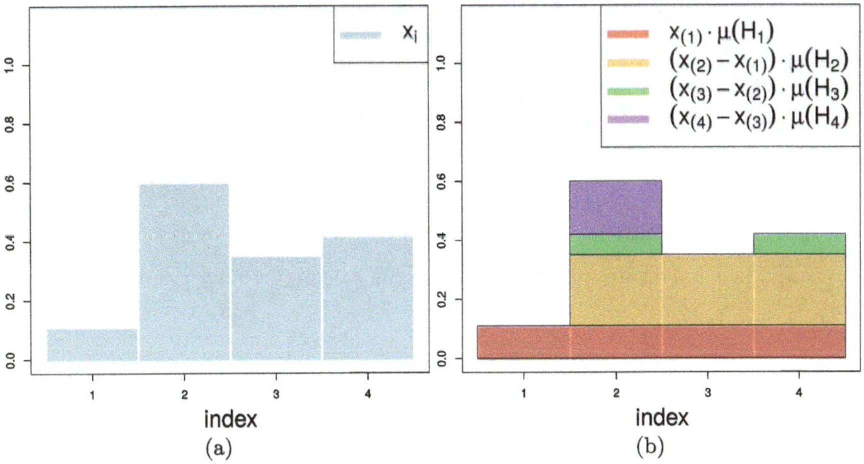

Fig. 5.3 Image in (**b**) shows the calculation breakdown for the vector in (**a**), $\mathbf{x} = (0.11, 0.6, 0.35, 0.42)$. When the capacity is additive, the effective "width" of each section would be the sum of the individual columns, which could be weighted but overall would add to 1. However, if the capacity is nonadditive, then the measure of each section may not be proportional to any visual representation and needs only to satisfy $\mu(H_1) \geqslant \mu(H_2) \geqslant \mu(H_3) \geqslant \mu(H_4)$

as a linear function with respect to all the values of μ, which is useful when learning capacities from data using linear programming and other methods.

For each $\mathcal{A} \subseteq \mathcal{N}$, let us define the function $g_{\mathcal{A}}$ such that

$$g_{\mathcal{A}}(\mathbf{x}) = \max(0, \min_{i \in \mathcal{A}} x_i - \max_{i \in \mathcal{N} \setminus \mathcal{A}} x_i), \tag{5.5}$$

The maximum and minimum over an empty set are taken as 0. Note that we will have $g_{\mathcal{A}}(\mathbf{x}) = 0$ unless $\mathcal{A}$ is the subset of indices of the k largest components of $\mathbf{x}$, in which case $g_{\mathcal{A}}(\mathbf{x}) = x_{\searrow(k)} - x_{\searrow(k+1)}$. Then it is a matter of simple calculation to show that

$$C_\mu(\mathbf{x}) = \sum_{\mathcal{A} \subseteq \mathcal{N}} \mu(\mathcal{A}) g_{\mathcal{A}}(\mathbf{x}). \tag{5.6}$$

The Choquet integral in Möbius representation also involves a linear summation involving all subsets:

$$C_\mu(\mathbf{x}) = \sum_{\mathcal{A} \subseteq \mathcal{N}} M(\mathcal{A}) \min_{i \in \mathcal{A}} x_i = \sum_{\mathcal{A} \subseteq \mathcal{N}} M(\mathcal{A}) h_{\mathcal{A}}(\mathbf{x}), \tag{5.7}$$

with $h_{\mathcal{A}}(\mathbf{x}) = \min_{i \in \mathcal{A}} x_i$. By using Definition 1.41, we obtain

$$C_\mu(\mathbf{x}) = \sum_{\mathcal{A} \subseteq \mathcal{N}} \mu(\mathcal{A}) \sum_{\mathcal{B}|\mathcal{A} \subseteq \mathcal{B}} (-1)^{|\mathcal{B}\setminus\mathcal{A}|} \min_{i \in \mathcal{B}} x_i, \tag{5.8}$$

which can be compared with (5.6) so that we have

$$g_{\mathcal{A}}(\mathbf{x}) = \max(0, \min_{i \in \mathcal{A}} x_i - \max_{i \in \mathcal{N}\setminus\mathcal{A}} x_i) = \sum_{\mathcal{B}|\mathcal{A} \subseteq \mathcal{B}} (-1)^{|\mathcal{B}\setminus\mathcal{A}|} h_{\mathcal{B}}(\mathbf{x}). \tag{5.9}$$

Using (5.6) and representing μ as the vector $\mathbf{v}$ using the binary indexing system from Sect. 4.3, we can write

$$C_\mu(\mathbf{x}) = \sum_{j=0}^{2^n-1} v_j g_j(\mathbf{x}) =< \mathbf{g}(\mathbf{x}), \mathbf{v} >, \tag{5.10}$$

where as earlier, functions g_j, $j = 0, \ldots, 2^n - 1$ are defined by

$$g_j(\mathbf{x}) = \max(0, \min_{i \in \mathcal{J}} x_i - \max_{i \in \mathcal{N}\setminus\mathcal{J}} x_i), \tag{5.11}$$

and the characteristic vector of the set $\mathcal{J} \subseteq \mathcal{N}$ corresponds to the binary representation of j.

5.2.2 *Main Properties*

In the following, we summarise some of the key properties known for the Choquet integral:

- The Choquet integral is a continuous piecewise linear idempotent aggregation function.
- An aggregation function is a Choquet integral if and only if it is homogeneous, shift-invariant, and *comonotone additive*, i.e. $C_\mu(\mathbf{x}+\mathbf{y}) = C_\mu(\mathbf{x}) + C_\mu(\mathbf{y})$ for all comonotone[2] $\mathbf{x}, \mathbf{y}$;
- The Choquet integral is uniquely defined by its values at the vertices of the unit cube $[0, 1]^n$, i.e. at the points $\mathbf{x}$, with coordinates $x_i \in \{0, 1\}$. Note that there are 2^n such points, the same as the number of values that determines the capacity μ.
- The Choquet integral is Lipschitz-continuous, with the Lipschitz constant 1 in any p-norm.

[2] Two vectors $\mathbf{x}, \mathbf{y} \in \mathbb{R}^n$ are called comonotone if there exists a common permutation P of $\{1, 2, \ldots, n\}$, such that $x_{P(1)} \leqslant x_{P(2)} \leqslant \cdots \leqslant x_{P(n)}$ and $y_{P(1)} \leqslant y_{P(2)} \leqslant \cdots \leqslant y_{P(n)}$. Equivalently, this condition is frequently expressed as $(x_i - x_j)(y_i - y_j) \geqslant 0$ for all $i, j \in \{1, \ldots, n\}$.

- The Choquet integral is a convex function if and only if the underlying capacity is submodular [Den94, Cho54] and if and only if it is a subadditive function, i.e. $C_\mu(\mathbf{x}+\mathbf{y}) \leqslant C_\mu(\mathbf{x}) + C_\mu(\mathbf{y})$.
- The class of Choquet integrals includes weighted means and OWA functions, as well as the minimum, maximum, and order statistics as special cases (see Chap. 6).
- A linear convex combination of Choquet integrals with respect to capacities μ_1 and μ_2, $\alpha C_{\mu_1} + (1-\alpha)C_{\mu_2}$, $\alpha \in [0, 1]$, is also a Choquet integral with respect to $\mu = \alpha\mu_1 + (1-\alpha)\mu_2$.[3]
- A pointwise maximum or minimum of Choquet integrals is not necessarily a Choquet integral (but it is in the bivariate case).
- The class of Choquet integrals is closed under duality.
- Choquet integrals have neutral and absorbent elements only in the limiting cases of min and max.

Certain types of capacity can lead to specific behaviour and properties for the Choquet integral. For example, the weighted arithmetic mean (WAM), ordered weighted averaging (OWA) operator, and weighted OWA (WOWA) are all obtained as special cases with respect to specific types of capacity (see Chap. 2). Some of these will be discussed later in Chap. 6.

5.2.3 Calculation

The following calculation is based on Eq. (5.4). We assume a set of vector pairs $((x_1, 1), (x_2, 2), \ldots, (x_n, n))$, where the second component of each pair is just the index i of x_i and helps us keep track of all permutations.

Calculation of $C_\mu(\mathbf{x})$.

1. Sort the components $((x_1, 1), (x_2, 2), \ldots, (x_n, n))$ with respect to the first component of each pair in non-decreasing order. We obtain $((x_{(1)}, i_1), (x_{(2)}, i_2), \ldots, (x_{(n)}, i_n))$, so that $x_{(j)} = x_{i_j}$ and $x_{(j)} \leqslant x_{(j+1)}$ for all i. Let $x_{(0)} = 0$.
2. Let $\mathcal{T} = \{1, \ldots, n\}$, and $\mathcal{S} = 0$.
3. For $j = 1, \ldots, n$ do
   a. $\mathcal{S} := \mathcal{S} + [x_{(j)} - x_{(j-1)}]\mu(\mathcal{T})$
   b. $\mathcal{T} := \mathcal{T} \setminus \{i_j\}$
4. Return $\mathcal{S}$.

[3] As a consequence, this property holds for a linear convex combination of any number of capacities.

Example 5.1 Let $n = 3$, and $\mathbf{x} = (0.8, 0.1, 0.6)$. In terms of μ, we will calculate as follows:
Step 1. We take $((0.8, 1), (0.1, 2), (0.6, 3))$.
Sort this vector of pairs to obtain $((0.1, 2), (0.6, 3), (0.8, 1))$.
Step 2. Take $\mathcal{T} = \{1, 2, 3\}$ and $\mathcal{S} = 0$.
Step 3. a) $\mathcal{S} := 0 + [0.1 - 0]\mu(\{1, 2, 3\}) = 0.1 \times 1 = 0.1$.
b) $\mathcal{T} = \{1, 2, 3\} \setminus \{2\} = \{1, 3\}$.
a) $\mathcal{S} := 0.1 + [0.6 - 0.1]\mu(\{1, 3\}) = 0.1 + 0.5\mu(\{1, 3\})$.
b)$\mathcal{T} := \{1, 3\} \setminus \{3\} = \{1\}$.
a) $\mathcal{S} := [0.1 + 0.5\mu(\{1, 3\})] + [0.8 - 0.6]\mu(\{1\})$.
Step 4. Therefore, $C_\mu(\mathbf{x}) = 0.1 + 0.5\mu(\{1, 3\}) + 0.2\mu(\{1\})$.

5.2.4 *Orness*

The degree of orness is a useful property defined by Dujmovic [Duj74, Duj73] (and rediscovered, e.g. in [Yag88, FR94]) for averaging aggregation functions. It gives an overall measure of how similar a function is to the maximum function, with an orness level of 1 corresponding with the maximum itself and 0 with the minimum.[4] It hence ranges between 0 and 1, with the definition as follows [FR94, FSM03, Duj74].

Definition 5.2 (Measure of orness) Let f be an averaging aggregation function on $\mathbb{I}^n = [0, 1]^n$. Then its measure of orness is

$$orness(f) = \frac{\int_{\mathbb{I}^n} f(\mathbf{x})\, d\mathbf{x} - \int_{\mathbb{I}^n} \min(\mathbf{x})\, d\mathbf{x}}{\int_{\mathbb{I}^n} \max(\mathbf{x})\, d\mathbf{x} - \int_{\mathbb{I}^n} \min(\mathbf{x})\, d\mathbf{x}}. \tag{5.12}$$

The integral calculations for the maximum and minimum were provided in [Duj73], resulting in simple expressions:

$$\int_{[0,1]^n} \max(\mathbf{x})\, d\mathbf{x} = \frac{n}{n+1} \text{ and } \int_{[0,1]^n} \min(\mathbf{x})\, d\mathbf{x} = \frac{1}{n+1}. \tag{5.13}$$

The orness of a Choquet integral can be calculated directly from the values of its capacity, μ.

Theorem 5.1 (Orness of Choquet Integral [Mar04]) *For any capacity μ, the orness of the Choquet integral with respect to μ is*

$$orness(C_\mu) = \frac{1}{n-1} \sum_{\mathcal{A} \subset \mathcal{N}} \frac{(n - |\mathcal{A}|)!\,|\mathcal{A}|!}{n!} \mu(\mathcal{A}). \tag{5.14}$$

[4] Recall that averaging aggregation functions are bound between the minimum and maximum (Definition 1.3).

In terms of the Möbius transform, the orness is calculated as follows:

$$orness(C_\mu) = \frac{1}{n-1} \sum_{\mathcal{A} \subseteq \mathcal{N}} \frac{n - |\mathcal{A}|}{|\mathcal{A}| + 1} M(\mathcal{A}), \tag{5.15}$$

where $M(\mathcal{A})$ is the Möbius representation of $\mu(\mathcal{A})$.

Example 5.2 For the following capacity with $n = 3$,

$$\begin{array}{ccc} & 1 & \\ 0.8 & 0.5 & 0.7 \\ 0.3 & 0.5 & 0.4 \\ & 0 & \end{array}$$

the orness (using Eq. (5.14)) is given by

$$\begin{aligned} orness &= \frac{1}{2}\left(\frac{\mu(\{1\}) + \mu(\{2\}) + \mu(\{3\})}{3} + \frac{\mu(\{1,2\}) + \mu(\{1,3\}) + \mu(\{2,3\})}{3}\right) \\ &= \frac{0.3 + 0.5 + 0.4 + 0.8 + 0.5 + 0.7}{6} = 0.533. \end{aligned}$$

Additive capacities always result in Choquet integrals with an orness of 0.5.

An alternative approach to calculation is provided by the following theorem.

Theorem 5.2 ([WB20]) *The orness of the Choquet integral C_μ is given as*

$$orness(C_\mu) = \frac{1}{n!} \sum_{\pi} C_\mu(\mathbf{r}_\pi),$$

where $\mathbf{r}_\pi$ are permutations of the vector $\mathbf{r} = (\frac{n-1}{n-1}, \frac{n-2}{n-1}, \ldots, 0)$ and the sum is over all $n!$ permutations.

5.3 Sugeno Integral

The Sugeno integral does not generalise the Lebesgue integral as the Choquet integral does, but rather approaches integration in a unique way that works for both ordinal and numeric inputs. Semantically, the Sugeno integral of an input vector $\mathbf{x}$ will give an output of at least y if there is a subset such that $\mu(\mathcal{A}) = y$ and for all members $i \in \mathcal{A}$, it holds that $x_i \geqslant y$. In this sense, it does not matter whether y is a number like 0.6 or an ordinal rating like "very good", in which case we would require all members of $\mathcal{A}$ to be at least *very good* and the value of $\mu(\mathcal{A})$ to be *very good*.

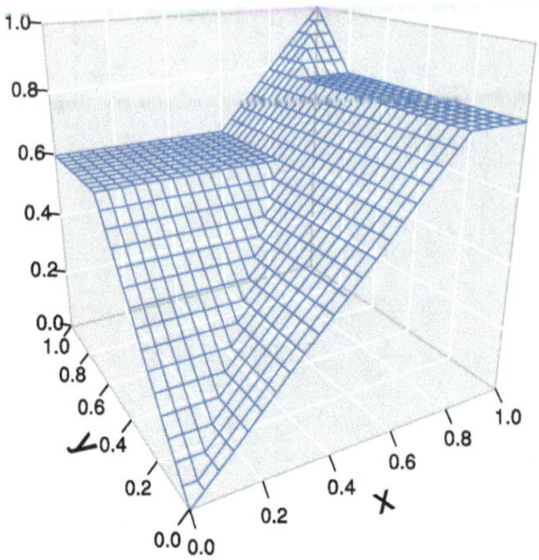

(a)

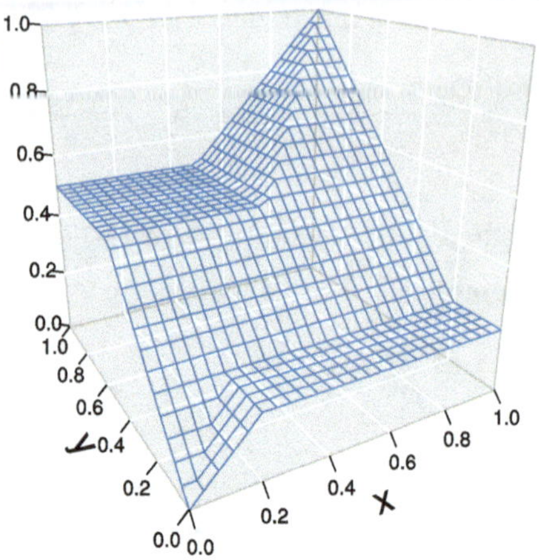

(b)

Fig. 5.4 3D plots of Sugeno integral functions with (**a**) $\mu(\{1\}) = 0.8$, $\mu(\{2\}) = 0.6$, $\mu(\{1, 2\}) = 1$ and (**b**) $\mu(\{1\}) = 0.2$, $\mu(\{2\}) = 0.5$, $\mu(\{1, 2\}) = 1$

The following definition is consistent with a number of more generalised nonadditive integral frameworks, expressing the calculation with respect to all subsets of $\mathcal{N}$.

Definition 5.3 (Discrete Sugeno Integral) The Sugeno integral with respect to a capacity μ is given by

$$S_\mu(\mathbf{x}) = \bigvee_{\mathcal{A} \subseteq \mathcal{N}} \left((\min_{i \in \mathcal{A}} x_i) \wedge \mu(\mathcal{A}) \right), \tag{5.16}$$

where $\vee$ is the maximum and $\wedge$ is the minimum.

Examples of Sugeno integrals when $n = 2$ are shown in Fig. 5.4. A distinguishing feature of these graphs are the level sections corresponding with the capacity values that take effect over specific parts of the domain.

5.3.1 *Alternative Expressions*

The following definition is given in terms of the same $\mu(H_i)$ decompositions as was used for the Choquet integral (also see Definition 1.43):

$$S_\mu(\mathbf{x}) = \max_{i=1,\dots,n} \min\{x_{(i)}, \mu(H_i)\}, \tag{5.17}$$

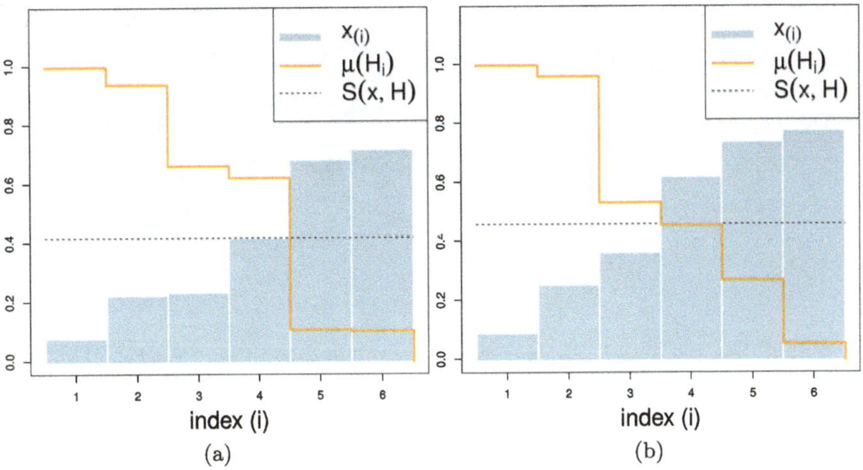

Fig. 5.5 Example input vectors (after ordering) and corresponding H_i sequences. The value of the Sugeno integral is the highest point of overlap. In (**a**), the output coincides with one of the inputs, while in (**b**) it coincides with one of the H_i

where $\mathbf{x}_{\nearrow} = (x_{(1)}, x_{(2)}, \ldots, x_{(n)})$ is a non-decreasing permutation of the input $\mathbf{x}$, and $H_i = \{(i), \ldots, (n)\}$.

This eliminates the need to consider many of the redundant subsets that are included in Eq. (5.16) and can be visualised as in Fig. 5.5, with the output being the highest point of overlap of the increasing input vector $x_{(1)}, x_{(2)}, \ldots, x_{(n)}$ and decreasing sequence $\mu(H_1), \mu(H_2), \ldots, \mu(H_n)$.

Whether expressed in terms of the decumulative distribution subsets or across all values of μ, only min and max operations are required, allowing the Sugeno integral to be applied in both numeric and ordinal contexts.

Example 5.3 For the ordinal labels {low, neutral, high, very high}, consider the capacity defined for $n = 3$ such that (also depicted in Fig. 5.6)

$$\mu(\{1, 2, 3\}) = \mu(\{1, 2\}) = \text{very high}, \mu(\{1, 3\}) = \text{neutral}, \mu(\{2, 3\}) = \text{high},$$

$$\mu(\{1\}) = \text{neutral}, \mu(\{2\}) = \text{high}, \mu(\{3\}) = \text{low}, \mu(\emptyset) = \text{low}.$$

The Sugeno integral of the inputs $\mathbf{x} =$ (neutral, very high, low) is given by (using Eq. (5.17))

$$\begin{aligned} S_\mu(\mathbf{x}) &= \max(\min\{\text{low}, \mu(\{1, 2, 3\})\}, \min\{\text{neutral}, \mu(\{1, 2\})\}, \\ &\quad \min\{\text{very high}, \mu(\{2\})\}) \\ &= \max(\min\{\text{low}, \text{very high}\}, \min\{\text{neutral}, \text{very high}\}, \min\{\text{very high}, \text{high}\}) \\ &= \max(\text{low}, \text{neutral}, \text{high}) = \text{high}. \end{aligned}$$

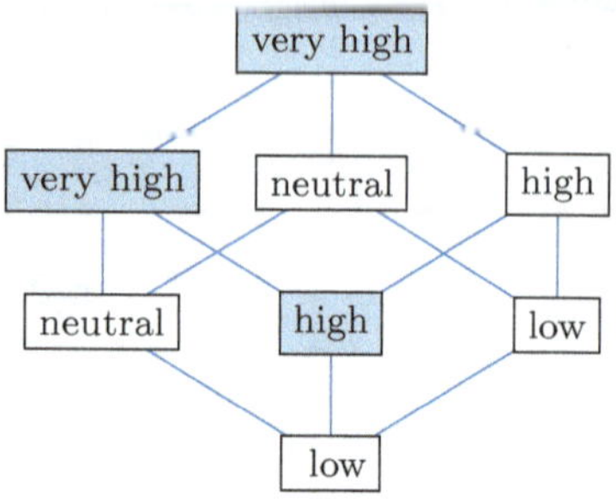

Fig. 5.6 Ordinal valued capacity for $n = 3$ from Example 5.3 with the values along the simplex used highlighted

Rather than the maximum over minima, we can instead perform calculation as the minimum over maxima, either as

$$S_\mu(\mathbf{x}) = \bigwedge_{\mathcal{A} \subseteq \mathcal{N}} \left((\max_{i \in \mathcal{A}} x_i) \vee \mu(\mathcal{N} \setminus \mathcal{A}) \right), \tag{5.18}$$

or in terms of H_i subsets as

$$S_\mu(\mathbf{x}) = \min_{i=1,\ldots,n} \max\{x_{(i)}, \mu(H_{i+1})\}, \tag{5.19}$$

with $H_{n+1} = \emptyset$.

The median also facilitates an equivalent expression, with

$$S_\mu(\mathbf{x}) = Med(x_1, \ldots, x_n, \mu(H_2), \mu(H_3), \ldots, \mu(H_n)). \tag{5.20}$$

Example 5.4 (Cont'd from Example 5.3) For the same capacity and inputs, calculation via Eq. (5.20) gives

$$S_\mu(\mathbf{x}) = Med(\underbrace{\text{neutral, very high, low}}_{x_1, x_2, x_3}, \underbrace{\text{very high, high}}_{\mu(H_2), \mu(H_3)}) = \text{high},$$

and using Eq. (5.19), we have

$$\begin{aligned} S_\mu(\mathbf{x}) &= \min(\max\{\text{low}, \mu(\{1,2\})\}, \max\{\text{neutral}, \mu(\{2\})\}, \\ &\quad \max\{\text{very high}, \mu(\emptyset)\}) \\ &= \min(\max\{\text{low, very high}\}, \max\{\text{neutral, high}\}, \max\{\text{very high, low}\}) \\ &= \min(\text{very high, high, very high}) = \text{high}. \end{aligned}$$

An alternative capacity representation that can lead to more efficient calculation for the Sugeno integral is the possibilistic Möbius transform M_p (Sect. 7.10). Since $\mu(\mathcal{A}) = \max_{\mathcal{B} \subset \mathcal{A}} \mu(\mathcal{B})$ suggests that there is a $\mathcal{B} \subset \mathcal{A}$ such that

$$\left((\min_{i \in \mathcal{B}} x_i) \wedge \mu(\mathcal{B})\right) \geqslant \left((\min_{i \in \mathcal{A}} x_i) \wedge \mu(\mathcal{A})\right),$$

in terms of M_p values, we simply have

$$S_\mu(\mathbf{x}) = \bigvee_{\mathcal{A} \subseteq \mathcal{N}} \left((\min_{i \in \mathcal{A}} x_i) \wedge M_p(\mathcal{A})\right), \tag{5.21}$$

noting that we can ignore any subsets such that $M_p(\mathcal{A}) = 0$ (see Sect. 7.6).

5.3.2 *Main Properties*

The main properties of the Sugeno integral can be summarised as follows:

- The Sugeno integral is a continuous idempotent aggregation function.
- An aggregation function is a Sugeno integral if and only if it is *min-homogeneous*, i.e. $S_\mu(x_1 \wedge r, \ldots, x_n \wedge r) = S_\mu(x_1, \ldots, x_n) \wedge r$ and *max-homogeneous*, i.e., $S_\mu(x_1 \vee r, \ldots, x_n \vee r) = S_\mu(x_1, \ldots, x_n) \vee r$ for all $\mathbf{x} \in [0, 1]^n$, $r \in [0, 1]$ (see [Mar00], Theorem 4.3. There are also alternative characterisations).
- The Sugeno integral is *comonotone maxitive* and *comonotone minimitive*, i.e. $S_\mu(\mathbf{x} \vee \mathbf{y}) = S_\mu(\mathbf{x}) \vee S_\mu(\mathbf{y})$ and $S_\mu(\mathbf{x} \wedge \mathbf{y}) = S_\mu(\mathbf{x}) \wedge S_\mu(\mathbf{y})$ for all comonotone[5] $\mathbf{x}, \mathbf{y} \in [0, 1]^n$.
- The Sugeno integral is Lipschitz-continuous (Definition 1.22), with the Lipschitz constant 1 in any p-norm, which means it is a kernel aggregation function.
- The class of Sugeno integrals is closed under duality.
- For symmetric capacities, $\mathsf{S}_\mu(\mathbf{x})$ is also referred to as the ordered weighted maximum (OWMax) operator (see Definition 6.30) [DPT88, DP96].

5.3.3 *Calculation*

The following calculation is given in terms of the ordered values $x_{(i)}$ and $\mu(H_i)$ decompositions, following the expression given in Eq. 5.17.

[5] See footnote 2 on p.112.

Calculation of $S_\mu(\mathbf{x})$.

1. Sort the components $((x_1, 1), (x_2, 2), \ldots, (x_n, n))$ with respect to the first component of each pair in non-decreasing order. We obtain $((x_{(1)}, i_1), (x_{(2)}, i_2), \ldots, (x_{(n)}, i_n))$, so that $x_{(j)} = x_{i_j}$ and $x_{(j)} \leqslant x_{(j+1)}$ for all i. Let $x_{(0)} = 0$.
2. Let $\mathcal{T} = \{1, \ldots, n\}$, and $\mathcal{S} = 0$.
3. For $j = 1, \ldots, n$ do
   a. $\mathcal{S} := \max(\mathcal{S}, \min(x_{(j)}, \mu(\mathcal{T})))$
   b. $\mathcal{T} := \mathcal{T} \setminus \{i_j\}$
4. Return $\mathcal{S}$.

Example 5.5 Let $n = 3$, $\mathbf{x} = (0.5, 0.9, 0.4)$, and the values of μ be given by

$$\mu(\emptyset) = 0, \mu(\mathcal{N}) = 1, \mu(\{1\}) = 0.5, \mu(\{2\}) = \mu(\{3\}) = 0.2$$

$$\mu(\{1, 2\}) = \mu(\{2, 3\}) = 0.6, \mu(\{1, 3\}) = 0.9$$

Step 1. We take $((0.5, 1), (0.9, 2), (0.4, 3))$.
Sort this vector of pairs to obtain $((0.4, 3), (0.5, 1), (0.9, 2))$.
Step 2. Take $\mathcal{T} = \{1, 2, 3\}$ and $\mathcal{S} = 0$.
Step 3. a) $\mathcal{S} := \max(0, \min(0.4, \mu(\{1, 2, 3\}))) = \max(0, \min(0.4, 1)) = 0.4$;
b) $\mathcal{T} = \{1, 2, 3\} \setminus \{3\} = \{1, 2\}$;
a) $\mathcal{S} := \max(0.4, \min(0.5, \mu(\{1, 2\}) = \max(0.4, \min(0.5, 0.6)) = 0.5$;
b)$\mathcal{T} := \{1, 2\} \setminus \{1\} = \{2\}$;
a) $\mathcal{S} := \max(0.5, \min(0.9, \mu(\{2\}))) = \max(0.5, \min(0.9, 0.2)) = 0.5$.
Step 4. Therefore, $S_\mu(\mathbf{x}) = 0.5$.

5.3.4 *Orness*

An orness calculation for the Sugeno integral is not as straightforward as for the Choquet integral. One can gather from the differences between their graphical depictions in the two-variate case (Figs. 5.1 and 5.4) that even for numeric inputs, we will not obtain simple direct calculations. In the ordinal case, the inputs and capacity values range over a lattice rather than the unit interval and so a different approach altogether is required.

A method for calculating orness for qualitative OWA functions was proposed in [Och+17]. It works by constructing a descending chain of equidistant ordinal values and then using this as an input vector.

For a lattice $L = \{a_1, \ldots, a_l\}$ where a_i are ordinal values, let $[b_l, \ldots, b_1]$ denote an ordered chain such that

$$
\begin{aligned}
b_1 &= a_1 \vee \cdots \vee a_l \\
b_2 &= \bigvee_{i \in \{1,\ldots,l-1\}} \bigvee_{j \in \{i+1,\ldots,l\}} (a_i \wedge a_j) \\
&\vdots \\
b_k &= \bigvee_{j_1 \in \{1,\ldots,l-k+1\}} \bigvee_{j_2 \in \{j_1+1,\ldots,l-k+2\}} \cdots \bigvee_{j_k \in \{j_{k-1}+1,\ldots,l\}} (a_{j_1} \wedge a_{j_2} \wedge \cdots \wedge a_{j_k}) \\
&\vdots \\
b_l &= a_1 \wedge \cdots \wedge a_l.
\end{aligned}
\tag{5.22}
$$

Note that if L is a chain with $a_l \prec \cdots \prec a_2 \prec a_1$, then we simply have $b_i = a_i$. The descending chain $[d_1, \ldots, d_n]$ is then constructed with $d_1 = b_1$ and for $j = 1, \ldots, n-1$:

$$
d_{j+1} = b_k, \; k = 1 + \left\lfloor \frac{jl-1}{n-1} \right\rfloor. \tag{5.23}
$$

The orness of f is then given by

$$
orness(f) = f(d_1, \ldots, d_n). \tag{5.24}
$$

However, this would only be appropriate in the case of symmetric capacities, since calculation of $S_\mu(d_1, \ldots, d_n)$ would only use one simplex. One possible way around this is to define an average capacity μ_{med} using the medians across each cardinality and then calculate $S_{\mu_{\text{med}}}(d_1, \ldots, d_n)$.

Example 5.6 Consider a Sugeno integral with $n = 3$ defined over nine ordinal labels, $\{0, 0.1, 0.3, 0.4, 0.5, 0.7, 0.8, 0.9, 1\}$ (which for consistency we have expressed over the unit interval):

$$
\begin{array}{ccc}
 & 1 & \\
0.8 & 0.5 & 0.7 \\
0.3 & 0.5 & 0.4 \\
 & 0 &
\end{array}
$$

Using Eq. (5.22), we have $b_1 = 1, b_2 = 0.9, b_3 = 0.8, \ldots, b_9 = 0$, and then from Eq. 5.23, we have $d_1 = 1$, $d_2 = b_{1+\lfloor 8/2 \rfloor} = b_5 = 0.5$, and $d_3 = b_{1+\lfloor 17/2 \rfloor} = b_9 = 0$.

The set of medians, leading to a kind of qualitative cardinality index for the capacity, is given by $(1, Med(0.8, 0.5, 0.7), Med(0.3, 0.5, 0.4) = (1, 0.7, 0.4)$, and hence an approximation for orness with this approach is given by

$$orness(S_\mu) = (1 \wedge 0) \vee (0.7 \wedge 0.5) \vee (0.4 \wedge 0) = 0.5.$$

5.4 Generalised and Other Fuzzy Integrals

The popularity and usefulness of both the Choquet integral and the Sugeno integral have led to many interesting generalisations and extensions. In some cases, these generalisations facilitate nonlinear behaviour, e.g. by transforming the inputs, transforming the capacity values, or generalising and replacing some of the operations involved. Researchers have also looked at different decompositions of the input set. Notable examples of more general integrals defined with respect to capacities or nonadditive measures include the Pan integral [Yan85], concave (upper/lower) integral [Wan+00, Leh09], decomposition integral [EL14], inclusion–exclusion integral [HO10], Choquet-like integrals [Mes+16], and binary tree-based integrals [Bel18] (see [BJW19] and the references therein for a brief overview). Many of the generalised constructions work towards unifying the approaches of Choquet and Sugeno integration, as well as connecting with previously defined functions such as the Shilkret integral [Shi71].

We present one straightforward generalisation that uses generating functions.

Definition 5.4 (Generalised Discrete Choquet Integral) Let g denote a continuous strictly monotone function $g : [0, 1] \to [-\infty, \infty]$. The generalised discrete Choquet integral with respect to a capacity μ is given by

$$C_{\mu,g}(\mathbf{x}) = g^{-1}(C_\mu(g(\mathbf{x}))), \tag{5.25}$$

where C_μ is the discrete Choquet integral with respect to μ and $g(\mathbf{x}) = (g(x_1), \ldots, g(x_n))$.

The averaging behaviour of the Choquet integral is not lost with this generalisation, as established in the following proposition from [CB10] (based on [BB87]).

Proposition 5.1 *Let f be an averaging aggregation function on $\mathbb{I}^n$, and g be a continuous strictly monotone function $\mathbb{I} \to \mathbb{R}$, called a generating function. Then $f_g(\mathbf{x}) = g^{-1}(f(g(\mathbf{x})))$ is also an averaging aggregation function on $\mathbb{I}^n$.*

The quasi-arithmetic means (of which the weighted power means are a special case when $g(t) = t^p$; see Definition 1.31) and Archimedean t-norms (Eq. 1.9) have a similar construction; however, in those cases, the function g is independent

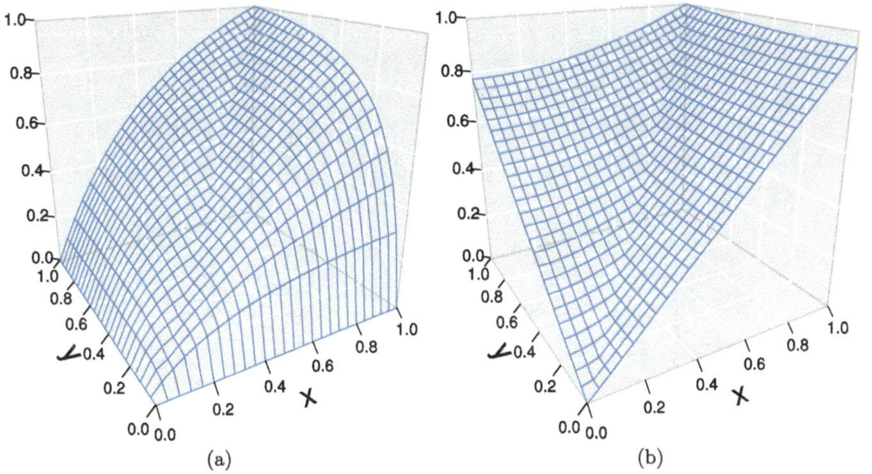

Fig. 5.7 3D plots of generalised Choquet integral functions with respect to the capacity $v(1) = 0.9$, $v(2) = 0.6$, $v(1, 2) = 1$. In (**a**) the generator used is $g(t) = t^{-1}$ (a harmonic Choquet integral), while (**b**) shows $g(t) = t^2$ (a quadratic Choquet integral)

of any scaling constant and so g and $-g$ result in the same function, which is not necessarily the case with Definition 5.4 since a decreasing generating function would reverse the ordering induced by $\mathbf{x}_\nearrow$. If this is problematic for implementation, we can require that g is increasing or fix the permutation of the inputs prior to applying the transformation, i.e. by calculating $\min_{i \in \mathcal{A}} x_i)$ first, in which case, we have

$$C_{\mu,g}(\mathbf{x}) = g^{-1}\left(\sum_{i=1}^{n} g(x_{(i)})[\mu(H_i) - \mu(H_{i+1})]\right) \quad (5.26)$$

$$= g^{-1}\left(\sum_{\mathcal{A} \subseteq \mathcal{N}} M(\mathcal{A}) g(\min_{i \in \mathcal{A}} x_i)\right).$$

This is the approach, taken in, e.g. [Yag04], for generalised OWA functions. It means that for additive capacities, we can recover any of the quasi-arithmetic means as special cases. For example, the generator $g(t) = t$ returns the arithmetic mean, $g(t) = \ln t$ leads to the geometric mean, and $g(t) = t^p$ results in power means. Behaviours stemming from g will be adopted by $C_{\mu,g}$. For example, if $g(0) = -\infty$, then $C_{\mu,g}$ will have an absorbing element at $a = 0$. The orness will depend both on the capacity and the choice of generator, which is apparent in Fig. 5.7, depicting generalised Choquet integrals defined with respect to different generating functions but the same capacity.

Table 5.1 Different expressions for the Choquet and Sugeno integral

Representation/Sets	Choquet integral $C_\mu(\mathbf{x})$	Sugeno integral $S_\mu(\mathbf{x})$
Capacity	$\sum_{i=1}^{n} x_{(i)}[\mu(\{j \mid x_j \geqslant x_{(i)}\}) - \mu(\{j \mid x_j \geqslant x_{(i+1)}\})]$	$\bigvee_{\mathcal{A}\subseteq\mathcal{N}} \left((\min_{i\in\mathcal{A}} x_i) \wedge \mu(\mathcal{A}) \right) = \bigwedge_{\mathcal{A}\subseteq\mathcal{N}} \left((\max_{i\in\mathcal{A}} x_i) \vee \mu(\mathcal{N} \setminus \mathcal{A}) \right)$
(In terms of H_i [a])	$\sum_{i=1}^{n} x_{(i)}[\mu(H_i) - \mu(H_{i+1})]$	$\max_{i=1,\ldots,n} \min\{x_{(i)}, \mu(H_i)\}$
	$\sum_{i=1}^{n} \left[x_{(i)} - x_{(i-1)}\right] \mu(H_i)$	$\min_{i=1,\ldots,n} \max\{x_{(i)}, \mu(H_{i+1})\}$
		$S_\mu(\mathbf{x}) = Med(x_1, \ldots, x_n, \mu(H_2), \mu(H_3), \ldots, \mu(H_n))$
(Using basis $g(\mathbf{x})$ [b])	$\sum_{\mathcal{A}\subseteq\mathcal{N}} \mu(\mathcal{A}) g_{\mathcal{A}}(\mathbf{x})$	
	$\sum_{\mathcal{A}\subseteq\mathcal{N}} \mu(\mathcal{A}) \sum_{\mathcal{B} \mid \mathcal{A}\subseteq\mathcal{B}} (-1)^{\lvert\mathcal{B}\setminus\mathcal{A}\rvert} \min_{i\in\mathcal{B}} x_i$	
Möbius	$\sum_{\mathcal{A}\subseteq\mathcal{N}} M(\mathcal{A}) \min_{i\in\mathcal{A}} x_i$	$\bigvee_{\mathcal{A}\subseteq\mathcal{N}} \left((\min_{i\in\mathcal{A}} x_i) \wedge M_p(\mathcal{A}) \right)$
co-Möbius	$\sum_{\mathcal{A}\subseteq\mathcal{N}} (-1)^{\lvert\mathcal{A}\rvert+1} \check{M}(\mathcal{A}) \max_{i\in\mathcal{A}} x_i$	
Marginal contributions	$\sum_{i=1}^{n} x_{(i)} \cdot \Delta_{(i)} \mu(H_i)$	

[a] $H_i = \{(i), \ldots, (n)\}$

[b] $g_{\mathcal{A}}(\mathbf{x}) = \max(0, \min_{i\in\mathcal{A}} x_i - \max_{i\in\mathcal{N}\setminus\mathcal{A}} x_i)$

References

[Bel18] G. Beliakov, A new type of fuzzy integrals for decision making based on bivariate symmetric means. Int. J. Intell. Syst. **33**(8), 1660–1671 (2018)

[BJ11] G. Beliakov, S. James, Citation-based journal ranks: The use of fuzzy measures. Fuzzy Sets Syst. **167**, 101–119 (2011)

[BJW19] G. Beliakov, S. James, J.-Z. Wu, *Discrete Fuzzy Measures: Computational Aspects* (Springer, Berlin, Heidelberg, 2019)

[BB87] J. Borwein, P. Borwein, *PI and the AGM: A Study in Analytic Number Theory and Computational Complexity* (Wiley, New York, 1987)

[CB10] T. Calvo, G. Beliakov, Aggregation functions based on penalties. Fuzzy Sets Syst. **161**(10), 1420–1436 (2010)

[CCT02] T. Chen, H. Chang, G. Tzeng, Using fuzzy measures and habitual domains to analyze the public attitude and apply to the gas taxi policy. Eur. J. Oper. Res. **137**, 145–161 (2002)

[Cho54] G. Choquet, Theory of capacities. Annales de l'institut Fourier **5**, 131–295 (1954)

[Den94] D. Denneberg, *Non-additive Measure and Integral* (Springer Science & Business Media, Dordrecht, 1994)

[DP96] D. Dubois, H. Prade, Semantics of quotient operators in fuzzy relational databases. Fuzzy Sets Syst. **78**(1), 89–93 (1996)

[DPT88] D. Dubois, H. Prade, C. Testemale, Weighted fuzzy pattern matching. Fuzzy Sets Syst. **28**, 313–331 (1988)

[Dub96] D. Dubois, Book Review: 'Non-additive measure and integral' by Dieter Denneberg. Int. J. Uncertainty Fuzziness Knowledge Based Syst. **4**(1), 95–97 (1996)

[Duj73] J. Dujmovic, Two integrals related to means. J. Univ. Belgrade EE Depart. Math. Phys. **Nos. 412–460**, 231–232 (1973)

[Duj74] J. Dujmovic, Weighted conjunctive and disjunctive means and their application in system evaluation. J. Univ. Belgrade EE Depart. Math. Phys. **483**, 147–158 (1974)

[EL14] Y. Even, E. Lehrer, Decomposition-integral: unifying Choquet and the concave integrals. Econ. Theory **56**, 33–58 (2014)

[FSM03] J. Fernändez Salido, S. Murakami, Extending Yager's orness concept for the OWA aggregators to other mean operators. Fuzzy Sets Syst. **139**, 515–542 (2003)

[FR94] J. Fodor, M. Roubens, *Fuzzy Preference Modelling and Multicriteria Decision Support* (Kluwer, Dordrecht, 1994)

[Gra16] M. Grabisch, *Set Functions, Games and Capacities in Decision Making* (Springer, Berlin, New York, 2016)

[GKM08] M. Grabisch, I. Kojadinovic, P. Meyer, A review of methods for capacity identification in Choquet integral based multi-attribute utility theory: Applications of the Kappalab R package. Eur. J. Oper. Res. **186**(2), 766–785 (2008)

[GMS00] M. Grabisch, T. Murofushi, M. Sugeno (eds.), *Fuzzy Measures and Integrals. Theory and Applications* (Physica-Verlag, Heidelberg, 2000)

[HO10] A. Honda, J. Okamoto, Inclusion-exclusion integral and its application to subjective video quality estimation, in *Information Processing and Management of Uncertainty in Knowledge-Based Systems. Theory and Methods. Vol. 80 of the series Communications in Computer and Information Science* (2010), pp. 480–489

[KCP05] K. Keun-Chang, W. Pedrycz, Face recognition: A study in information fusion using fuzzy integral. Pattern Recognit. Lett. **26**, 719–733 (2005)

[Leh09] E. Lehrer, A new integral for capacities. Econ. Theory **39**, 157–176 (2009)

[LO05] D. Liginlal, T. Ow, On policy capturing with fuzzy measures. Eur. J. Oper. Res. **167**, 461–474 (2005)

[LO06] D. Liginlal, T. Ow, Modeling attitude to risk in human decision processes: An application of fuzzy measures. Fuzzy Sets Syst. **157**, 3040–3054 (2006)

[Mar00] J.-L. Marichal, On Choquet and Sugeno integrals as aggregation functions. Fuzzy measures and integrals, in *Theory and Applications*, eds. by M. Grabisch, T. Murofushi, M. Sugeno (Physica-Verlag, Heidelberg, 2000), pp. 247–272

[Mar04] J.-L. Marichal, Tolerant or intolerant character of interacting criteria in aggregation by the Choquet integral. Eur. J. Oper. Res. **155**, 771–791 (2004)

[Mes+16] R. Mesiar, A. Kolesärovä, H. Bustince, G. Dimuro, B. Bedregal, Fusion functions based discrete Choquet-like integrals. Eur. J. Oper. Res. **252**, 601–609 (2016)

[MS89] T. Murofushi, M. Sugeno, An interpretation of fuzzy measures and the Choquet integral as an integral with respect to a fuzzy measure. Fuzzy Sets Syst. **29**(2), 201–227 (1989)

[NT05] Y. Narukawa, V. Torra, Fuzzy measure and probability distributions: distorted probabilities. IEEE Trans. Fuzzy Syst. **13**, 617–629 (2005)

[NT07] Y. Narukawa, V. Torra, Fuzzy measures and integrals in evaluation of strategies. Inf. Sci. **177**, 4686–4695 (2007)

[Och+17] G. Ochoa, I. Lizasoain, D. Paternain, H. Bustince, N.R. Pal, From quantitative to qualitative orness for lattice OWA operators. Int. J. General Syst. **46**(6), 640–669 (2017)

[RSMG09] A. Rico, O. Strauss, D. Mariano-Goulart, Choquet integrals as projection operators for quantified tomographic re-construction. Fuzzy Sets Syst. **160**, 198–211 (2009)

[Sch86] D. Schmeidler, Integral representation without additivity. Proc. Amer. Math. Soc. **97**, 255–261 (1986)

[Shi71] N. Shilkret, Maxitive measure and integration. Indagationes Mathematicae (Proceedings) **74**, 109–116 (1971)

[Sug74] M. Sugeno, Theory of Fuzzy Integrals and Applications. PhD Thesis. Tokyo Inst. of Technology, 1974

[Vit97] G. Vitali, On the definition of integral of functions of one variable. Rivista di matematica per le scienze economiche e sociali **20**(2), 159–168 (1997). Originally published in Italian in 1925, pp. 159–168

[WK09] Z. Wang, G. Klir, *Generalized Measure Theory* (Springer US, New York, 2009)

[Wan+00] Z. Wang, K.-S. Leung, M.-L. Wong, J. Fang, A new type of nonlinear integrals and the computation algorithm. Fuzzy Sets Syst. **112**, 223–231 (2000)

[Web84] S. Weber, ⊥-decomposable measures and integrals for Archimedean t-conorms ⊥. J. Math. Anal. Appl. **101**(1), 114–138 (1984)

[WB20] J.-Z. Wu, G. Beliakov, Marginal contribution representation of capacity based multicriteria decision making. Int. J. Intell. Syst. **35**(3), 373–400 (2020)

[Yag88] R. Yager, On ordered weighted averaging aggregation operators in multicriteria decision making. IEEE Trans. Syst. Man Cybern. **18**, 183–190 (1988)

[Yag04] R. Yager, Generalized OWA aggregation operators. Fuzzy Optim. Decis. Making **3**, 93–107 (2004)

[Yan85] Q. Yang, The Pan-integral on fuzzy measure space. Fuzzy Math. **3**, 107–114 (1985)

[Yan+08] R. Yang, Z. Wang, P. Heng, K. Leung, Fuzzified Choquet integral with a fuzzy-valued integrand and its application on temperature prediction. IEEE Trans. Syst. Man Cybern. Part B Cybern. **38**(2), 367–380 (2008)

Chapter 6
Symmetric Capacities: OWA

Abstract The Choquet integral with respect to a symmetric capacity results in aggregation equivalent to the ordered weighted averaging (OWA) operator. The use of the OWA in research and practical applications is extensive, and so in this chapter we focus on important results and properties relating to OWA functions. Special extensions such as the weighted OWA and induced OWA (WOWA and IOWA) are presented, as well as useful methods for learning weights from data.

6.1 Ordered Weighted Averaging

Symmetric capacities have the same value assigned to subsets with the same cardinality. Interaction can still be considered to occur as elements are added to or removed from subsets, but these effects are independent of which elements or subsets are involved. Let us recall the definition that expresses this property in terms of an underlying monotone function.

Definition 6.1 (Symmetric Capacity) A capacity on $\mathcal{N}$ is symmetric if its values depend only on the cardinality of the subsets:

$$\mu(\mathcal{A}) = Q\left(\frac{|\mathcal{A}|}{n}\right)$$

for $n = |\mathcal{N}|$ and some monotone increasing function $Q : [0, 1] \to [0, 1]$, $Q(0) = 0$, $Q(1) = 1$.

More simply, a capacity is symmetric if $|\mathcal{A}| = |\mathcal{B}| \to \mu(\mathcal{A}) = \mu(\mathcal{B})$. It has been well established that when the Choquet integral is defined with respect to a symmetric capacity, the resulting aggregation function is equivalent to the ordered weighted averaging (OWA) operator. OWA operators were introduced by Yager in [Yag88] and have since received a great deal of attention in the fuzzy sets and decision-making communities [YK97, YKB11, EM14]. They can be defined without referring to capacities, but rather with respect to a weighting vector $\mathbf{w}$ and `sort()` operation.

G. Beliakov et al., *Choquet Capacities and Fuzzy Integrals*, Theory and Applications of Computability, https://doi.org/10.1007/978-3-031-97070-2_6

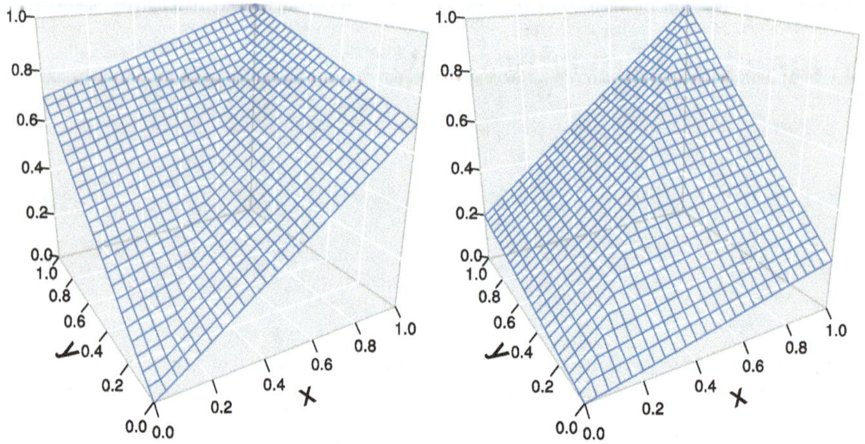

Fig. 6.1 3D plots of OWA functions $OWA_{(0.7,0.3)}$ and $OWA_{(0.2,0.8)}$

Definition 6.2 (OWA) For a given weighting vector $\mathbf{w}$, $w_i \geqslant 0$, $\sum w_i = 1$, the OWA function is given by

$$OWA_{\mathbf{w}}(\mathbf{x}) = \sum_{i=1}^{n} w_i x_{(i)} =< \mathbf{w}, \mathbf{x}_{\searrow} >, \tag{6.1}$$

where $\mathbf{x}_{\searrow} = x_{(1)} \geqslant x_{(2)} \geqslant \ldots \geqslant x_{(n)}$ denotes the vector obtained from $\mathbf{x}$ by arranging its components in *non-increasing* order.

Surface plots for $n = 2$ (the same as depicted in Fig. 1.6) are shown in Fig. 6.1.

The symmetry of OWA functions means that all inputs are treated equally or considered equally important—for instance, calculating the Shapley or Banzhaf indices (see Chap. 3) for a symmetric capacity will result in equal values. The weights are applied depending on the relative size of the input, so that high, low, or medium values can be emphasised. This aggregation approach accommodates a number of existing functions, including the median and trimmed means (see special cases below), but it is also a natural fit for many applications. For example, the average amount we spend on wine at a local restaurant might be closer to the cost of the "second-cheapest wine" than the average across the whole menu. The orness value (mentioned in Chap. 5) of the weighting vector has also found useful interpretations, e.g. as the degree of optimism or pessimism, risk tolerance or risk aversion when making decisions depending on whether one leans more heavily towards high or low values.

While this chapter will focus on results relating to the OWA as defined with inputs in increasing order, we mention the reverse OWA, which uses the same ordering as is common for the Choquet integral.

Definition 6.3 (Reverse OWA) Given an OWA function $OWA_{\mathbf{w}}$, the reverse OWA is $OWA_{\mathbf{w}_d}$ with the weighting vector $\mathbf{w}_d = (w_n, w_{n-1}, \ldots, w_1)$.

For the remainder of the chapter, we will be looking at OWA functions in the context of aggregation functions rather than capacities, presenting results and properties in terms of $\mathbf{w}$ rather than μ. Any of the results can be established in one or the other setting by noting that the weights of $\mathbf{w}$ can be calculated from the values of a symmetric capacity using $w_i = \mu(\mathcal{A}) - \mu(\mathcal{A} \setminus \{j\})$ for any subset $\mathcal{A}$ of cardinality i and $j \in \mathcal{A}$, and conversely, from a weighting vector, we can define a capacity such that for all $\mathcal{A}$, we have

$$\mu(\mathcal{A}) = \sum_{i=1}^{|\mathcal{A}|} w_i.$$

6.1.1 *Main Properties*

Key properties relating to the OWA can be summarised as follows:

- As with all averaging aggregation functions, OWA are increasing (strictly increasing if all weights are positive) and idempotent.
- The dual of an OWA function is the *reverse* OWA, with the vector of weights $\mathbf{w}_d = (w_n, w_{n-1}, \ldots, w_1)$.
- OWA functions are continuous, symmetric, homogeneous, and shift-invariant.
- OWA functions do not have neutral or absorbing elements, except for the special cases min and max.

6.1.2 *Special Cases*

- If all weights are equal, $w_i = \frac{1}{n}$, the OWA becomes the arithmetic mean $OWA_{\mathbf{w}}(\mathbf{x}) = AM(\mathbf{x})$.
- If $\mathbf{w} = (1, 0, \ldots, 0)$, then $OWA_{\mathbf{w}}(\mathbf{x}) = \max(\mathbf{x})$;
- If $\mathbf{w} = (0, \ldots, 0, 1)$, then $OWA_{\mathbf{w}}(\mathbf{x}) = \min(\mathbf{x})$.
- If $\mathbf{w} = (\alpha, 0, \ldots, 0, 1 - \alpha)$, then the OWA becomes the *Hurwizc aggregation function*, $OWA_{\mathbf{w}}(\mathbf{x}) = \alpha \max(\mathbf{x}) + (1 - \alpha) \min(\mathbf{x})$; for $\alpha = \frac{1}{2}$, it is called the half-range function.
- If $w_i = 0$ for all i except the k-th, and $w_k = 1$, then the OWA becomes a k-th-order statistic, $OWA_{\mathbf{w}}(\mathbf{x}) = x_{(k)}$.
- Weighting vectors can be defined so that $OWA_{\mathbf{w}}(\mathbf{x}) = Med(\mathbf{x})$, the *median* (see Sect. 6.8). When n is odd, we will have $\mathbf{w} = (0, \ldots, 0, 1, 0, \ldots, 0)$, and for even n, we have $\mathbf{w} = (0, \ldots, 0, \frac{1}{2}, \frac{1}{2}, 0, \ldots, 0)$.

- If we take weighting vectors with several non-zero values assigned to the middle inputs, for example, $\mathbf{w} = (0, \ldots, 0, \frac{1}{6}, \frac{1}{3}, \frac{1}{3}, \frac{1}{6}, 0, \ldots, 0)$, we obtain the *central OWA* function proposed in [Yag07a] and later investigated in [ZSA08]. In [Yag07a], the weights were also symmetric ($w_j = w_{n+1-j}$), strongly decaying ($w_i < w_j$ if either $i < j \leqslant (n+1)/2$ or $i > j \geqslant (n+1)/2$), and inclusive ($w_j > 0$).
- One special case of centred OWA is the *α-trimmed mean* ($0 \leqslant \alpha \leqslant \frac{1}{2}$), which can be written as an OWA function with the weights

$$\mathbf{w} = \left(0, \ldots, 0, \frac{1}{n - 2\lfloor n\alpha \rfloor}, \ldots, \frac{1}{n - 2\lfloor n\alpha \rfloor}, 0, \ldots, 0\right),$$

where $\lfloor n\alpha \rfloor$ is the nearest integer no greater than $n\alpha$.
- The "Olympic OWA" has the weighting vector $\mathbf{w} = (0, \frac{1}{n-2}, \ldots, \frac{1}{n-2}, 0)$ and is a special case of trimmed means.
- Another interesting case is the *Winsorised mean*, in which extreme low and high values are replaced. An $\alpha\%$ Winsorised mean can be expressed as an OWA function with the weights

$$\mathbf{w} = (0, \ldots, 0, (\lfloor n\alpha \rfloor + 1)a, a, \ldots, a, (\lfloor n\alpha \rfloor + 1)a, 0, \ldots, 0),$$

with $a = \frac{1}{n}$.

As special cases of OWA, all the mentioned functions are symmetric, homogeneous, and shift-invariant aggregation functions.

Note 6.1 Calculation of α-trimmed and Winsorised means can be achieved without sorting the arguments $\mathbf{x}$ (which has complexity $O(n \log n)$ and could be time consuming for repetitive calculations and large n). The Winsorised mean can be written as

$$WM(\mathbf{x}) = \sum_{i=1}^{n} \min(\max(x_i, x_{(\lfloor n\alpha \rfloor+1)}), x_{(n-\lfloor n\alpha \rfloor)}),$$

where $x_{([n\alpha]+1)}, x_{(n-[n\alpha])}$ are order statistics that are computed in $O(n)$ operations (using the quickselect algorithm). Then the α-trimmed mean is

$$TM(\mathbf{x}) = \frac{1}{n - 2\lfloor n\alpha \rfloor} \left(nWM(\mathbf{x}) - \lfloor n\alpha \rfloor (x_{\lfloor n\alpha \rfloor+1)} + x_{(n-\lfloor n\alpha \rfloor)})\right).$$

6.2 Orness and Entropy

We have addressed entropy for general capacities in Sect. 3.8 and the orness of fuzzy integrals throughout Chap. 5. For the special case of symmetric capacities and the OWA function, we have a number of simplified results.

6.2.1 *Orness*

The general expression for the measure of orness, given in (5.12), translates to the following simple formula:

$$orness(OWA_{\mathbf{w}}) = \sum_{i=1}^{n} w_i \frac{n-i}{n-1} = OWA_{\mathbf{w}}\left(1, \frac{n-2}{n-1}, \ldots, \frac{1}{n-1}, 0\right). \quad (6.2)$$

We now summarise some additional properties involving the orness value:

- The orness of OWA and its dual are related by

$$orness(OWA_{\mathbf{w}}) = 1 - orness(OWA_{\mathbf{w}_d}).$$

- An OWA function is self-dual if and only if $orness(OWA_{\mathbf{w}}) = \frac{1}{2}$.
- For the special cases of the maximum, minimum, and arithmetic mean (AM), we have $orness(\max) = 1$, $orness(\min) = 0$, and $orness(AM) = \frac{1}{2}$. Furthermore, the orness of an OWA is 1 only if it is the max function and 0 only if it is the min function. However, orness can be $\frac{1}{2}$ for an OWA different from the arithmetic mean, which is nevertheless self-dual, e.g. the Olympic OWA mentioned above.
- If the weighting vector is non-decreasing, i.e. $w_i \leqslant w_{i+1}, i = 1, \ldots, n-1$, then $orness(OWA_{\mathbf{w}}) \in [0, \frac{1}{2}]$. If the weighting vector is non-increasing, then $orness(OWA_{\mathbf{w}}) \in [\frac{1}{2}, 1]$. Such weighting vectors are referred to as being *antibuoyant* and *buoyant*, respectively (see Sect. 2.10).
- If two OWA functions with weighting vectors $\mathbf{w}_1, \mathbf{w}_2$ have their respective orness values O_1, O_2, and if $\mathbf{w}_3 = a\mathbf{w}_1 + (1-a)\mathbf{w}_2$, $a \in [0, 1]$, then an OWA function with the weighting vector $\mathbf{w}_3$ has orness value [Ahn06]

$$orness(OWA_{\mathbf{w}_3}) = aO_1 + (1-a)O_2.$$

Example 6.1 The measure of orness for some special weighting vectors has been precalculated in [Ahn06]:

$$w_i = \frac{1}{n}\sum_{j=i}^{n}\frac{1}{j}, \quad orness(OWA_{\mathbf{w}}) = \frac{3}{4},$$

$$w_i = \frac{2(n+1-i)}{n(n+1)}, \quad orness(OWA_{\mathbf{w}}) = \frac{2}{3}.$$

Note 6.2 The classes of recursive and iterative OWA functions, which have the same orness value for any given n, were investigated in [TY05b].

Yager has defined andlike and orlike S-OWA functions [Yag93] using parameter $-1 \leqslant \alpha \leqslant 1$. The weighting vectors of S-OWA functions are given by

$$w_1 = \frac{1-\alpha}{n} + \alpha, w_i = \frac{1-\alpha}{n}, i = 2, \ldots, n,$$

where orlike OWA will have a positive α and andlike OWA will have a negative α. The orness of these functions is $orness(OWA_{\mathbf{w}}) = \alpha + \frac{1}{2}(1-\alpha)$.

The concept of the orness measure has been revisited in [KSP14]. Here the authors proposed an axiomatic framework for orness measures based on the following axioms:

O1 Fuzzy OR: $orness(OWA_{\mathbf{w}}) = 1$ if and only if $\mathbf{w} = \mathbf{w}^* = (1, 0, \ldots, 0)$.
O2 Fuzzy AND: $orness(OWA_{\mathbf{w}}) = 0$ if and only if $\mathbf{w} = \mathbf{w}_* = (0, \ldots, 0, 1)$.
O3 If $\mathbf{w} = (w_1, w_w, \ldots, w_n)$ and $\mathbf{w}' = (w_1, \ldots, w_j - \varepsilon, \ldots, w_k + \varepsilon, \ldots, w_n)$, where $\varepsilon > 0$ and $j < k$, then $orness(OWA_{\mathbf{w}}) > orness(OWA_{\mathbf{w}'})$.
O4 $orness(OWA_{\mathbf{w}}) = \frac{1}{2}$ if and only if $\sum_{i=1}^{n'} c_i w_i = -\sum_{i=n'+1}^{n} c_i w_i$ for $c_i = n' + 1 - i - \lfloor\frac{2i-1}{n}\rfloor - \lfloor\frac{2n'-1}{n}\rfloor\lfloor\frac{i-1}{n'}\rfloor$ and $n' = \lfloor\frac{n+1}{2}\rfloor$.

The axiom O4 is a generalisation of the following simpler axiom:
O4a. The arithmetic mean has orness $\frac{1}{2}$: $orness(\frac{1}{n}, \ldots, \frac{1}{n}) = \frac{1}{2}$.
The modified orness measure is defined as follows.

Definition 6.4 (Modified OWA orness Value) Let $\mathbf{w}$ be a weighting vector. Then the modified orness value is

$$orness(OWA_{\mathbf{w}}) = \sum_{i=1}^{n} d_i w_i, \tag{6.3}$$

where $d_i = \frac{1}{2n'}\left(2n' + 1 - i - \lfloor\frac{2i-1}{n}\rfloor - \lfloor\frac{2n'-1}{n}\rfloor\lfloor\frac{i-1}{n'}\rfloor\right)$ and $n' = \lfloor\frac{n+1}{2}\rfloor$.

The vector $\mathbf{d}$ has the following properties:

- For an even n, $d_i = \frac{1}{n}(n + 1 - i - \lfloor\frac{2i-1}{n}\rfloor)$.
- For an odd n, $d_i = \frac{1}{n+1}(n + 2 - i - \lfloor\frac{2i-1}{n}\rfloor - \lfloor\frac{2i-2}{n+1}\rfloor)$.
- The values d_i form a monotonically decreasing sequence, with $d_1 = 1$ and $d_n = 0$.
- $d_i + d_{n-i+1} = 1$.
- $\sum d_i = \frac{n}{2}$.

The modified measure of orness in Definition 6.4 satisfies the four axioms O1–O4, as shown in [KSP14], where an interpretation of this quantity in terms of the normalised sum of moments is presented. Furthermore, the following results hold:

- $orness(OWA_{\mathbf{w}}) = 1 - orness(OWA_{\mathbf{w}_d})$.
- If $\mathbf{w}$ is buoyant , i.e. $w_i \geqslant w_j$ for all $i < j$, then $orness(OWA_{\mathbf{w}}) \geqslant \frac{1}{2}$.
- An OWA is defined with respect to a buoyant weighting vector if and only if it is a convex function, and if and only if it is a subadditive function $OWA(\mathbf{x}+\mathbf{y}) \leqslant OWA(\mathbf{x}) + OWA(\mathbf{y})$.

6.2.2 Entropy

The following is the most popularly adopted measure of dispersion or entropy for OWA weights. It corresponds with the Shannon entropy (Definition 3.24).

Definition 6.5 (Weight Dispersion (Entropy)) For a given non-negative weighting vector $\mathbf{w}$, the weight dispersion, or the entropy, is defined by

$$Disp(\mathbf{w}) = -\sum_{i=1}^{n} w_i \log w_i.$$

This quantity measures the degree to which all the information (i.e. each of the inputs) is used in the aggregation process. The entropy is used to define the weights with the maximal entropy (functions called MEOWA), subject to a predefined orness value:

- If the orness is not specified, the maximum of $Disp$ is achieved at $w_i = \frac{1}{n}$, i.e. the arithmetic mean, and $Disp(\frac{1}{n}, \ldots, \frac{1}{n}) = \log n$.
- The minimum value of $Disp$, 0, is achieved if and only if $w_i = 0, i \neq k$, and $w_k = 1$, i.e. the order statistic (see Sect. 6.8).
- The entropy of an OWA and its dual (reverse OWA) coincide, $Disp(\mathbf{w}) = Disp(\mathbf{w}_d)$.

We note that there are other types of entropy used to quantify weight dispersion, e.g. Rényi entropy (Definition 3.25), which for OWA weights is calculated for a given $\alpha > 0, \alpha \neq 1$, as

$$H_\alpha(\mathbf{w}) = \frac{1}{1-\alpha} \ln \left(\sum_{i=1}^{n} w_i^\alpha \right). \tag{6.4}$$

A number of other measures have been proposed, e.g. [Yag95]. One such measure of dispersion was presented in [TY05a] and is calculated as

$$\rho(\mathbf{w}) = \frac{\sum_{i=1}^{n} \frac{i-1}{n-1}(w_{(i)} - w_{(i+1)})}{\sum_{i=1}^{n}(w_{(i)} - w_{(i+1)})} = \frac{1}{n-1} \cdot \frac{1 - w_{(1)}}{w_{(1)}}, \tag{6.5}$$

where $w_{(i)}$ denotes the i-th largest weight. A related measure of weight dispersion is $\tilde{\rho}(\mathbf{w}) = 1 - w_{(1)}$ [Yag93]. Another useful measure is weight variance [FM03]:

$$D^2(\mathbf{w}) = \frac{1}{n}\sum_{i=1}^{n}(w_i - M(\mathbf{w}))^2 = \frac{1}{n}\sum_{i=1}^{n} w_i^2 - \frac{1}{n^2}, \tag{6.6}$$

where $M(\mathbf{w})$ is the arithmetic mean of $\mathbf{w}$.

6.3 Special Types of OWA Functions

Here we outline a number of special OWA generalisations. In some cases, the generalisations correspond with particular cases of the Choquet integral or of generalised Choquet integral models, while others are not usually considered in the context of capacities.

6.3.1 Neat OWA

OWA functions have been generalised to functions whose weights depend on the aggregated inputs.

Definition 6.6 (Neat OWA) An OWA function whose weights are defined by

$$w_i = \frac{x_{(i)}^p}{\sum_{i=1}^{n} x_{(i)}^p},$$

with $p \in]-\infty, \infty[$ is called a neat OWA.

Note 6.3 Neat OWA functions are counter-harmonic means [BBC16]. They are not monotone (hence not aggregation functions).

Example 6.2 When $p = 1$ and $\mathbf{x}$ is a probability distribution, the neat OWA coincides with Simpson's diversity index used in ecology [Sim49].

6.3.2 Generalised OWA

Similar to quasi-arithmetic means and as a special case of Definition 5.4, OWA functions have been generalised with the help of generating functions $g : [0, 1] \to [-\infty, \infty]$.

Definition 6.7 (Generalised OWA) Let $g : [0, 1] \to [-\infty, \infty]$ be a continuous strictly monotone function and let $\mathbf{w}$ be a weighting vector. The function

$$GenOWA_{\mathbf{w},g}(\mathbf{x}) = g^{-1}\left(\sum_{i=1}^{n} w_i g(x_{(i)})\right) \tag{6.7}$$

is called a generalised OWA (also known as an ordered weighted quasi-arithmetic mean [Cal+02]). As for OWA, $x_{(i)}$ denotes the i-th largest value of $\mathbf{x}$.

The ordered weighted geometric function was studied in [HHV03, XD02].

Definition 6.8 (Ordered Weighted Geometric (OWG) Function) For a given weighting vector $\mathbf{w}$, the OWG function is

$$OWG_{\mathbf{w}}(\mathbf{x}) = \prod_{i=1}^{n} x_{(i)}^{w_i}. \tag{6.8}$$

Note 6.4 Similar to the weighted geometric mean, OWG is a special case of (6.7) with the generating function $g(t) = \log t$.

Other special cases that have received attention include the ordered weighted harmonic (OWH) function and power-based OWA.

Definition 6.9 (Ordered Weighted Harmonic (OWH)) Function For a given weighting vector $\mathbf{w}$, the OWH function is

$$OWH_{\mathbf{w}}(\mathbf{x}) = \left(\sum_{i=1}^{n} \frac{w_i}{x_{(i)}}\right)^{-1}. \tag{6.9}$$

A large family of generalised OWA functions is based on power functions, similar to weighted power means [Yag04b]. Let g_r denote the family of power functions:

$$g_r(t) = \begin{cases} t^r, & \text{if } r \neq 0, \\ \log(t), & \text{if } r = 0. \end{cases}$$

Definition 6.10 (Power-Based Generalised OWA) For a given weighting vector $\mathbf{w}$, and a value $r \in \mathbb{R}$, the function,

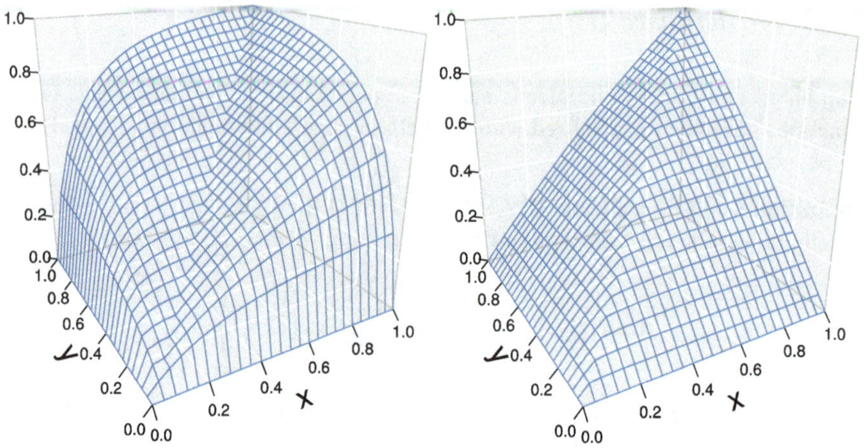

Fig. 6.2 3D plots of OWH functions $OWH_{(0.9,0.1)}$ and $OWH_{(0.2,0.8)}$

$$GenOWA_{\mathbf{w},[r]}(\mathbf{x}) = \left(\sum_{i=1}^{n} w_i x_{(i)}^r\right)^{1/r}, \tag{6.10}$$

for $r \neq 0$, and $GenOWA_{\mathbf{w},[r]}(\mathbf{x}) = OWG_{\mathbf{w}}(\mathbf{x})$ if $r = 0$, is called a power-based generalised OWA.

Of course, both OWG and OWH functions are special cases of power-based OWA with $r = 0$ and $r = -1$, respectively. The usual OWA corresponds with $r = 1$. Another special case is that of the quadratic OWA, $r = 2$, given by

$$OWQ_{\mathbf{w}}(\mathbf{x}) = \sqrt{\sum_{i=1}^{n} w_i x_{(i)}^2}.$$

Plots depicting OWH and OWQ with respect to different weighting vectors for $n = 2$ are shown in Figs. 6.2 and 6.3. Note that with the ordered weighted harmonic function, the absorbent element of 0 is preserved, even though the weighting vector would define a function similar to the maximum for the standard OWA.

Other generating functions can also be used to define generalised OWA functions.

Example 6.3 (Trigonometric OWA) Let $g_1(t) = \sin(\frac{\pi}{2}t)$, $g_2(t) = \cos(\frac{\pi}{2}t)$, and $g_3(t) = \tan(\frac{\pi}{2}t)$ denote generating functions. The trigonometric OWA functions are given by

$$OWAS_{\mathbf{w}}(\mathbf{x}) = \frac{2}{\pi} \arcsin\left(\sum_{i=1}^{n} w_i \sin\left(\frac{\pi}{2} x_{(i)}\right)\right),$$

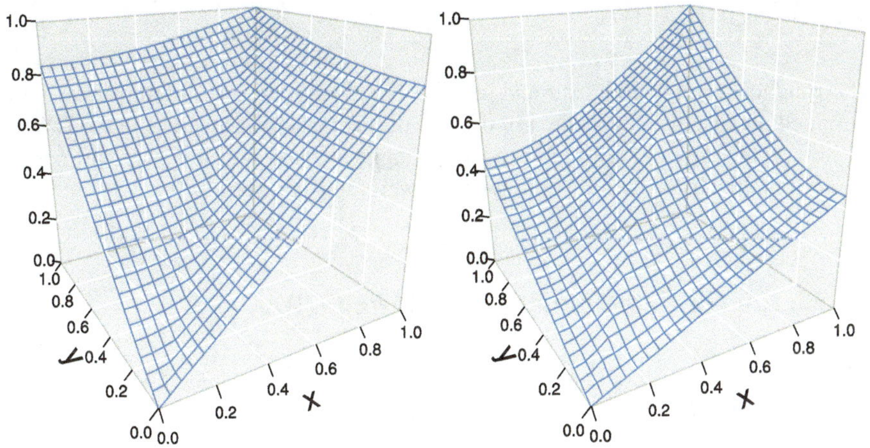

Fig. 6.3 3D plots of quadratic OWA functions $OWQ_{(0.7,0.3)}$ and $OWQ_{(0.2,0.8)}$

$$OWAC_{\mathbf{w}}(\mathbf{x}) = \frac{2}{\pi} \arccos(\sum_{i=1}^{n} w_i \cos(\frac{\pi}{2} x_{(i)})), \text{ and}$$

$$OWAT_{\mathbf{w}}(\mathbf{x}) = \frac{2}{\pi} \arctan(\sum_{i=1}^{n} w_i \tan(\frac{\pi}{2} x_{(i)})).$$

Example 6.4 (Exponential OWA) Consider the generating function:

$$g(t) = \begin{cases} \gamma^t, & \text{if } \gamma \neq 1, \\ t, & \text{if } \gamma = 1 \end{cases}$$

The exponential OWA is the function

$$OWAE_{\mathbf{w},\gamma}(\mathbf{x}) = \begin{cases} \log_\gamma \left(\sum_{i=1}^{n} w_i \gamma^{x_{(i)}} \right), & \text{if } \gamma \neq 1, \\ OWA_{\mathbf{w}}(\mathbf{x}), & \text{if } \gamma = 1. \end{cases}$$

Example 6.5 (Radical OWA) Let $\gamma > 0$, $\gamma \neq 1$ for the following generating function:

$$g(t) = \gamma^{1/t}.$$

The radical OWA is given by

$$OWAR_{\mathbf{w},\gamma}(\mathbf{x}) = \left(\log_\gamma (\sum_{i=1}^{n} w_i \gamma^{1/x_{(i)}}) \right)^{-1}.$$

6.4 Weighted OWA

The weights in weighted means and in OWA functions represent different things. In weighted means, w_i reflects the importance of the i-th input, whereas in the OWA, w_i reflects the importance of the i-th largest input. We now consider some proposed frameworks for incorporating both types of weighting schemes.

6.4.1 Convex Combination of WAM and OWA

A straightforward way of combining aspects of both the WAM and OWA is to take their weighted arithmetic mean. This function is called OWAWA in [Mer09].

Definition 6.11 (OWAWA) Let $\mathbf{w}$, $\mathbf{u}$ be weighting vectors (i.e. both have non-negative components and add to one), and $\alpha \in [0, 1]$. Then

$$OWAWA_{\mathbf{w},\mathbf{u},\alpha}(\mathbf{x}) = (1-\alpha)WAM_{\mathbf{u}}(\mathbf{x}) + \alpha OWA_{\mathbf{w}}(\mathbf{x}). \tag{6.11}$$

It is clear that OWAWA is an idempotent aggregation function, whose limiting cases $\alpha = 0$ and $\alpha = 1$ are the WAM and OWA, respectively. It is also not difficult to show that the OWAWA function is the discrete Choquet integral with respect to the capacity $\mu = (1-\alpha)\mu_1 + \alpha\mu_2$, where μ_1 is the additive capacity corresponding with the weights of the WAM, and μ_2 is the symmetric capacity corresponding with the weights of the OWA. For this, note the properties of the Choquet integral on p. 113.

An extension of this that combines three weighting vectors (probabilities, WAM weights, and OWA weights) called POWAWA [Mer10] is given by the WAM of three respective functions. Of course, other combinations of WAM and OWA functions are possible, for example, the weighted quasi-arithmetic mean of WAM and OWA, in which case the resulting function is averaging and monotone, but no longer corresponds to the Choquet integral with respect to some capacity.

6.4.2 Weighted OWA Approach by Torra

In [Tor97], Torra proposed a generalisation of both weighted means and OWA, called WOWA. This aggregation function has two sets of weights $\mathbf{w}$ and $\mathbf{p}$. The vector $\mathbf{p}$ plays the same role as the weighting vector in weighted means, and $\mathbf{w}$ plays the role of the weighting vector in OWA functions.

Consider the following motivation. A robot needs to combine information coming from n different sensors, which provide distances to various obstacles. The reliability of the sensors is known (i.e. we have weights $\mathbf{p}$). However, independent of their reliability, the distances to the nearest obstacles are more important, so irrespective of the reliability of each sensor, their inputs are also weighted according to their numerical value; hence, we have another weighting vector $\mathbf{w}$. Thus, both factors, the size of the inputs and the reliability of the inputs, need to be taken into account. The WOWA provides exactly this type of aggregation function.

Definition 6.12 (Weighted OWA) Let $\mathbf{w}$, $\mathbf{p}$ be two weighting vectors, $w_i, p_i \geqslant 0$, $\sum w_i = \sum p_i = 1$. The following function is called the weighted OWA function:

$$WOWA_{\mathbf{w},\mathbf{p}}(\mathbf{x}) = \sum_{i=1}^{n} u_i x_{(i)},$$

where $x_{(i)}$ is the i-th largest component of $\mathbf{x}$, and the weights u_i are defined as

$$u_i = g\left(\sum_{j \in H_i} p_j\right) - g\left(\sum_{j \in H_{i-1}} p_j\right),$$

where the set $H_i = \{j | x_j \geqslant x_i\}$ is the set of indices of the i largest elements of $\mathbf{x}$, and g is a monotone non-decreasing function with two properties:

1. $g(i/n) = \sum_{j \leqslant i} w_j, i = 0, \ldots, n$ (of course $g(0) = 0$).
2. g is linear if the points $(i/n, \sum_{j \leqslant i} w_j)$ lie on a straight line.

Note 6.5 The second condition is in the form formulated in [Tor97]. It can be reformulated as follows: 2a. g is linear if all w_i are equal.

Thus, computation of WOWA involves a very similar procedure as that of the OWA (i.e. sorting components of $\mathbf{x}$ and then computing their weighted sum), but the weights u_i are defined by using both vectors $\mathbf{w}$, $\mathbf{p}$, a special monotone function g, and depend on the components of $\mathbf{x}$ as well. One can see WOWA as an OWA function with the weights $\mathbf{u}$.

In [Tor97, Tor00], the weights were introduced through an auxiliary interpolation function. It was later shown that such a WOWA function is a discrete Choquet integral with respect to a distorted probability measure [Tor98, NT05]. It allows one to operate with two weighting vectors, one vector $\mathbf{p}$ related to the inputs' magnitude, another, $\mathbf{w}$, related to the inputs themselves. Another work in this direction was done by Yager in [Yag98b, YA14a, YA16]. In [Yag98b] a transformation function that combines the inputs of the OWA with their importances was built using fuzzy modelling techniques from partial knowledge about that function in the prototypical examples.

Let us list some of the properties of the WOWA function:

- The weighting vector $\mathbf{u}$ satisfies $u_i \geqslant 0$, $\sum u_i = 1$.
- If $w_i = \frac{1}{n}$, then $WOWA_{\mathbf{w},\mathbf{p}}(\mathbf{x}) = WAM_{\mathbf{p}}(\mathbf{x})$, the weighted arithmetic mean.
- If $p_i = \frac{1}{n}$, $WOWA_{\mathbf{w},\mathbf{p}}(\mathbf{x}) = OWA_{\mathbf{w}}(\mathbf{x})$.
- The WOWA is an idempotent aggregation function.

As noted, the weights $\mathbf{u}$ also depend on the generating function g. This function can be chosen as a linear spline (i.e. a broken line interpolant), interpolating the points $(i/n, \sum_{j\leqslant i} w_j)$ (in which case, it automatically becomes a linear function if these points are on a straight line), or as a monotone quadratic spline, as was suggested in [Tor97, Tor00]; see also [Bel01] where Schumaker's quadratic spline algorithm was used [Sch83], which automatically satisfies the straight line condition when needed.

It turns out that WOWA belongs to a class of Choquet integral-based aggregation functions, with respect to *distorted probabilities*; see Definition 7.4 [Tor98, NT05, Tor07]. It is a piecewise linear function whose linear segments are defined on the simplicial partition of the unit cube $[0, 1]^n$: $\mathcal{S}_i = \{\mathbf{x} \in [0, 1]^n | x_{p(j)} \geqslant x_{p(j+1)}\}$, where p is a permutation of the set $\{1, \ldots, n\}$. Note that there are exactly $n!$ possible permutations, the union of all $\mathcal{S}_i$ is $[0, 1]^n$, and the intersection of the interiors of $\mathcal{S}_i \cap \mathcal{S}_j = \emptyset, i \neq j$.

There have been a number of alternative attempts to incorporate weights into OWA, some of which turned out to be mathematically equivalent. We refer the reader to the following publications [Lla15, YA14a, Mer11, LJ14]. Unfortunately, these generalisations of OWA do not produce idempotent functions except in a few special cases.

6.4.3 Interpolation of the RIM Quantifier Function

Let us now consider an alternative approach based on interpolating a RIM quantifier function. Here we use a method from [CMY04], in which the weights of a function are computed by repeating the inputs a suitable number of times. Consider an auxiliary vector of arguments $\mathbf{X} = (x_1, \ldots, x_1, x_2, \ldots, x_2, \ldots, x_n, \ldots, x_n)$, so that x_1 is taken k_1 times and x_2 is taken k_2 times, and so on, so that $\frac{k_1}{M} = p_1$, $\frac{k_2}{M} = p_2, \ldots$, and $k_1 + k_2 + \ldots + k_n = M$. We assume the weights p_i are rational numbers, which is not a strong restriction if we look at computer implementation of the method in finite precision arithmetics.

The approach from [CMY04] consists in using the auxiliary vector $\mathbf{X}$ in a strongly idempotent symmetric function (such as OWA induced by a quantifier) whose output will be a weighted function of the inputs $\mathbf{x}$.

In the case of OWA, in order to apply it to a larger dimensional auxiliary input vector $\mathbf{X}$, we need to produce the weighting vector $\mathbf{w}$ of the corresponding dimension M, denoted here by $\mathbf{u}$. We apply a similar approach to Torra's construction, that is, we construct a generating function g by interpolating the data $g(i/n) = \sum_{j\leqslant i} w_j, i = 0, \ldots, n$ and $g(0) = 0$, and the straight line condition, that is, using the two conditions in Definition 6.12. The latter is necessary to obtain the standard weighted mean in case all $w_i = \frac{1}{n}$.

Hence, we can use a piecewise linear or piecewise quadratic interpolation as in [Tor97, Tor00, Bel01] to construct the RIM quantifier g. Now, in a way that differs

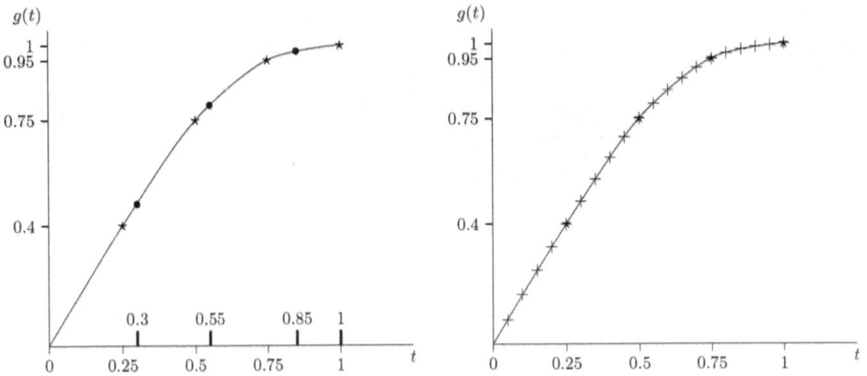

Fig. 6.4 Quantifier function g interpolated from the OWA weights $\mathbf{w}$ (represented by stars). The weights u_i for Torra's method are found from the points marked with circles (left), and the weights for the vector X ($M = 20$ in this example) are found from the points marked with crosses (right). The abscissae of those points are marked above the horizontal axis

to Torra's approach, we calculate the weighted OWA as

$$WOWA_{\mathbf{w},\mathbf{p}}(\mathbf{x}) = OWA_{\mathbf{u}}(\mathbf{X}) = \sum_{i=1}^{n} u_i X_{(i)},$$

where the weights u_i are defined as

$$u_i = g\left(\frac{i}{M}\right) - g\left(\frac{i-1}{M}\right), \quad i = 1, \dots, M.$$

Figure 6.4 illustrates the two approaches.

If we compare both methods based on the generating RIM quantifier function, we can see that both produce exactly the same WOWA function. Indeed, since the vector $\mathbf{X}$ has groups of repeated arguments, we can break the expression for WOWA into individual sums:

$$OWA_{\mathbf{u}}(\mathbf{X}) = \sum_{i=1}^{n} u_i X_{(i)} = \sum_{j=1}^{n} \sum_{i=1}^{k_j} x_j \left(g\left(\frac{i + k_{j-1}}{M}\right) - g\left(\frac{i + k_{j-1} - 1}{M}\right)\right),$$

with $k_0 = 0$. The inner sums collapse and we have

$$OWA_{\mathbf{u}}(\mathbf{X}) = \sum_{j=1}^{n} x_j \left(g\left(\frac{k_j + k_{j-1}}{M}\right) - g\left(\frac{k_{j-1}}{M}\right)\right),$$

with $g(0) = 0$ as usual. From the definition of $k_j = Mp_j$, we conclude that the above expression matches the one in Definition 6.12. We conclude that Torra's formula for the weights of the WOWA can be seen as an instance of the approach from [CMY04] based on replicating the inputs, although it obviously predates the work.

The next section introduces an alternative and generic construction to incorporate weights into any symmetric averaging function. In particular, it will work for OWA and will not have the somewhat unclear issue of selecting the function g as in Torra's WOWA.

6.4.4 *n*-Ary Tree Construction for OWA by Dujmovic and Beliakov

We apply the method of incorporating weights into any symmetric function by using binary trees [BD16, DB17]. This method can also be applied to fuzzy integrals.

To introduce the weights into a symmetric function, we use the approach from [CMY04], where each argument x_i is replicated a suitable number of times. We consider an auxiliary vector of arguments $\mathbf{X} = (x_1, \ldots, x_1, x_2, \ldots, x_2)$, so that x_1 is taken k_1 times and x_2 is taken k_2 times, so that $\frac{k_1}{2^L} \approx p_1$, $\frac{k_2}{2^L} \approx p_2$, and $k_1 + k_2 = M = 2^L$. Here, M is a power of two and $L \geqslant 1$ is a specified number of levels of the binary tree shown in Fig. 6.6. One way of doing so is to take $k_1 = \lfloor p_1 2^L + \frac{1}{2} \rfloor$ and $k_2 = 2^L - k_1$. The vector $\mathbf{X}$ needs to be sorted into increasing or decreasing order.

Next, let us build a binary tree as presented in Fig. 6.5, where at each node, a value is produced by aggregating the values of two children nodes with the given bivariate symmetric averaging function f (with weights equal to $\frac{1}{2}$). We start from the leaves of the tree, which contain the elements of the vector $\mathbf{X}$. In this example, we took $p_1 = \frac{5}{8}$ and $p_2 = \frac{3}{8}$. The value y at the root node will be the desired output of the n-variate weighted function.

An efficient algorithm based on pruning the binary tree was presented in [BD16]. The pruning of the binary tree is done by using the idempotency of f; see Fig. 6.5, right. No invocation of f is necessary if both of its arguments are equal. A branch

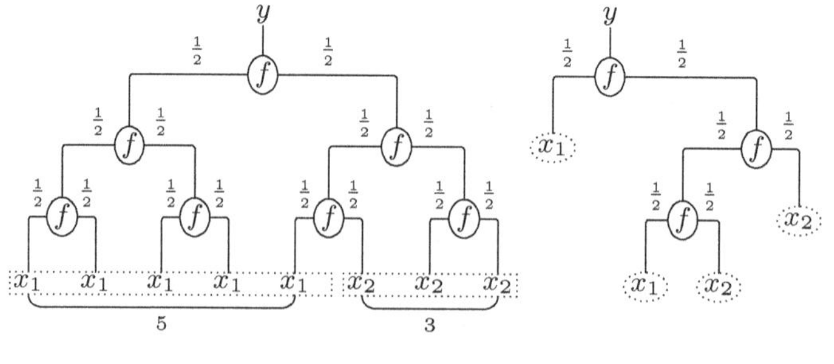

Fig. 6.5 Representation of a weighted arithmetic mean in a binary tree construction. The tree on the right is pruned by using idempotency

is pruned if it is clear that all its leaves have exactly the same value, and by idempotency this is the value of the root node of that branch. The algorithm is a recursive depth-first traversing of the binary tree, given as Algorithm 6.1. It has worst-case complexity $O(L)$, which makes it practically applicable for larger L.

Algorithm 6.1: Pruned tree aggregation (PTA) algorithm

function $node(m, N, K, x)$

1. If $N[K] \geqslant 2^m$ then do:

   a. $N[K] := N[K] - 2^m$;
   b. $y := x[K]$;
   c. If $N[K] = 0$ then $K := K + 1$;
   d. return y;

   else
2. return $f(node(m-1, N, K, x), node(m-1, N, K, x))$.

function $f\_weighted(p, x, L)$

1. create the array $N := (k_1, k_2)$ by using $k_1 := \lfloor p_1 2^L + \frac{1}{2} \rfloor$, and $k_2 := 2^L - k_1$;
2. $K := 1$;
3. return $node(L, N, K, x)$.

In this algorithm, the array N serves as a counter of how many copies of each of $x[K]$ remains. If there are more than 2^m copies, they belong to a branch that can be pruned, so the function $node$ returns $x[K]$ and never visits the nodes of that branch. If $N[K] = 1$, then the last remaining copy of $x[K]$ is returned and the value of K is incremented. Every time a branch whose leaves contain identical arguments is encountered (which is detected by the counter $N[K] \geqslant 2^m$), this branch is pruned.

Following, the same algorithm PTA can be used to extend a symmetric bivariate mean to a weighted n-variate mean (as was done in [BD16]). For this, we construct an auxiliary weighting vector:

$$\mathbf{X} = (x_1, \ldots, x_1, x_2, \ldots, x_2, \ldots, x_n, \ldots, x_n)$$

where each argument x_i is replicated a suitable number of times. So x_1 is taken k_1 times, x_2 is taken k_2 times, and so on; then $\frac{k_1}{2^L} \approx p_1, \frac{k_2}{2^L} \approx p_2, \ldots$, and $\sum k_i = 2^L$, where $L \geqslant 1$ is a specified number of levels of the binary tree shown in Fig. 6.5. One way of doing so is to take $k_i = \lfloor p_i 2^L + \frac{1}{2} \rfloor, i = 1, \ldots, n-1$, and $k_n = 2^L - k_1 - k_2 - \ldots - k_{n-1}$.

The algorithm for computing a weighted n-variate mean f_n is shown as Algorithm 6.2.

To see the complexity of this algorithm, note that f is never executed (nor is the corresponding node of the tree visited) if its arguments are the same. There is exactly one node at each level of the tree where the child nodes contain distinct

Algorithm 6.2: n-variate extension of a bivariate idempotent function

function $f_n(n, p, x, L)$

1. create the array $N := (k_1, k_2, \ldots, k_n)$ by using $k_i := \lfloor p_i n^L + \frac{1}{n} \rfloor, i = 1, \ldots, n-1$, and $k_n := n^L - k_1 - \ldots - k_{n-1}$;
2. $K := 1$;
3. return $node(L, N, K, x)$.

arguments; hence, f is executed exactly L times. Also note that both N and K are input–output parameters, so that the two arguments of f at step 2 are different as N and K change from one invocation of the function $node$ to another; however, the order of execution of the calls to $node$ does not matter as the lists of formal parameters are identical.

The properties of the binary tree construction include preservation of idempotency, monotonicity, continuity, convexity (concavity), homogeneity, and shift invariance, due to preservation of these properties in function composition. Furthermore, when the weights are given in a finite binary representation (as is always the case in machine arithmetic), the sequence of the outputs of the PTA algorithm with increasing $L = 2, 3, \ldots$,etc., converges to a weighted mean with the specified weights, and in fact L need not exceed the number of bits in the mantissa of the weights p_i to match these weights exactly. When f is a quasi-arithmetic mean, f_p is a weighted quasi-arithmetic mean with the same generator.

The big advantage of the binary tree construction is its universality and transparency. It is applicable to any bivariate idempotent function f without modification, and the role of the weights as the respective multiplicities of the arguments as argued in [CMY04] is very clear. The availability of a fast and uncomplicated algorithm for computing the output makes this method immediately applicable.

We now extend the algorithm PTA to n-variate OWA functions following [Bel17a, Bel18]. Our goal here is to incorporate a vector $\mathbf{p}$ of non-negative weights (which add to one) into a symmetric n-variate function, by replicating the arguments a suitable number of times. As in the binary tree construction, we build an n-ary tree with L levels, as shown in Fig. 6.6. As the base symmetric aggregator f, we take an OWA function $OWA_{\mathbf{w}}$ with specified weights $\mathbf{w}$ (although the origins of f are not important for the algorithm).

Let us create an auxiliary vector $\mathbf{X} = (x_1, \ldots, x_1, x_2, \ldots, x_2, \ldots, x_n, \ldots, x_n)$, so that x_1 is taken k_1 times, x_2 is taken k_2 times, and so on, with $\frac{k_1}{n^L} \approx p_1, \frac{k_2}{n^L} \approx p_2$, …, and $\sum k_i = n^L$, where $L \geqslant 1$ is a specified number of levels of the tree shown in Fig. 6.6. One way of doing this is to take $k_i = \lfloor p_i n^L + \frac{1}{n} \rfloor, i = 1, \ldots, n-1$, and $k_n = n^L - k_1 - k_2 - \ldots - k_{n-1}$.

The PnTA algorithm (Algorithm 6.3) works in the same way as the PTA algorithm for binary trees. The vector of counters N helps determine whether there are more than n^m identical elements of the auxiliary array $\mathbf{X}$, in which case they are the leaves of a branch of the tree with m levels. This branch is pruned. The function

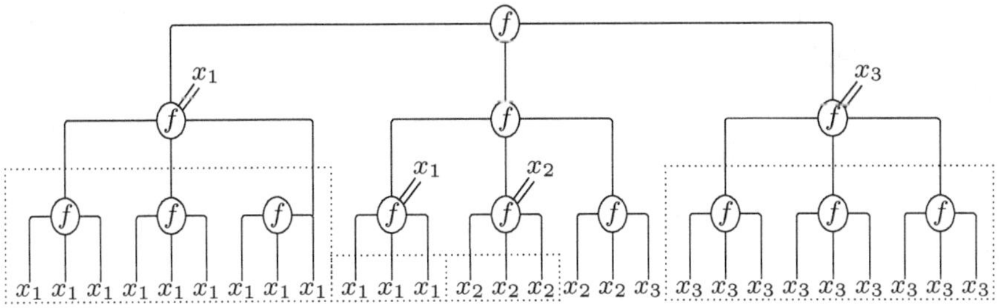

Fig. 6.6 Representation of a weighted tri-variate function f in a ternary tree construction. The weights are chosen as $\mathbf{p} = (\frac{12}{27}, \frac{5}{27}, \frac{10}{27})$ and $L = 3$. The framed branches are pruned by the algorithm

f is executed only when some of its arguments are distinct, and since the elements of $\mathbf{X}$ are ordered, there are at most $n - 1$ such possibilities at each level of the tree; hence, the complexity of the algorithm is $O((n - 1)L)$.

Note that the complexity is linear in terms of L, as is the case for the PTA algorithm, which means that the dimension of the base aggregator f does not matter in this respect. Of course, nominally the n-ary tree is larger than the binary tree, but since we only track the multiplicities of the arguments, never creating the array $\mathbf{X}$ explicitly, memorywise the complexity of the PnTA algorithm is the same as that of PTA.

Algorithm 6.3: Pruned n-tree aggregation (PnTA) algorithm

---

function $node(n, m, N, K, x)$

1. If $N[K] \geqslant n^m$ then do:

   a. $N[K] := N[K] - n^m$;
   b. $y := x[K]$;
   c. If $N[K] = 0$ then $K := K + 1$;
   d. return y;

   else
2. for $i := 1, \ldots, n$ do
   $z[i] := node(n, m - 1, N, K, x)$
3. return $f(\mathbf{z})$.

function $f\_n(n, x, p, L)$

1. create the array $N := (k_1, k_2, \ldots, k_n)$ by using
   $k_i := \lfloor p_i n^L + \frac{1}{n} \rfloor$, $i = 1, \ldots, n - 1$, and $k_n := n^L - k_1 - \ldots - k_{n-1}$;
2. $K := 1$;
3. return $node(n, L, N, K, x)$.

The vector $\mathbf{X}$ needs to be sorted, which is equivalent to sorting the inputs $\mathbf{x}$ jointly with the multiplicities of the inputs N (i.e. using the components of $\mathbf{x}$ as the key), so the complexity of the sort operation is the same $O(n \log n)$ as for OWA functions.

We list some useful properties of the function f_p generated by the PnTA algorithm established in [Bel17a].

Theorem 6.1 (The Inheritance Theorem) *The weighted extension f_p of a function f by the PnTA algorithm preserves the intrinsic properties of the parent function f as follows:*

1. *f_p is idempotent since f is idempotent.*
2. *If f is monotone increasing, then f_p is monotone increasing.*
3. *If f is continuous, then f_p is continuous.*
4. *If f is convex (resp. concave), then f_p is convex (resp. concave).*
5. *If f is homogeneous, then f_p is homogeneous.*
6. *If f is shift-invariant, then f_p is shift-invariant.*
7. *f_p has the same absorbing element as f (if any).*
8. *If f generates f_p, then a φ-transform of f generates the corresponding φ-transform of f_p.*

Proof The proof easily follows from the properties of composition of the respective functions and idempotency of f. For the φ-transform, notice that at each inner level of the tree, the composition $\varphi^{-1} \circ \varphi = Id$, while φ is applied to the leaves of the tree and φ^{-1} is applied to the root. □

The next results are applicable when an OWA function is taken as the base aggregator f. For proofs, see [Bel17a, Bel18].

Theorem 6.2 *Let $f = OWA_{\mathbf{w}}$. Then the PnTA algorithm generates the weighted function f_p, which is a discrete Choquet integral (and is hence homogeneous and shift-invariant).*

For special cases of the Choquet integral, we have the following.

Theorem 6.3 *Let $f = OWA_{\mathbf{w}}$. Then the PnTA algorithm generates the weighted function f_p with the following properties:*

1. *For the weights $w_i = \frac{1}{n}$, f_p is the WAM with weights $\mathbf{p}$.*
2. *For the weights $p_i = \frac{1}{n}$, f_p is $OWA_{\mathbf{w}}$.*
3. *When $f = OWA_{\mathbf{w}} = \min$ (or $= \max$), and $p_i > 0$ for all i, f_p is also the $\min$ (respectively, $\max$).*
4. *When $f = OWA_{\mathbf{w}} = Med(\mathbf{x})$ and n is odd, f_p is the weighted median.*
5. *If $OWA_{\mathbf{w}}$ generates f_p, then the dual $OWA^d_{\mathbf{w}}$ generates the dual f^d_p, and in particular an OWA with the reverse weights generates the respective weighted OWA with the reverse weights.*

Theorem 6.4 *Let $f = OWA_{\mathbf{w}}$ and let the weighting vector be decreasing (increasing). Then the PnTA algorithm generates a Choquet integral with respect to a submodular (supermodular) capacity.*

This last result is useful when constructing weighted norms from OWA with decreasing weights; see [BJL11, Yag10].

Note 6.6 On the technical side, we note that we do not need to sort the arguments in each OWA function in the n-ary tree, as the vector $\mathbf{x}$ is already sorted; hence, only one sort of operation for the inputs is required. Another note is that when the weights $\mathbf{p}$ are specified to m digits in base n, $L = m$ levels of the n-ary tree are sufficient to match these weights exactly. For example, if $\mathbf{p}$ are specified to three decimal places and $n = 10$, we only need to take $L = 3$. Therefore, to match the weights to machine precision (e.g. 53 bits for data type `double`) n^L need not exceed the largest 64-bit integer, and hence the algorithm PnTA can be implemented with 64-bit data types. The source code in C++ is presented in Fig. 6.7 and the software can be downloaded from [Bel17b].

Finally, we can introduce weights into generalised OWA functions in the same way as for OWA functions, by using the n-ary tree construction. This can be done by using a φ-transform of a weighted OWA with $\varphi = g$, that is, by applying g and g^{-1} only to the leaves and to the root node of the tree, relying on the preservation of φ-transforms. This method is computationally efficient as functions g and g^{-1} need not be used in the middle of the tree, where they cancel each other.

This way we also obtain the special cases of ordered weighted geometric and ordered weighted power-based generalised OWA functions.

6.4.5 WOWA Based on Implicit Averaging

A different approach to introducing weights into averaging functions was recently presented in [BCF17] under the name of implicit averaging. It follows an analogy with weighted arithmetic means, which can be written in this form:

$$y \cdot \frac{\sum_{j=1}^{n} p_j}{n} = \frac{\sum_{i=1}^{n} p_i x_i}{n}. \tag{6.12}$$

with $y = WAM(x_1, \ldots, x_n)$ with respect to the weighting vector $\mathbf{p}$. We use the following equation to compute the values of a weighted function f_p from a symmetric mean M:

$$C(M(p_1, \ldots, p_n), f_p(\mathbf{x})) = M(C(p_1, x_1), \ldots, C(p_n, x_n)). \tag{6.13}$$

Here, M is a mean and C is a suitable bivariate operation such as a t-norm. The motivation behind the study in [BCF17] is to produce alternative ways of incorporating weights $\mathbf{p}$ by replacing the product with another suitable operation and replacing the arithmetic mean with an arbitrary mean M. The function f_p is given implicitly through solution to the algebraic equation (6.13). This equation can

```
double OWA(int n, double x[],double w[])
{ /* no sorting is needed when used in the tree */
   double z=0;
   for(int i=0;i<n;i++) z+=x[i]*w[i];
   return z;
}
double node(int n, double x[], long int N[], long int C, int & k,
     double w[], double(*F)(int, double [],double[]), double* z)
{
  /* recursive function in the n-ary tree processing
  Parameters: x - input vector, N vector of multiplicities of  x
  m current level of recursion counted from L (root node)  to 0
  k - input-output parameter, the index of x being processed */
  if(N[k]==0) k++;
  if(N[k]>= C) {  /* we use idempotency to prune the tree */
      N[k]  -= C;
      if(N[k]<=0) return x[k++]; else return x[k];
  }
   C /= n;
  /* tree not pruned, process the children nodes */
  for(int i=0;i<n;i++) z[i]=node(n,x,N,C,k,w,F,z+n);
  return F(n,z,w);
}
double weightedf(double x[], double p[], double w[], int n,
          double(*F)(int, double[],double[]), int L)
/*
 Function F is the symmetric base aggregator.
 p[] = array of weights of inputs x[],
 w[] = array of weights for OWA, n = the dimension of x, p, w.
 the weights must add to one and be non-negative.
 L = number of binary tree levels. Run time = O[(n-1)L]   */
{
  long int t=0, C=1;
  int k=0;
  for(int i=0;i<L;i++) C*=n;   /* C=n^L */
  sortpairs(x, x+n, p);
  long int N[n]; /* multiplicities of x based on the weights */
  for(int i=0;i<n-1;i++)  { N[i]=p[i]*C+1./n;   t+=N[i]; }
  N[n-1]=C-t;
  double z[n*L]; /* working memory */
  return node(n,x,N,C,k,w,F,z);
}
/* example: calling the function */
int n=4, L=4;
double x[4]={0.2,0.2,0.4,0.8};
double w[4]={0.1,0.2,0.3,0.4};
double p[4]={0.3,0.2,0.1,0.4};
double y=weightedf(x,p,w,n,&OWA,L);
```

Fig. 6.7 A C++ implementation of the pruned n-ary tree algorithm PnTA. The function `sortpairs` (not shown) implements sorting of an array of pairs (x_i, p_i) in the order of decreasing x_i

be written in a compact form as

$$C(\bar{p}, \bar{x}_p) = \overline{C(p_i, x_i)},$$

where $\bar{p}$ denotes the (unweighted) average weight, $\overline{C(p_i, x_i)}$ denotes the average value of C, and $y = \bar{f}_p(\mathbf{x}) = \bar{x}_p$ is the weighted average of the x_i.

The work [BCF17] established a number of useful theoretical properties of the implicit averages, which also apply to the case of $M = OWA$. Instantiating Eq. (6.13) with C being the product and M being an OWA function with weights $\mathbf{w}$, we can resolve it explicitly and obtain

$$WOWA_{\mathbf{w},\mathbf{p}}(\mathbf{x}) = \frac{OWA_{\mathbf{w}}(p_1x_1, \ldots, p_nx_n)}{OWA_{\mathbf{w}}(p_1, \ldots, p_n)} = \frac{OWA_{\mathbf{w}}(\mathbf{px})}{OWA_{\mathbf{w}}(\mathbf{p})}. \tag{6.14}$$

Note that the weights p_i are otherwise unrestricted (i.e. they need not add to one) as the denominator in (6.14) will produce the required normalising factor to ensure idempotency.

Special care should be taken when the denominator vanishes, as this WOWA may be discontinuous or not well defined if we allow 0 weights. For strictly positive weighting vectors $\mathbf{w}$, the proposed WOWA is well defined and continuous. It is a piecewise linear, increasing, idempotent, and homogeneous function. However, unlike the other mentioned WOWA, this function *is not* a discrete Choquet integral. This is evident from Fig. 6.8 (note the set of non-differentiability).

It is not difficult to see that the special case of equal weights $p_i = \frac{1}{n}$ corresponds to the unweighted OWA and $w_i = \frac{1}{n}$ corresponds with the WAM. However, reversing the weights of the OWA does not produce the dual of the original function. The implicit WOWA in (6.14) is a valid alternative to the existing WOWA where the arguments are weighted both according to their position and magnitude.

Example 6.6 Consider the Hurwitz operator $M = H(\mathbf{x}) = a\max(\mathbf{x}) + (1 - a)\min(\mathbf{x})$, $a \in]0, 1]$ and $C = \prod$. We have

$$WOWA_{(a,1-a)}(\mathbf{x}) = \frac{a\max(\mathbf{px}) + (1-a)\min(\mathbf{px})}{a\max(\mathbf{p}) + (1-a)\min(\mathbf{p})}.$$

Let us consider a particular case where $\mathbf{x} = (x_1, x_2)$, $a = \frac{9}{10}$, $\mathbf{p} = (p_1, p_2) = (\frac{1}{3}, \frac{2}{3})$; then

$$\begin{aligned} WOWA_{\mathbf{w},\mathbf{p}}(\mathbf{x}) &= \frac{\frac{9}{10}\max(\frac{1}{3}x_1, \frac{2}{3}x_2) + \frac{1}{10}\min(\frac{1}{3}x_1, \frac{2}{3}x_2)}{\frac{9}{10}\max(\frac{1}{3}, \frac{2}{3}) + \frac{1}{10}\min(\frac{1}{3}, \frac{2}{3})} \\ &= \frac{9\max(x_1, 2x_2) + \min(x_1, 2x_2)}{19}. \end{aligned}$$

F is plotted in Fig. 6.8.

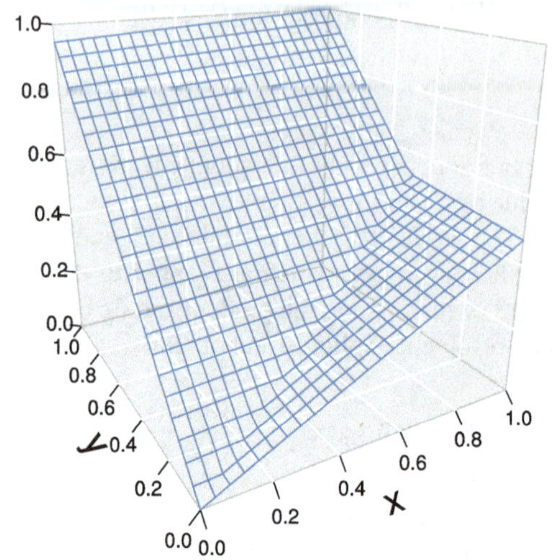

Fig. 6.8 Function $WOWA$ in Example 6.6 for $C = \prod$ and M the Hurwitz operator H

If we replace the products $p_i x_i$ with a more general function C, strictly increasing on $\mathbb{R}_{++} \times]0, 1]$, we obtain a generalisation of the WOWA function.

Example 6.7 When C is replaced by the square of the product function in Example 6.6, we have

$$GenWOWA_{\mathbf{w},\mathbf{p}}(\mathbf{x}) = \left[\frac{\frac{9}{10}\max(\frac{1}{9}x_1^2, \frac{4}{9}x_2^2) + \frac{1}{10}\min(\frac{1}{9}x_1^2, \frac{4}{9}x_2^2)}{\frac{9}{10}\max(\frac{1}{9}, \frac{4}{9}) + \frac{1}{10}\min(\frac{1}{9}, \frac{4}{9})} \right]^{1/2}$$

Therefore,

$$GenWOWA_{\mathbf{w},\mathbf{p}}(\mathbf{x}) = \left[\frac{9\max(x_1^2, 4x_2^2) + \min(x_1^2, 4x_2^2)}{37} \right]^{1/2}.$$

Unlike Example 6.6, $GenWOWA$ is not piecewise linear, yet it is strictly increasing and idempotent.

6.4.6 *Illustrative Example*

We now illustrate the behaviour of the three WOWA functions using Example 6.8.

Example 6.8 An academic department wishes to hire faculty that can contribute significantly to research, teaching, industry engagement, and service and leadership. The department advertised for a junior to mid-level position and therefore does not anticipate the candidates with significant past roles in service and leadership. The research is given marginal priority over teaching, and industry engagement is also welcome. The selection committee decides to assess the

Table 6.1 The applicants specified by their input vectors (R, T, I, S) in Example 6.8. The overall aggregated score computed using three types of WOWA and the WAM

Candidate	Satisfaction scores	PnTA	Torra	Implicit	WAM
A	(1,0,1,1)	0.95	0.95	0.90	0.75
B	(1,1,0,1)	0.94	0.93	0.84	0.70
C	(0,1,1,1)	0.94	0.93	0.84	0.70
D	(1,1,1,0)	0.99	0.98	0.97	0.85
E	(1,0,1,0)	0.89	0.84	0.79	0.60
F	(1,0,0,1)	0.74	0.69	0.61	0.45

applicants on their strongest two points out of the four. The weightings given to research, teaching, industry engagement, and service will be denoted by $\mathbf{p} = (p_r, p_t, p_i, p_s) = (0.3, 0.25, 0.3, 0.15)$. The OWA weighting vector is given by $\mathbf{w} = (0.4, 0.35, 0.2, 0.05)$, so that the two strongest inputs are the most important, while the weakest input is only of marginal importance.

We consider a number of model applicants presented in Table 6.1. For simplicity, we take perfect satisfaction scores with respect to our criteria.

In this case, all the methods including the WAM give a reasonable ranking for the first six candidates, $D \succ A \succ B \sim C \succ E \succ F$, although the outputs of the methods differ in numerical values. For instance, the PnTA and Torra's method give very close scores to the candidates A, B, C, D, which somewhat agrees with our intuition.

If we now change the OWA weights to $\mathbf{w} = (0.45, 0.45, 0.05, 0.05)$, we get different rankings, namely, $D \succ A \succ B \sim C \sim E \succ F$ for PnTA but $D \succ A \succ B \sim C \succ E \succ F$ for Torra's and WAM and $D \succ A \succ E \succ B \sim C \succ F$ for implicit OWA. We see here that the implicit OWA placed E over B, C, discounting the third largest criterion (service) but favouring higher weighted research and industry engagement.

If we bring another candidate $G = (0.8, 0.5, 0, 0.8)$, all WOWA methods will place him last (as they discounted the third largest input 0.5), but the WAM will rank $G \succ F$.

Another useful illustration is to compare the preferences generated by the WOWA and the weighted arithmetic mean. Consider two applicants $A = (0.8, 0.8, 0, 0.2)$ and $B = (0.5, 0.5, 0.5, 0.5)$, so that A shows excellence in research and teaching and some service capacity, whereas B is all-round but mediocre in all components. The WAM with the same weighting vector $\mathbf{p}$ will give the scores $WAM(A) = 0.47$, $WAM(B) = 0.5$, so that B is preferred to A. In contrast, $WOWA(A) = 0.68$ and $WOWA(B) = 0.5$, so that A is preferred to B, and despite lack of industry experience and little service, A would be considered stronger than B based on her strongest points. Similarly, the candidates with strong research and industry experience and no teaching experience, or excellence in teaching and industry engagement and no research, would still be preferred to the mediocre B. We believe that WOWA better reflects the desire of the department as it is unreasonable to expect excellence in all the selection criteria from junior academics. The WOWA function allows one to model "k out of n" criteria, whereas the WAM discriminates against the candidates who may not have had an opportunity to excel in one or two areas.

A very similar situation occurs in academic promotion committees, where the assessment is done on the four mentioned criteria. As some applicants come from research-only or teaching-only positions, they obviously lack experience in one or another criterion. While the WAM pulls their overall score down due to zero score in that missing criterion, the WOWA ignores the zero score and allows one to compare candidates with very different skillsets based on their strongest points, which matches the desires of decision-makers.

Of course, OWA weights could be adjusted as to emphasise the strongest points more and disregard the weaknesses. And to the contrary, OWA weights can be chosen to discard the biggest inputs and emphasise the smallest or middle-sized inputs, which is useful when filtering outliers.

Let us now compare the three mentioned types of WOWA functions. All three methods produce averaging piecewise linear continuous functions and match the OWA and WAM functions in the limiting cases. WOWA based on interpolating the quantifier function and WOWA based on n-nary tree are both computationally efficient and preserve several useful properties of the original OWA function. Both result in a discrete Choquet integral. However, both methods are computer-oriented: while the algorithms employed are not too complicated, it is cumbersome to use these methods for quick pen and paper calculations. In contrast, the WOWA based on implicit averaging is quite simple for manual calculations, or to be used in a spreadsheet formula. Care should be taken to ensure the weights are positive in this case to avoid division by zero.

When the OWA function is max or min, the methods based on the quantifier function also produce the max and min function irrespective of the weights $\mathbf{p}$, as long as $p_i > 0$. However, the third method produces a different function, similar to the weighted maximum and minimum functions. This may be advantageous, as in this case, small weights p_i ensure that the aggregated value depends on the corresponding variables to a lesser degree (unlike the max and min functions that are symmetric).

From numerical evidence, it appears that the implicit WOWA provides scores somewhat in between the other two WOWA and the WAM, although this behaviour is not universal (e.g. candidate D in Table 6.1).

The implementation of all the mentioned WOWA methods is available in [Bel17b].

6.5 OWA-Based Metrics

The following results are also relevant to Choquet integral-based metrics; however, we focus on the symmetric case.

The choice of a suitable distance function, or metric, is very important in data analysis. In particular, the performance of many supervised and unsupervised classification methods can be highly dependent on whether the notion of proximity or similarity captured by the metric is actually representative of the data.

We recall the definition of a norm.

Definition 6.13 (Norm) A norm is a function $s : \mathbb{R}^n \to [0, \infty)$ that satisfies the following conditions:

$$S1)\ s(\mathbf{x}) = 0 \Longleftrightarrow x_i = 0, \qquad \forall\, i;$$

$$S2)\ s(a\mathbf{x}) = |a| s(\mathbf{x}), \qquad \forall \mathbf{x} \in \mathbb{R}^n, \forall a \in \mathbb{R};$$

$$S3)\ s(\mathbf{x}_1) + s(\mathbf{x}_2) \geqslant s(\mathbf{x}_1 + \mathbf{y}_2), \qquad \forall \mathbf{x}_1, \mathbf{x}_2 \in \mathbb{R}^n.$$

Norms are useful, among other things, for defining distance metrics.

Definition 6.14 (Metric) A metric is a function $d : \mathbb{R}^n \times \mathbb{R}^n \to [0, \infty)$ that satisfies the following conditions $\forall \mathbf{x}_1, \mathbf{x}_2, \mathbf{x}_3 \in \mathbb{R}^n$:

$$D1)\ d(\mathbf{x}_1, \mathbf{x}_2) = 0 \Longleftrightarrow \mathbf{x}_1 = \mathbf{x}_2;$$

$$D2)\ d(\mathbf{x}_1, \mathbf{x}_2) = d(\mathbf{x}_2, \mathbf{x}_1);$$

$$D3)\ d(\mathbf{x}_1, \mathbf{x}_2) + d(\mathbf{x}_2, \mathbf{x}_3) \geqslant d(\mathbf{x}_1, \mathbf{x}_3).$$

The third condition is well-known as the triangle inequality. If s is a norm, then $d(\mathbf{x}_1, \mathbf{x}_2) = s(\mathbf{x}_1 - \mathbf{x}_2)$ is a metric.

Various authors have proposed the use of metrics defined by the discrete Choquet integral [Nar07, BGW08] and identified the conditions under which it satisfies the properties of a metric. Similar investigations concern the OWA function [Yag10]. In this section, we outline the conditions under which the metric based on OWA functions satisfies the axioms of a metric. This is useful in particular when learning a metric from data in applications such as semi-supervised clustering [BJL11].

The following definitions are useful in this context.

Definition 6.15 (Star-Shaped Set) A closed set $\mathcal{S}$ is called star-shaped, if it includes the origin in its interior, and any ray originating from the origin is either contained in $\mathcal{S}$ or crosses its boundary at exactly one point.

It is clear that if $\mathcal{S}$ is star-shaped and $\mathbf{x} \in \mathcal{S}$, then for all $\lambda \in [0, 1]$, $\lambda\mathbf{x} \in \mathcal{S}$.

Definition 6.16 (Minkowski Gauge) Let $\mathcal{S}$ be a star-shaped set in $\mathbb{R}^n$. The Minkowski gauge is the function $M_{\mathcal{S}}(\mathbf{x}) = \inf\{\lambda > 0 : \mathbf{x} \in \lambda\mathcal{S}\}$.

The Minkowski gauge enjoys several interesting properties, as reported in [Rub00, DR95]. In particular, it is defined on $\mathbb{R}^n$ and is non-negative and positively homogeneous of degree one. The set $\mathcal{S}$ is convex if and only if $M_{\mathcal{S}}$ is convex and hence sublinear. The function $M_{\mathcal{S}}$ is a norm if and only if $\mathcal{S}$ is bounded, convex, and symmetric with respect to the origin (the triangle inequality holds if and only if $\mathcal{S}$ is convex, whereas symmetry is needed in order to obtain $M_{\mathcal{S}}(\mathbf{x}) = M_{\mathcal{S}}(-\mathbf{x})$, in which case $M_{\mathcal{S}}$ is homogeneous). All norms are special cases of the Minkowski gauge with a bounded convex S, which is interpreted as the unit ball.

Let us summarise and relate various concepts.

Definition 6.17 (Choquet Integral-Based Norm) A Choquet integral-based norm s_{Ch} is defined as

$$s_{Ch}(\mathbf{x}) = C_{\mu}(|x_1|, |x_2|, \ldots, |x_n|), \tag{6.15}$$

provided that s_{Ch} satisfies conditions S1) and S3).

Proposition 6.1 *([Cho54, Den94]). The Choquet integral C_{μ} is convex if and only if the capacity μ is submodular.*

Proposition 6.2 *The Choquet integral C_{μ} defines a norm in (6.15) if and only if μ is submodular.*

Corollary 6.1 *If μ is a Sugeno λ-capacity, the Choquet integral C_{μ} defines a norm if and only if $-1 < \lambda \leqslant 0$.*

Definition 6.18 (Buoyancy of Weights) OWA weights satisfy the buoyancy property if $w_i \geqslant w_k, \forall i < k$ (i.e. the weighting vector is non-increasing).

Proposition 6.3 *An OWA function has the buoyancy property if and only if it is the Choquet integral with respect to a symmetric submodular capacity.*

Corollary 6.2 *An OWA function defines a norm by $s_{OWA}(\mathbf{x}) = OWA_{\mathbf{w}}(|\mathbf{x}|)$ if and only if the weights have the buoyancy property.*

Corollary 6.3 *A power-based OWA function, $p \geqslant 1$, defines a norm by*

$$s_{OWA,p}(\mathbf{x}) = OWA_{p,\mathbf{w}}(|\mathbf{x}|) = \left(OWA_{\mathbf{w}}(|x_1|^p, \ldots, |x_n|^p)\right)^{\frac{1}{p}}$$

if and only if the weights have the buoyancy property.

6.6 Learning OWA Weights

The problem of identification of weights of OWA functions has been studied by several authors [YK97, FY98, FM01, Xu05, YA14b, Bel07, Bel03, BJL11]. Here, we summarise several of the proposed approaches.

6.6.1 Methods Based on Data

A common feature of all methods is to eliminate nonlinearity due to reordering of the components of $\mathbf{x}$ by restricting the domain of this function to the simplex $S \subset \mathbb{I}^n$ defined by the inequalities $x_1 \geqslant x_2 \geqslant \ldots \geqslant x_n$. On that domain, the OWA function is a linear function (it coincides with the arithmetic mean). Once the coefficients of this function are found, the OWA function can be computed on the whole $\mathbb{I}^n$ by using its symmetry. Algorithmically, it amounts to using an auxiliary dataset $\{(\mathbf{z}^j, y_j)\}$, where vectors $\mathbf{z}^j = \mathbf{x}^j_{\searrow}$. Thus, identification of weights of OWA functions is a very similar problem to identification of weights of arithmetic means. Depending on whether we use the least squares or least absolute deviation criterion, we solve it by using either quadratic or linear programming techniques. In the first case, we have the problem

$$\text{minimise} \sum_{j=1}^{J} \left(\sum_{i=1}^{n} w_i z_i^j - y_j \right)^2 \tag{6.16}$$

$$\text{s.t.} \sum_{i=1}^{n} w_i = 1, w_i \geqslant 0, i = 1, \ldots, n.$$

In the second case, we have

$$\text{minimise} \sum_{j=1}^{J} \left| \sum_{i=1}^{n} w_i z_i^j - y_j \right| \tag{6.17}$$

$$\text{s.t.} \sum_{i=1}^{n} w_i = 1, w_i \geqslant 0, i = 1, \ldots, n,$$

which converts to a linear programming problem.

Filev and Yager [FY98] proposed a nonlinear change in variables to obtain an unrestricted minimisation problem, which they proposed to solve using nonlinear local optimisation methods. Unfortunately, the resulting nonlinear optimisation problem is difficult due to a large number of local minimisers and traditional optimisation methods become stuck in local minima.

An approach relying on quadratic programming was used in [YF94, Tor02, Tor04, Bel03, Bel05], and it was shown to be numerically efficient and stable with respect to rank deficiency (e.g. when $J < n$, or the data are linearly dependent).

Often an additional requirement is imposed, which is the desired value of orness, $orness(f) = \alpha \in [0, 1]$. This requirement is easily incorporated into a QP or LP problem as an additional linear equality constraint, namely,

$$\sum_{i-1}^{n} w_i \frac{n-i}{n-1} = \alpha.$$

6.6.2 Preservation of Ordering of the Outputs

We may also require that the ordering of the outputs is preserved, i.e. if $y_j \leqslant y_k$, then we expect $f(\mathbf{x}^j) \leqslant f(\mathbf{x}^k)$. We arrange the data, so that the outputs are in non-decreasing order, $y_j \leqslant y_{j+1}$, $j = 1, \ldots, J-1$. Then we define the additional linear constraints:

$$< \mathbf{z}^{j+1} - \mathbf{z}^j, \mathbf{w} >= \sum_{i=1}^{n} w_i (z_i^{j+1} - z_i^j) \geqslant 0,$$

$j = 1, \ldots, J-1$. We add the above constraints to problem (6.16) or (6.17) and solve it. The addition of an extra $J-1$ constraints neither changes the structure of the optimisation problem nor does it drastically affect its complexity.

Additional constraints could be related to the buoyancy property as discussed in Sect. 6.5, in which case the OWA functions define a metric. The methods related to metric learning are discussed in [BJL11].

6.6.3 Methods Based on a Measure of Dispersion

Rather than the regression approach where differences between observed and predicted outputs are minimised, some researchers have focused on defining weights based on measures such as dispersion and variance.

Maximum Entropy OWA

One such approach to choosing OWA weights was proposed in [O'H88] and followed in [FM01]. It does not use any empirical data, but various measures of weight entropy or dispersion. The measure of weight dispersion is defined as in Definition 6.5, i.e.

$$Disp(\mathbf{w}) = -\sum_{i=1}^{n} w_i \log w_i. \tag{6.18}$$

The idea is to choose, for a given n, such a vector of weights that maximises the dispersion $Disp(\mathbf{w})$.

It is formulated as an optimisation problem:

$$\text{minimise} \sum_{i=1}^{n} w_i \log w_i \tag{6.19}$$

$$\text{s.t.} \sum_{i=1}^{n} w_i = 1$$

$$\sum_{i=1}^{n} w_i \frac{n-i}{n-1} = \alpha$$

$$w_i \geqslant 0, i = 1, \ldots, n.$$

The solution is provided in [FM01] and is called maximum entropy OWA (MEOWA). Using the method of Lagrange multipliers, the authors obtain the following expressions for w_i:

$$w_i = (w_1^{n-i} w_n^{i-1})^{\frac{1}{n-1}}, i = 2, \ldots, n-1, \tag{6.20}$$

$$w_n = \frac{((n-1)\alpha - n)w_1 + 1}{(n-1)\alpha + 1 - nw_1},$$

and w_1 being the unique solution to the equation

$$w_1[(n-1)\alpha + 1 - nw_1]^n = ((n-1)\alpha)^{n-1}[((n-1)\alpha - n)w_1 + 1] \tag{6.21}$$

on the interval $(0, \frac{1}{n})$.

Note 6.7 For $n = 3$, we obtain $w_2 = \sqrt{w_1 w_3}$ independently of the value of α.

A different representation of the same solution was given in [CMM97]. Let t be the (unique) positive solution to the equation

$$dt^{n-1} + (d+1)t^{n-2} + \ldots + (d+n-2)t + (d+n-1) = 0, \tag{6.22}$$

with $d = -\alpha(n-1)$. Then the MEOWA weights are identified from

$$w_i = \frac{t^i}{T}, \quad i = 1, \ldots, n, \quad \text{where } T = \sum_{k=1}^{n} t^k. \tag{6.23}$$

Note 6.8 It is not difficult to check that both (6.20) and (6.23) represent the same set of weights, noting that $t = \sqrt[n-1]{\frac{w_n}{w_1}} = -\frac{1-d-nw_1}{d}$, or $w_1 = \frac{1+td-d}{n}$, and that substituting w_1 into (6.21) yields

$$1 - t^n = \frac{n(1-t)}{1-d(1-t)},$$

which translates into

$$\frac{1-t^n}{1-t} - d(1-t^n) - n = 0,$$

and then into

$$dt^n + t^{n-1} + t^{n-2} + \ldots + t + (1-d-n) = 0.$$

After factoring out $(t-1)$, we obtain (6.22).

Minimum Variance OWA

Another popular characteristic of weighting vectors is weight variance (as given in (6.6)):

$$D^2(\mathbf{w}) = \frac{1}{n}\sum_{i=1}^{n}(w_i - AM(\mathbf{w}))^2 = \frac{1}{n}\sum_{i=1}^{n} w_i^2 - \frac{1}{n^2}, \tag{6.24}$$

where $AM(\mathbf{w})$ is the arithmetic mean of $\mathbf{w}$.

Here, one minimises $D^2(\mathbf{w})$ subject to a given orness measure. The resulting OWA function is called minimum variance OWA (MVOWA). Since adding a constant to the objective function does not change the minimiser, this is equivalent to the problem:

$$\begin{aligned} \text{minimise} \quad & \sum_{i=1}^{n} w_i^2 \\ \text{s.t.} \quad & \sum_{i=1}^{n} w_i \frac{n-i}{n-1} = \alpha \\ & \sum_{i=1}^{n} w_i = 1, w_i \geqslant 0, i = 1, \ldots, n. \end{aligned} \tag{6.25}$$

For $\alpha = \frac{1}{2}$, the optimal solution is always $w_i = \frac{1}{n}, i = 1, \ldots, n$. It is also worth noting that the optimal solution to (6.25) for $\alpha > \frac{1}{2}$, $\mathbf{w}^*$, is related to the optimal solution for $\alpha < \frac{1}{2}$, $\mathbf{w}$, by $w_i^* = w_{n-i+1}$, i.e. it gives the reverse OWA. Thus, it is sufficient to establish the optimal solution in the case $\alpha < \frac{1}{2}$.

The optimal solution [FM03, Liu07] for $\alpha < \frac{1}{2}$ is given as the vector $\mathbf{w} = (0, 0, \dots, 0, w_r, \dots, w_n)$, i.e. $w_i = 0$ if $i < r$, and

$$w_r = \frac{6(n-1)\alpha - 2(n-r-1)}{(n-r+1)(n-r+2)},$$

$$w_n = \frac{2(2n-2r+1) - 6(n-1)\alpha)}{(n-r+1)(n-r+2)},$$

and

$$w_i = w_r + \frac{i-r}{n-r}(w_n - w_r),\ r < i < n.$$

The index r depends on the value of α and is found from the inequalities:

$$n - 3(n-1)\alpha - 1 < r \leqslant n - 3(n-1)\alpha.$$

It was also established in [Liu07] that the solution to the minimum variance OWA weight problem is equivalent to that of minimax disparity [WP05], i.e. the solution to

$$\text{minimise} \left\{ \max_{i=1,\dots,n-1} |w_i - w_{i-1}| \right\} \tag{6.26}$$

$$\text{s.t.} \sum_{i=1}^{n} w_i \frac{n-i}{n-1} = \alpha$$

$$\sum_{i=1}^{n} w_i - 1,\ w_i \geqslant 0,\ i = 1, \dots, n.$$

We reiterate that the weights of OWA functions obtained as solutions to maximum entropy or minimum variance problems are fixed for any given n and orness measure and can be precomputed. However, both criteria are also useful for data-driven weight identification (in Sect. 6.6.1), if there are multiple optimal solutions. Then the solution maximising $Disp(\mathbf{w})$ or minimising $D(\mathbf{w})$ is chosen. Torra [Tor02] proposes to solve an auxiliary univariate optimisation problem to maximise weight dispersion, subject to a given value of (6.18). On the other hand, one can fit the orness value α of MEOWA or MVOWA to empirical data, using a univariate nonlinear optimisation method, in which at each iteration, the vector $\mathbf{w}$ is computed using analytical solutions to problems (6.19) and (6.25).

Furthermore, it is possible to include both criteria directly into problem (6.16). It is especially convenient for the minimum variance criterion, as it yields a modified quadratic programming problem,

$$\text{minimise} \quad \sum_{j=1}^{J}\left(\sum_{i=1}^{n} w_i z_i^j - y_j\right)^2 + \lambda \sum_{i=1}^{n} w_i^2 \tag{6.27}$$

$$\text{s.t.} \quad \sum_{i=1}^{n} w_i \frac{n-i}{n-1} = \alpha$$

$$\sum_{i=1}^{n} w_i = 1, w_i \geqslant 0, i = 1, \ldots, n,$$

where $\lambda \geqslant 0$ is a user-specified parameter controlling the balance between the criterion of fitting the data and that of obtaining minimum variance weights.

6.6.4 Methods Based on Weight-Generating Functions

Yager [Yag91, Yag96] has proposed to use monotone continuous functions $Q : [0, 1] \to [0, 1]$, $Q(0) = 0$, $Q(1) = 1$, called basic unit-interval monotone (BUM) functions, or regular increasing monotone (RIM) quantifiers [Yag91]. These functions generate OWA weights for any n using

$$w_i = Q\left(\frac{i}{n}\right) - Q\left(\frac{i-1}{n}\right). \tag{6.28}$$

RIM quantifiers are fuzzy linguistic quantifiers[1] that express the concept of fuzzy majority. Yager defined such quantifiers for fuzzy sets "for all", "there exists", "identity", "most", "at least half", and "as many as possible" as follows:

- "for all": $Q_{forall}(t) = 0$ for all $t \in [0, 1)$ and $Q_{forall}(1) = 1$.
- "there exists": $Q_{exists}(t) = 1$ for all $t \in (0, 1]$ and $Q_{exists}(0) = 0$.
- "identity": $Q_{Id}(t) = t$.

Other mentioned quantifiers are expressed by

$$Q_{a,b}(t) = \begin{cases} 0, & \text{if } t \leqslant a, \\ \frac{t-a}{b-a} & \text{if } a < t < b, \\ 1, & \text{if } t \geqslant b. \end{cases} \tag{6.29}$$

Then we can choose pairs $(a, b) = (0.3, 0.8)$ for "most", $(a, b) = (0, 0.5)$ for "at least half", and $(a, b) = (0.5, 1)$ for "as many as possible".

Calculation of weights results in the following OWA:

[1] That is, Q is a monotone increasing function $[0, 1] \to [0, 1]$, $Q(0) = 0$, $Q(1) = 1$ whose value $Q(t)$ represents the degree to which t satisfies the fuzzy concept represented by the quantifier.

- "for all": $\mathbf{w} = (0, 0, \ldots, 0, 1)$, $OWA_{\mathbf{w}} = \min$.
- "there exists": $\mathbf{w} = (1, 0, 0, \ldots, 0)$, $OWA_{\mathbf{w}} = \max$.
- "identity": $\mathbf{w} = (\frac{1}{n}, \ldots, \frac{1}{n})$, $OWA_{\mathbf{w}} = AM$.

Example 6.9 Consider the linguistic quantifier "most", given by (6.29) with $(a, b) = (0.3, 0.8)$ and $n = 5$. The weighting vector is then $(0, 0.2, 0.4, 0.4, 0)$.

Weight-generating functions are applied to generate weights of both quasi-arithmetic means and OWA functions. They allow one to compute the degree of orness of an OWA function in the limiting case:

$$\lim_{n\to\infty} orness(f_n) = orness(Q) = \int_0^1 Q(t)\, dt.$$

Entropy and other characteristics can also be computed based on Q; see [TY06].

Yager [Yag07b] has proposed using generating, or stress, functions (see also [Liu05]), defined by

Definition 6.19 (Generating Function of RIM Quantifiers) Let $q : [0, 1] \to [0, \infty]$ be an (integrable) function. It is a generating function of RIM quantifier Q, if

$$Q(t) = \frac{1}{K} \int_0^t q(u)\, du,$$

where $K = \int_0^1 q(u)du$ is the normalisation constant. The normalised generating function will be referred to as $\tilde{q}(t) = \frac{q(t)}{K}$.

Note 6.9 The generating function has the properties of a density function (e.g. a probability distribution density, although Q is not necessarily interpreted as a probability). If Q is differentiable, we may put $q(t) = Q'(t)$. Of course, for a given Q, if a generating function exists, it is not unique.

Note 6.10 In general, Q needs not be continuous to have a generating function. For example, it may be generated by Dirac's delta function,[2]

$$\delta(t) = \begin{cases} \infty, & \text{if } t = 0, \\ 0 & \text{otherwise,} \end{cases}$$

constrained by $\int_{-\infty}^{\infty} \delta(t)\, dt = 1$.

By using the generating function, we generate the weights as

[2] This is an informal definition. The proper definition involves the concepts of distributions and measures; see, e.g. [Rud91].

$$\tilde{w}_i = \frac{1}{K} q\left(\frac{i}{n}\right)\frac{1}{n}.$$

Note that these weights provide an approximation to the weights generated by (6.28) and that they do not necessarily sum to one. To ensure the latter, we shall use the weights:

$$w_i = \frac{\frac{1}{K} q\left(\frac{i}{n}\right)\frac{1}{n}}{\sum_{j=1}^{n} \frac{1}{K} q\left(\frac{j}{n}\right)\frac{1}{n}} = \frac{q\left(\frac{i}{n}\right)}{\sum_{j=1}^{n} q\left(\frac{j}{n}\right)}. \tag{6.30}$$

Equation (6.30) provides an alternative method for OWA weight generation, independent of Q, while at the same time, it gives an approximation to the weights provided by (6.28).

Example 6.10 Various interpretations of generating functions are provided in [Yag07b], from which we quote just a few examples:

- A constant generating function $q(t) = 1$ generates weights $w_i = \frac{1}{n}$, i.e. the arithmetic mean.
- A constant in range function $q(t) = 1$ for $t \leqslant \beta$ and 0 otherwise emphasises the larger arguments and generates the weights $w_i = \frac{1}{r}, i = 1, \ldots, r$ and $w_i = 0, i = r+1, \ldots, n$, where r is the largest integer less than or equal to βn.
- Generating function $q(t) = 1$, for $\alpha \leqslant t \leqslant \beta$ and 0 otherwise, emphasises the "middle" arguments and generates the weights $w_i = \frac{1}{p-r}, i = r+1, \ldots, p$ and 0 otherwise, with (for simplicity) $\alpha n = r$ and $\beta n = p$.
- A generating function with two tails $q(t) = 1$ if $t \in [0, \alpha]$ or $t \in [\beta, 1]$ and 0 otherwise emphasises both large and small arguments and yields $w_i = \frac{1}{r_1+r_2}$ for $i = 1, \ldots, r_1$ and $i = n+1-r_2, \ldots, n$, and $w_i = 0, i = r_1+1, \ldots, n-r_2$, with $r_1 = \alpha n, r_2 = \beta n$ integers.
- A linear stress function $q(t) = t$ generates weights $w_i = \frac{i}{\sum_{j=1}^{n} i} = \frac{2i}{n(n+1)}$, which gives orness value $\frac{1}{3}$, which can be compared to Example 6.1. It emphasises smaller arguments.

Of course, by using the same approach (i.e. $Q(t)$ or $q(t)$), one can generate the weights for generalised OWA and weighted quasi-arithmetic means. However, the interpretation and limiting cases for the means will be different. For example, the weighting vector $\mathbf{w} = (1, 0, \ldots, 0)$ results not in the *max* function, but in the projection to the first coordinate $f(\mathbf{x}) = x_1$.

6.6.5 Fitting Weight-Generating Functions

Weight-generating functions allow one to compute weighting vectors of OWA and weighted means for any number of arguments, i.e. to obtain extended aggregation

Table 6.2 A dataset with inputs of varying dimension

k	n_k	x_1	x_2	x_3	x_4	x_5	y
1	3	x_{11}	x_{21}	x_{31}			y_1
2	2	x_{12}	x_{22}				y_2
3	3	x_{13}	x_{23}	x_{33}			y_3
4	5	x_{14}	x_{24}	x_{34}	x_{44}	x_{54}	y_4
5	4	x_{15}	x_{25}	x_{35}	x_{45}		y_5
$\vdots$							

functions in the sense of Definition 1.2. This is very convenient when the number of arguments is not known a priori. Next, we pose the question as to whether it is possible to learn weight-generating functions from empirical data, similar to determining weighting vectors of aggregation functions of a fixed dimension.

A positive answer was provided in [BMV04, Bel05]. The method consists in representing a weight-generating function with a spline or polynomial and fitting its coefficients by solving a least squares or least absolute deviation problem subject to a number of linear constraints. Consider a dataset $(\mathbf{x}^j, y_j)$, $j = 1, \ldots, J$, where vectors $\mathbf{x}^j \in [0, 1]^{n_j}$ need not have the same dimension (see Table 6.2). This is because we are dealing with an extended aggregation function—a family of n-ary aggregation functions.

First, let us use the method of monotone splines, discussed in [Bel00]. We write

$$Q(t) = \sum_{k=1}^{K} c_k B_k(t), \; t \in (0, 1) \text{ and } Q(0) = 0, \; Q(1) = 1,$$

where functions $B_k(t)$ constitute a convenient basis for polynomial splines [Bel00], in which the condition of monotonicity of Q is expressed as $c_k \geqslant 0, k = 1, \ldots, K$. We do not require Q to be continuous on $[0, 1]$ but only on $]0, 1[$. We also have two linear constraints:

$$\sum_{k=1}^{K} c_k B_k(0) \geqslant 0, \quad \sum_{k=1}^{K} c_k B_k(1) \leqslant 1,$$

which convert to equalities if we want Q to be continuous on $[0, 1]$. Next, put this expression in (6.28) to get

$$f(\mathbf{x}^j) = \sum_{i=1}^{n_j} z_i^j \left(Q\left(\frac{i}{n_j}\right) - Q\left(\frac{i-1}{n_j}\right)\right)$$

$$= \sum_{i=2}^{n_j - 1} z_i^j \left(\sum_{k=1}^{K} c_k \left[B_k\left(\frac{i}{n_j}\right) - B_k\left(\frac{i-1}{n_j}\right)\right]\right)$$

$$
\begin{aligned}
&+z_1^j\left[\sum_{k=1}^{K} c_k B_k\left(\frac{1}{n_j}\right) - 0\right] + z_{n_j}^j\left[1 - \sum_{k=1}^{K} c_k B_k\left(\frac{n_j-1}{n_j}\right)\right]\\
&= \sum_{k=1}^{K} c_k \left(\sum_{i=2}^{n_j-1} z_i^j\left[B_k\left(\frac{i}{n_j}\right) - B_k\left(\frac{i-1}{n_j}\right)\right]\right.\\
&\quad \left. + z_1^j B_k\left(\frac{1}{n_j}\right) - z_{n_j}^j B_k\left(\frac{n_j-1}{n_j}\right)\right) + z_{n_j}^j\\
&= \sum_{k=1}^{K} c_k A_k(\mathbf{x}^j) + z_{n_j}^j.
\end{aligned}
$$

The vectors $\mathbf{z}^j$ stand for $\mathbf{x}^j$ when we treat weighted arithmetic means, or $\mathbf{x}^j_{\searrow}$ when we deal with OWA functions. The entries $A_k(\mathbf{x}^j)$ are computed from z_i^j using the expression in the brackets. Note that if Q is continuous on [0, 1], the expression simplifies to

$$
f(\mathbf{x}^j) = \sum_{j=1}^{K} c_k \left(\sum_{i=1}^{n_j} z_i^j\left[B_k\left(\frac{i}{n_j}\right) - B_k\left(\frac{i-1}{n_j}\right)\right]\right).
$$

Consider now the least squares approximation of empirical data. We obtain a quadratic programming problem:

$$
\begin{aligned}
\text{minimise} \quad & \sum_{j=1}^{J}\left(\sum_{k=1}^{K} c_k A_k(\mathbf{x}^j) + z_{n_j}^j - y_j\right)^2 \qquad (6.31)\\
\text{s.t.} \quad & \sum_{k=1}^{K} c_k B_k(0) \geqslant 0, \quad \sum_{k=1}^{K} c_k B_k(1) \leqslant 1\\
& c_k \geqslant 0.
\end{aligned}
$$

The solution is performed by QP programming methods.

OWA aggregation functions and weighted arithmetic means are special cases of Choquet integral-based aggregation functions. There are special classes of capacities called k-additive measures (see Definition 7.5), which we will discuss in detail in Chap. 7, and in the remainder of this section, we will present a method for identifying weight-generating functions that correspond to symmetric two- and three-additive capacities. These capacities lead to OWA functions with special weight distributions.

Proposition 6.4 ([BMV04]) *The discrete Choquet integral with respect to a symmetric two-additive capacity is an OWA function whose weight-generating* *function is given by*

$$Q(t) = at^2 + (1-a)t \text{ for some } a \in [-1, 1].$$

Furthermore, such an OWA weighting vector is equidistant (i.e. $w_{i+1} - w_i = const$ *for all* $i = 1, \ldots, n-1$*).*

The discrete Choquet integral with respect to a symmetric three-additive capacity is an OWA function whose weight-generating function is given by

$$Q(t) = at^3 + bt^2 + (1-a-b)t \text{ for some } a \in [-2, 4],$$

such that:

- *If* $a \in [-2, 1]$, *then* $b \in [-2a-1, 1-a]$
- *If* $a \in]1, 4]$, *then* $b \in [-3a/2 - \sqrt{3a(4-a)/4}, -3a/2 + \sqrt{3a(4-a)/4}]$

Proposition 6.4 provides two parametric classes of OWA functions that correspond to two- and three-additive symmetric fuzzy measures. In these cases, rather than fitting a general monotone non-decreasing function, we fit a quadratic or cubic function, identified by parameters a and b.

Interestingly, in the case of two-additive symmetric capacities, we obtain the following formula, a linear combination of OWA and the arithmetic mean (see Sect. 6.4.1):

$$f(\mathbf{x}) = aOWA_{\mathbf{w}}(\mathbf{x}) + (1-a)AM(\mathbf{x}),$$

with $\mathbf{w} = (\frac{1}{n^2}, \frac{3}{n^2}, \ldots, \frac{2n-1}{n^2})$. In this case, the solution is explicit, and the optimal a is given by [BMV04]

$$a = \max\left\{-1, \min\left\{1, \frac{\sum_{j=1}^{J}(y_j - U_j)V_j}{\sum_{j=1}^{J} V_j^2}\right\}\right\},$$

where

$$U_j = \sum_{i=1}^{n_j} \frac{z_i^j}{n_j}, \text{ and}$$

$$V_j = \frac{1}{n_j}\sum_{i=1}^{n_j}\left(\frac{2i-1}{n_j} - 1\right) z_i^j.$$

For three-additive symmetric capacities, the solution is found numerically by solving a convex optimisation problem in the feasible domain D in Proposition 6.4, which is the intersection of a polytope and an ellipse. Details are provided in [BMV04].

6.6.6 Choosing Parameters for Generalised OWA

We now consider optimisation approaches for determining the parameters for generalised OWA functions.

Choosing Weight-Generating Functions

Consider the case of generalised OWA functions, where a given generating function g is given. As earlier, we have a dataset $(\mathbf{x}^j, y_j)$, $j = 1, \ldots, J$, and we are interested in finding the weighting vector $\mathbf{w}$ that fits the data best. When we use the least squares, we have the following optimisation problem:

$$\text{minimise} \quad \sum_{j=1}^{J} \left(g^{-1} \left(\sum_{i=1}^{n} w_i g(z_i^j) \right) - y_j \right)^2 \tag{6.32}$$

$$\text{s.t.} \quad \sum_{i=1}^{n} w_i = 1, w_i \geqslant 0, i = 1, \ldots, n,$$

where $\mathbf{z}^j = \mathbf{x}^j_{\searrow}$ (see Sect. 6.6). This problem is converted to a QP problem similar to the case of weighted quasi-arithmetic means:

$$\text{minimise} \quad \sum_{j=1}^{J} \left(\sum_{i=1}^{n} w_i g(z_i^j) - g(y_j) \right)^2 \tag{6.33}$$

$$\text{s.t.} \quad \sum_{i=1}^{n} w_i = 1, w_i \geqslant 0, i = 1, \ldots, n.$$

This is a standard convex QP problem. This approach is presented in [Bel05].

If one uses the least absolute deviation (LAD) criterion, we obtain a different optimisation problem:

$$\text{minimise} \quad \sum_{j=1}^{J} \left| \sum_{i=1}^{n} w_i g(z_i^j) - g(y_j) \right| \tag{6.34}$$

$$\text{s.t.} \sum_{i=1}^{n} w_i = 1, w_i \geqslant 0, i = 1, \dots, n.$$

This problem is subsequently converted into a linear programming problem.

As in the case of weighted quasi-arithmetic means, in the presence of multiple optimal solutions, one can use an additional criterion of the dispersion of weights [Tor02].

Preservation of Ordering of the Outputs

If we require that the ordering of the outputs be preserved, i.e. if $y_j \leqslant y_k$, then we expect $f(\mathbf{x}^j) \leqslant f(\mathbf{x}^k)$, we can arrange the data so that the outputs are in a non-decreasing order, $y_j \leqslant y_{j+1}, j = 1, \dots, J-1$. Then we define the additional linear constraints:

$$< \mathbf{g}(z^{j+1}) - g(\mathbf{z}^j), \mathbf{w} >= \sum_{i=1}^{n} w_i (g(z_i^{j+1}) - g(z_i^j)) \geqslant 0,$$

$j = 1, \dots, J-1$. We add the above constraints to problem (6.33) or (6.34) and solve the modified problem.

Choosing Generating Functions

Consider now the case where the generating function g is also unknown, and hence it has to be found based on the data. We study two cases: (a) when g is given algebraically, with one or more unknown parameters to estimate (e.g. $g_r(t) = t^r$, r unknown), and (b) when no specific algebraic form of g is given.

In the first case, we solve the problem:

$$\min_{r,\mathbf{w}} \sum_{j=1}^{J} \left(\sum_{i=1}^{n} w_i g_r(z_i^j) - g_r(y_j) \right)^2 \tag{6.35}$$

$$\text{s.t.} \sum_{i=1}^{n} w_i = 1, w_i \geqslant 0, i = 1, \dots, n$$

additional conditions on r.

While this general optimisation problem is non-convex and nonlinear (i.e. difficult to solve), we can convert it to a bi-level optimisation problem:

$$\min_{r} \left[\min_{\mathbf{w}} \sum_{j=1}^{J} \left(\sum_{i=1}^{n} w_i g_r(z_i^j) - g_r(y_j) \right)^2 \right] \tag{6.36}$$

$$\text{s.t.} \quad \sum_{i=1}^{n} w_i = 1, w_i \geqslant 0, i = 1, \ldots, n$$

additional conditions on r.

The problem at the inner level is the same as (6.33) with a fixed g_r, which is a QP problem. At the outer level, we have a global optimisation problem with respect to a single parameter r. It is solved by using one of the methods of global optimisation. We recommend the deterministic Pijavski-Shubert method.

Example 6.11 To determine the weights and the generating function of a family of generalised OWA based on the power function, subject to a given measure of orness α. We have $g_r(t) = t^r$ and hence solve the bi-level optimisation problem:

$$\min_{r\in[-\infty,\infty]} \left[\min_{\mathbf{w}} \sum_{j=1}^{J} \left(\sum_{i=1}^{n} w_i z_i^{jr} - y_j^r \right)^2 \right] \tag{6.37}$$

$$\text{s.t.} \quad \sum_{i=1}^{n} w_i = 1, w_i \geqslant 0, i = 1, \ldots, n$$

$$\sum_{i=1}^{n} w_i \left(\frac{n-i}{n-1} \right)^r = \alpha.$$

Of course, for numerical purposes, we need to limit the range of r to a finite interval and treat all the limiting cases $r \to \pm\infty$, $r \to 0$, and $r \to -1$.

A different situation arises when the parametric form of g is not given. The approach proposed in [Bel03, Bel05] is based on approximation of g with a monotone linear spline, as

$$g(t) = \sum_{k=1}^{K} c_k B_k(t), \tag{6.38}$$

where B_k are appropriately chosen basis functions and c_k are spline coefficients. The monotonicity of g is ensured by imposing linear restrictions on spline coefficients, in particular non-negativity, as in [Bel00]. Further, since the generating function is defined up to an arbitrary linear transformation, one has to fix a particular g by specifying two interpolation conditions, like $g(a) = 0$, $g(b) = 1$, $a, b \in]0, 1[$, and if necessary, properly model asymptotic behaviour if $g(0)$ or $g(1)$ are infinite.

After rearranging the terms of the sum, the problem of identification becomes (subject to linear conditions on **c**, **w**)

$$\min_{\mathbf{c},\mathbf{w}} \sum_{j=1}^{J} \left(\sum_{k=1}^{K} c_k \left[\sum_{i=1}^{n} w_i B_k(z_i^j) - B_k(y_j) \right] \right)^2 . \tag{6.39}$$

For a fixed **c** (i.e. fixed g), we have a quadratic programming problem to find **w**, and for a fixed **w**, we have a quadratic programming problem to find **c**. However, if we consider both **c**, **w** as variables, we obtain a difficult global optimisation problem. We convert it to a bi-level optimisation problem:

$$\min_{\mathbf{c}} \min_{\mathbf{w}} \sum_{j=1}^{J} \left(\sum_{k=1}^{K} c_k \left[\sum_{i=1}^{n} w_i B_k(z_i^j) - B_k(y_j) \right] \right)^2 ,$$

where at the inner level, we have a QP problem and at the outer level, we have a nonlinear problem with multiple local minima. When the number of spline coefficients K is not very large (< 10), this problem can be efficiently solved by using deterministic global optimisation methods. If the number of variables is small and K is large, then reversing the order of minimisation (i.e. using $\min_{\mathbf{w}}, \min_{\mathbf{c}}$) is more efficient.

Choosing Generating Functions and Weight-Generating Functions

We recall the definition of the generalised OWA (Definition 6.7). Consider the case of generating function $g(t) = t^r$, in which case

$$GenOWA_{\mathbf{w},[r]}(\mathbf{x}) = \left(\sum_{i=1}^{n} w_i z_i^r \right)^{1/r} .$$

Consider first a fixed r. To find a weight-generating function $Q(t)$, we first linearise the least squares problem to get

$$\min_{\mathbf{c}} \sum_{j=1}^{J} \left(\sum_{k=1}^{K} c_k \sum_{i=1}^{n_j} \left[B_k\left(\frac{i}{n_j}\right) - B_k\left(\frac{i-1}{n_j}\right) \right] (z_i^j)^r - y_j^r \right)^2$$

$$\text{s.t.} \sum_{k=1}^{K} c_k B_k(0) = 0, \quad \sum_{k=1}^{K} c_k B_k(1) = 1$$

$$c_k \geqslant 0.$$

This is a standard QP, which differs from (6.31) because z_i^j and y_j are raised to power r (we considered a simpler case of $Q(t)$ continuous on [0, 1]).

Now, if the parameter r is also unknown, we determine it from data by setting a bi-level optimisation problem:

$$\min_{r} \min_{\mathbf{c}} \sum_{j=1}^{J} \left(\sum_{k=1}^{K} c_k \sum_{i=1}^{n_j} \left[B_k \left(\frac{i}{n_j} \right) - B_k \left(\frac{i-1}{n_j} \right) \right] (z_i^j)^r - y_j^r \right)^2$$

$$\text{s.t.} \quad \sum_{k=1}^{K} c_k B_k(0) = 0, \quad \sum_{k=1}^{K} c_k B_k(1) = 1$$

$$c_k \geqslant 0,$$

in which at the inner level, we solve a QP problem with a fixed r and at the outer level, we optimise with respect to a single nonlinear parameter r, in the same way we did in Example 6.11.

For the more complicated case of both generating functions g and Q given non-parametrically (as splines), we refer to [Bel05].

6.7 Induced OWA

Yager and Filev proposed the induced OWA (IOWA) in [YF99], whereby the inputs are sorted according to the values of an auxiliary variable referred to as the *inducing* variable, rather than by their relative size. Some earlier studies in the field of image compression had also incorporated such a scheme with OWA functions [ME97, ME98].

Nearest-neighbour and time-series models can be expressed in this framework too, where weight is not necessarily applied according to the source or the value, but rather the relative distance. It is worth noting that ties in the inducing variable require special attention.

Definition 6.20 (Induced OWA) Given a weighting vector $\mathbf{w}$ and an inducing variable $\mathbf{z}$, the induced ordered weighted averaging (IOWA) function is

$$IOWA_{\mathbf{w}}(\langle x_1, z_1 \rangle, \ldots, \langle x_n, z_n \rangle) = \sum_{i=1}^{n} w_i x_{(i)}, \tag{6.40}$$

where the (.) notation denotes the inputs $\langle x_1, z_1 \rangle$ reordered such that $z_{(1)} \geqslant z_{(2)} \geqslant \ldots \geqslant z_{(n)}$ and the convention that if q of the $z_{(i)}$ are tied, i.e. $z_{(i)} = z_{(i+1)} = \ldots = z_{(i+q-1)}$,

$$x_{(i)} = \frac{1}{q+1} \sum_{j=(i)}^{(i+q-1)} x_j.$$

The input pairs $\langle x_i, z_i \rangle$ may be two independent features of the same input or can be related by some function, i.e. $z_i = f_i(x_i)$.

6.7.1 *Main Properties*

Although the properties and behaviour of the IOWA will be largely dependent on the inducing variable $\mathbf{z}$, in most cases, it will exhibit characteristics that have shown to hold for the standard OWA, namely, monotonicity, idempotency, and symmetry. Clearly, the IOWA still applies a weighted sum of the inputs, and hence the value obtained will be equivalent to an OWA or WAM whose weights are permuted [BJ11]. However, there are certain choices for $\mathbf{z}$ such that symmetry and even monotonicity may be violated:

- For a fixed vector $\mathbf{z}$, monotonicity with respect to $\mathbf{x}$ will hold, i.e.

$$\mathbf{x} \leqslant \mathbf{y} \Rightarrow IOWA_{\mathbf{w}}(\langle \mathbf{x}, \mathbf{z} \rangle) \leqslant IOWA_{\mathbf{w}}(\langle \mathbf{y}, \mathbf{z} \rangle).$$

  All values of the weighting vector $\mathbf{w}$ satisfy $w_i \geqslant 0$; hence, an increase to any of the x_i cannot decrease the overall output. If we have $w_i > 0, \forall i$, the IOWA will be strictly monotone increasing.
- As a corollary of the previous statement, comonotone pairs of the auxiliary variable $\mathbf{z}$ associated with $\mathbf{x}$ and $\mathbf{y}$, respectively, will also preserve monotonicity, since they will induce the same ordering.

  Note 6.11 In some situations, the inducing variable $\mathbf{z}$ may change when $\mathbf{x}$ changes. For example, suppose we have input pairs $\langle x_i, z_i \rangle_{i=1,\ldots,3}$ obtained from three observation stations. The value of z_i is the reliability of the reading x_i, based partially on the value of x as well as external information. We want to aggregate the x_i, giving preference to the most reliable readings. We hence use an IOWA that orders the observations from most to least reliable in accordance with the values z_i and define a weighting vector with decreasing weights. For a given reading, the input vector has an induced order of $z_2 \geqslant z_1 \geqslant z_3$, so the value x_2 is allocated the largest weight w_1. Now suppose the value of x_2 increases to a value that is unusual at Station 2 and is hence less reliable. The value of z_2 decreases and the induced order is now $z_1 \geqslant z_2 \geqslant z_3$. This results in a smaller weight being given to the input x_2 and monotonicity is violated.

- The positive and normalised (i.e. $\sum_{i=1}^{n} w_i = 1$) values of the weighting vector $\mathbf{w}$ ensure that the IOWA will be both averaging and idempotent, i.e.

$$\min(\mathbf{x}) \leqslant \text{IOWA}_{\mathbf{w}}(\langle \mathbf{x}, \mathbf{z} \rangle) \leqslant \max(\mathbf{x}),$$

$$\text{IOWA}_{\mathbf{w}}(\langle t, z_1 \rangle, \langle t, z_2 \rangle, ..., \langle t, z_n \rangle) = t.$$

Note 6.12 In the study of aggregation functions, averaging behaviour and idempotency are usually considered to be equivalent. Averaging behaviour necessarily implies idempotency; however, the property of idempotency is not sufficient for averaging behaviour without monotonicity. As discussed previously in Note 6.11, an IOWA-type function may not necessarily be monotone; however, this does not cause either of these properties to be lost.

- For inducing variables that are independent of the input vector $\mathbf{x}$, the IOWA will be symmetric, since it is the value of the inducing variable rather than the initial indexing that is important. However, there are certain choices of $\mathbf{z}$ that could result in loss of the symmetry property. For example, the z_i may be constant or a function that is somewhat dependent on the initial indexing, e.g. $z_i = f(i)$. Inducing variables that give a fixed calculation order result in weighted arithmetic means (see below). Such instances are unlikely to arise in contexts that call for the IOWA to be used since it would make more sense just to use a WAM.
- Whether the IOWA is shift-invariant or homogeneous will depend on the choice of inducing variable. Given any input pair $\langle x_i, z_i \rangle$, let $z_{i,+\lambda}$ be the value associated with the input $x_i + \lambda$ and $z_{i,\times\lambda}$ be the value associated with the input $\lambda \cdot x_i$.

  The induced OWA will be shift-invariant if

$$z_i < z_j \Leftrightarrow z_{i,+\lambda} < z_{j,+\lambda}, \forall i, j.$$

  The induced OWA will be homogeneous if

$$z_i < z_j \Leftrightarrow z_{i,\times\lambda} < z_{j,\times\lambda}, \forall i, j\,(\lambda > 0).$$

  The z_i may be specifically associated with each x_i, but need not be a function of the actual values, i.e. $x_i = x_j$ does not necessarily imply $z_i = z_j$. In other words, provided λ does not change the relative ordering, the IOWA will be stable for translations and homogenous. It should be noted that equalities should also be preserved, which follows from the above equations.
- As with the standard OWA, the dual IOWA can be defined as the IOWA with respect to the reverse weighting vector, $\mathbf{w}_d = (w_n, w_{n-1}, ..., w_1)$. It is also possible to define the dual function of an IOWA by the inducing variable $\mathbf{z}_d$, which induces the reverse ordering to $\mathbf{z}$.
- The inducing variable $\mathbf{z}$ can hence be chosen such that the reverse OWA is obtained, e.g. $z_i = 1 - x_i$ results in a function such that the inputs are aggregated in non-decreasing order $x_{(1)} \leqslant ... \leqslant x_{(n)}$.
- The induced OWA includes the standard OWA as a special case whenever the ordering of the inducing variable corresponds exactly with the order of the input

variable, e.g. $z_i = x_i$. Clearly, the special cases of the OWA, the minimum, maximum, and median, can be obtained through this choice and the appropriate selection of the weighting vector **w**.

- The IOWA reduces to the WAM for any **z** that corresponds with the original ordering of the inputs, e.g. $z_i = n + 1 - i$ will result in the weighted arithmetic mean $\sum_{i=1}^{n} w_i x_i$.

6.7.2 *Induced Generalised OWA*

Following from the definition of the generalised OWA (Definition 6.3.2), Chiclana et al. proposed a natural extension of the IOWA to the induced ordered weighted geometric (IOWG) function [Chi+04]. Of course, this corresponds to a special case of the induced generalised OWA, which was studied in [MGL09, BJ11]. The I-GenOWA function allows the application of quasi-arithmetic means with respect to an order-inducing variable.

Definition 6.21 (Induced Generalised OWA) Given a weighting vector **w**, an inducing variable **z**, and a continuous strictly monotone function $g : \mathbb{I} \to [-\infty, \infty]$, the induced generalised OWA (I-GenOWA) function is

$$I - GenOWA_{\mathbf{w},g}(\mathbf{x}) = g^{-1}\left(\sum_{i=1}^{n} w_i g(x_{(i)})\right) \tag{6.41}$$

As for IOWA, the inputs $\langle x_1, z_1 \rangle$ are reordered such that $z_{(1)} \geqslant z_{(2)} \geqslant \ldots \geqslant z_{(n)}$ and the convention that if q of the $z_{(i)}$ are tied, i.e. $z_{(i)} = z_{(i+1)} = \ldots = z_{(i+q-1)}$,

$$x_{(i)} = \frac{1}{q+1} \sum_{j=(i)}^{(i+q-1)} x_j.$$

Special cases of the quasi-arithmetic mean include power means and harmonic means; however, we will provide only the geometric mean, since many studies have investigated its use and properties.

Definition 6.22 (Induced Ordered Weighted Geometric Function) For a given weighting vector **w** and an inducing variable **z**, the IOWG function is

$$\mathrm{IOWG}_{\mathbf{w}}(\langle \mathbf{x}, \mathbf{z} \rangle) = \prod_{i-1}^{n} x_{(i)}^{w_i}. \tag{6.42}$$

where the inputs $\langle x_1, z_1 \rangle$ are reordered such that $z_{(1)} \geqslant z_{(2)} \geqslant \ldots \geqslant z_{(n)}$. The same convention is employed for ties as with IOWA and I GenOWA.

In [Chi+04], IOWG was used to aggregate the preferences of experts, where the inducing variable was given by either the importance, consistency, or preference of a given expert.

6.7.3 Choices for the Inducing Variable

As one might expect, the calculation of $\mathbf{z}$ and hence the process by which the arguments are reordered is the fundamental concern when considering induced aggregation functions. In some cases, we may wish to give weight to observations that are more similar or "closer" to other values in $\mathbf{x}$, so we will let z_i represent some function of distance to the mean, i.e. $z_i = f(|x_i - AM(\mathbf{x})|)$. In other cases, the z_i may be some representation of the reliability of the value x_i. If the z_i are constant, this essentially models a weighted arithmetic mean; however, in many cases, the accuracy of an observation may fluctuate. We consider some typical examples where induced aggregation functions can be useful, with particular focus on the choice of $\mathbf{z}$.

Standard Auxiliary Ordering

The inducing variable may simply be an attribute associated with the input $\mathbf{x}$ that is not considered in the actual aggregation process, but is informative about the object itself. For instance, consider the peer-review process for some journal and conference papers. Each reviewer allocates a score to each criterion, e.g. originality, relevance, etc. Sometimes the reviewer also provides an evaluation of their own familiarity with the topic, e.g. very familiar and only marginally familiar. This last input, of course, is not taken into account when aggregating the scores for submission, but could be taken into account in the weight allocation process. To give an overall score for each criterion, we can then use an IOWA where x_i is the score allocated by the i-th reviewer, z_i is the familiarity of the reviewer with the topic, and $\mathbf{w}$ is a weighting vector with non-increasing weights such that the heavier weight is given to experts with more expertise on the given paper.

This allocation of weighting is different to providing the expert a weight based on their expertise, as the variability of $\mathbf{z}$ suggests that this may fluctuate depending on the paper they are marking.

Example 6.12 An editor for a journal considers two papers that have been evaluated by the same three reviewers.

	Paper 1		Paper 2	
	Score (x)	Expertise (z)	Score (x)	Expertise (z)
Reviewer 1	70	8	92	4
Reviewer 2	85	7	62	8
Reviewer 3	76	4	86	5

Using the weighting vector $\mathbf{w} = (0.5, 0.3, 0.2)$, the score for the first paper is

$$IOWA_{\mathbf{w}}(\langle \mathbf{x}, \mathbf{z} \rangle) = 0.5(70) + 0.3(85) + 0.2(76) = 75.7\,.$$

The score for the second paper is

$$IOWA_{\mathbf{w}}(\langle \mathbf{x}, \mathbf{z} \rangle) = 0.5(62) + 0.3(86) + 0.2(92) = 75.2\,.$$

Nearest-Neighbour Rules

Nearest-neighbour methods and their variants have been popularly applied to classification and function approximation problems. The underlying assumption is that objects described similarly by their features will belong to the same class or have the same function output. For instance, consider a classification problem that requires an object $\mathbf{x} = (x_1, x_2, ..., x_n)$ to be assigned to a class Y_1 or Y_2. Given a number of training data $\mathcal{D} = \{(\mathbf{x}^1, y_1), \ldots, (\mathbf{x}^J, y_J)\}$, we can identify the object $\mathbf{x}^j$ most similar to $\mathbf{x}$ and allocate the same label. Extensions can be made such that the class labels of a number of the objects in $\mathcal{D}$ are aggregated to determine the class of $\mathbf{x}$. Induced aggregation functions can be used to model this situation, with the weights and inducing variable often reflecting the similarity of each of the training data. Of course, the way similarity is calculated becomes very important.

The nearest-neighbour approach can also be used for function approximation. Suppose $\mathbf{x}^j \in \mathbb{I}^n$ and $y_j \in \mathbb{I}$. We assume the training data are generated by some function $f(\mathbf{x}^j) + \epsilon_j = y_j$ where ϵ_j are random errors and then approximate an unknown y by aggregating the y_j of the most similar $\mathbf{x}^j$. The induced OWA and induced aggregation functions in general can be used in this context, where the input vector comprises the y_j values taken from $\mathcal{D}$ and $\mathbf{z}$ represents a similarity or distance function. For example, a standard representation of the nearest-neighbour model is

$$y(\mathbf{x}) = IOWA_{\mathbf{w}}(\langle \mathbf{y}, \mathbf{z} \rangle) = \sum_{j=1}^{J} w_j y_{\eta(j)},$$

with $\mathbf{w} = (1, 0, ..., 0)$, and $z_j = \frac{1}{||\mathbf{x}^j - \mathbf{x}||}$, the reciprocal of Euclidean distance between $\mathbf{x}$ and $\mathbf{x}^j$ associated with y_j.

A function that approximates y based on the k nearest-neighbours (known as the kNN method) could be calculated using the same function, with $w_j = \frac{1}{k}, j = 1, ..., k$, $w_j = 0$ otherwise. One can also consider weighted versions where the weights gradually decay, i.e. $w_1 \geqslant w_2 \geqslant ... \geqslant w_k$, e.g. $w_j = \frac{2(k+1-j)}{k(k+1)}$.

There exist alternative choices for the distance function, in particular there are many suitable metrics depending on the spatial distribution of the data. Yager has studied the use of IOWA for modelling these situations in [Yag02a, Yag02b, YK03, Yag04a] as well as in [YF99] with Filev.

Time-series smoothing can also be handled within the framework of induced aggregation functions. It is similar to a nearest-neighbour problem with a single-dimension variable, t. In time-series smoothing, we want to estimate or smooth the value y_t at a point in time t based on the values y_i obtained at previous times t_i. Induced OWA functions allow a simple framework for modelling this smoothing process. For instance, the 3-day simple moving average on a dataset $\mathcal{D} = \{\langle t_1, y_1\rangle, \ldots, \langle t_J, y_J\rangle\}$ can be modelled by $IOWA_{\mathbf{w}}(t)$, with $\mathbf{w} = (\frac{1}{3}, \frac{1}{3}, \frac{1}{3}, 0, 0, ..., 0)$, $z_j = \frac{1}{t - t_j}$.

Best-Yesterday Models

Extrapolation problems involve predicting future values based on previous observations. Weather prediction is an important example. In [YF99], Yager and Filev present the best-yesterday model for predicting stock market prices based on the opinions of multiple experts. We could aggregate their scores using a weighted mean, allocating a weight to the experts who seem more reliable; however, an alternative is to use an IOWA function, inducing the input vector based on the most accurate predictions from previous days. We consider an adapted example from [YF99] here.

Example 6.13 We have four experts who, daily, predict the next day's opening share price of the FUZ Company. Our data then consists of the predictions each day for each expert $i = 1, ..., 4$, $x_i(t)$ and the actual stock price of the FUZ Company each day $y(t)$. Our aggregated prediction could be the value obtained from the induced OWA:

$$IOWA_{\mathbf{w}}(t) = \sum_{i=1}^{4} w_i x_{\eta(i)}(t),$$

where $z_i = -|x_i(t-1) - y(t-1)|$ and the weights w_i are non-increasing. This allocates more importance to the expert whose predictions were closest to the actual price yesterday.

Of course, we could order our experts by their accuracy for the past 2, 3, etc., days or determine the weights using optimisation techniques. In the Yager and Filev example, for instance, the fitted weighting vector was $\mathbf{w} = (0.2, 0.12, 0.08, 0.6)$, i.e. the best fitting weighted model gave more influence to the expert who was closest to the mark the previous day (associated with the lowest value of z_i).

Group Decision-Making

IOWA functions can be useful in group decision-making (GDM) problems for modelling concepts such as consensus. In [Chi+07], the use of different inducing variables was considered for group decisions based on pair-wise preference matrices modelling multiple alternatives. The usual approach to varying weights is to have these reflect the importance of each expert. In the context of preference matrices, however, it might also make sense to allocate importance to each input based on the consistency of each expert, inferred from how well their preferences satisfy transitivity, etc. Another of the inducing variables presented took into account the overall preference for a particular alternative expressed by each expert.

The standard OWA is able to model majority concepts such as "most" or "80%" using weighting vectors based on linguistic quantifiers. In [PY06], it was proposed that consensus might be better achieved with inducing variables reflecting the *support* for each individual score. Consider the evaluations of five experts, $\mathbf{x} = (0.3, 0.1, 0.7, 0.9, 0.8)$. It makes sense that the score given by expert 5 is more representative of the group than say, expert 2. The support for evaluation x_i from x_j can be modelled by simply using

$$\mathrm{Sup}(x_i, x_j) = \begin{cases} 1, & \text{if } |x_i - x_j| < \alpha; \\ 0, & \text{otherwise,} \end{cases}$$

where α is a desired threshold. The inducing variable based on support can then be given by

$$z_i = \sum_{j=1, j\neq i}^{n} \mathrm{Sup}(x_i, x_j).$$

In turn, weighting vectors with non-increasing weights can be specified such that experts with more support are allocated higher importance.

Multiple Inducing Variables

In [MS00], a generalisation to multiple inducing variables was considered. Suppose we have N priorities and each of the inputs x_i are associated with N ratings or degrees of satisfaction with respect to these priorities. In this context, the order can be induced by some aggregation of these N scores, and a single inducing variable $\mathbf{z}^*$ can be considered as the vector of aggregated results.

Example 6.14 We want to measure the pollution levels at a beach, and we have multiple mobile sensors that report to a central computer for analysis. The reliability of the sensor readings depends somewhat on the time since they were transmitted, as well as the distance travelled in transmission and the local conditions as they were sent—for instance, varying water pressure, presence of animals, etc. We hence decide to aggregate the pollution levels with an $IOWA_{\mathbf{w}}(\langle \mathbf{x}, \mathbf{z}^* \rangle)$ where $\mathbf{z}_i^* = f(z_{i1}, z_{i2}, z_{i3})$ is the aggregated inducing input associated with each pollution level input x_i.

6.8 Medians and Order Statistics

OWA functions and symmetric capacities can be used for modelling median-like behaviour and order-statistic-based aggregation. Here, we focus on some of the common frameworks.

In statistics, the median of a sample is a number dividing the higher half of a sample from the lower half. The median of a finite list of numbers can be found by arranging all the numbers in increasing or decreasing order and picking the middle one. If the number of inputs is even, one takes the mean of the two middle values.

The median is a type of average that is more representative of a "typical" value than the mean. It essentially discards very high and very low values (outliers). For example, the median price of houses is often reported in the real estate market, because the mean can be influenced by just one or a few very expensive houses and will not represent the cost of a "typical" house in the area.

Definition 6.23 (Median) The median is the function

$$Med(\mathbf{x}) = \begin{cases} \frac{1}{2}(x_{(k)} + x_{(k+1)}), & \text{if } n = 2k \text{ is even} \\ x_{(k)}, & \text{if } n = 2k - 1 \text{ is odd,} \end{cases}$$

where $x_{(k)}$ is the k-th largest (or smallest) component of $\mathbf{x}$.

An attractive property of medians is that they are applicable to inputs given on an ordinal scale, i.e. when only the ordering, rather than the numerical values matter. For example, one can use medians for aggregation of inputs like labels of fuzzy sets, such as *very high*, *high*, *medium*, *low*, and *very low*.

Note 6.13 The median can be conveniently expressed as an OWA function with a special weighting vector. For an odd n, let $w_{\frac{n+1}{2}} = 1$ and all other $w_i = 0$, and for an even n, let $w_{\frac{n}{2}} = w_{\frac{n}{2}+1} = \frac{1}{2}$, and all other $w_i = 0$. Then $Med(\mathbf{x}) = OWA_{\mathbf{w}}(\mathbf{x})$.

One interesting median-based averaging function is the Hodges–Lehmann estimator, which is the median of all pairwise means of the components of $\mathbf{x}$,

$$Med(\mathbf{z}) \text{ with } z_k = \frac{x_i + x_j}{2}, 1 \leqslant i < j \leqslant n, k = 1, \ldots, \frac{n(n-1)}{2}.$$

Definition 6.24 (a-Median) Given a value $a \in [0, 1]$, the a-median is the function

$$Med_a(\mathbf{x}) = Med(x_1, \ldots, x_n, \overbrace{a, \ldots, a}^{n-1\ times}).$$

Note 6.14 a-medians are also the limiting cases of idempotent nullnorms. They have an absorbing element a and are continuous, symmetric, and associative (and, hence, bisymmetric). They can be expressed as

$$Med_a(\mathbf{x}) = \begin{cases} \max(\mathbf{x}), & \text{if } \mathbf{x} \in [0, a]^n, \\ \min(\mathbf{x}), & \text{if } \mathbf{x} \in [a, 1]^n, \\ a & \text{otherwise.} \end{cases}$$

Note 6.15 The Sugeno integral with respect to a capacity satisfying $\mu(\mathcal{A}) = a$ for all $\mathcal{A}$ such that $\emptyset \neq \mathcal{A} \subset \mathcal{N}$ results in the a-median.

The following construction involves a generating function, similar to quasi-arithmetic means. Since the median either involves calculation based on a single input or the average of two inputs, the generating function will only come into play for even n.

Definition 6.25 (Quasi-median) Given a continuous strictly monotone function $g : \mathbb{I} \to [-\infty, \infty]$, the quasi-median is the function

$$QMed_g(\mathbf{x}) = g^{-1}\left(Med(g(x_1), \ldots, g(x_n))\right).$$

For odd values of n, the outputs of the quasi-median and the median coincide.

The concept of the weighted median was treated in detail in [Yag98a].

Definition 6.26 (Weighted Median) Let $\mathbf{w}$ be a weighting vector, and let $\mathbf{u}$ denote the vector obtained from $\mathbf{w}$ by arranging its components in the order induced by the components of the input vector $\mathbf{x}$, such that $u_k = w_i$ if $x_i = x_{(k)}$ is the k-th largest component of $\mathbf{x}$. The lower weighted median is the function

$$Med_{\mathbf{w}}(\mathbf{x}) = x_{(k)}, \tag{6.43}$$

where k is the index obtained from the condition

$$\sum_{j=1}^{k-1} u_j < \frac{1}{2} \text{ and } \sum_{j=1}^{k} u_j \geqslant \frac{1}{2}. \tag{6.44}$$

The upper weighted median is the function (6.43) where k is the index obtained from the condition

$$\sum_{j=1}^{k-1} u_j \leqslant \frac{1}{2} \text{ and } \sum_{j=1}^{k} u_j > \frac{1}{2}.$$

Note 6.16 It is convenient to describe calculation of $Med_{\mathbf{w}}(\mathbf{x})$ using the following procedure. Take the vector of pairs $((x_1, w_1), (x_2, w_2), \ldots, (x_n, w_n))$ and sort them in the order of decreasing x. We obtain $((x_{(1)}, u_1), (x_{(2)}, u_2), \ldots, (x_{(n)}, u_n))$. Calculate the index k from condition (6.44) and return $x_{(k)}$.

The properties of the weighted median are consistent with averaging functions:

- The weighted median is a continuous idempotent aggregation function.
- If all the weights are given by $w_i = \frac{1}{n}$, the weighted median becomes the ordinary median Med.
- If any weight $w_i = 0$, then

  $$Med_{\mathbf{w}}(\mathbf{x}) = Med_{(w_1,\ldots,w_{i-1},w_{i+1},\ldots,w_n)}(x_1, \ldots, x_{i-1}, x_{i+1}, \ldots, x_n),$$

  i.e. the input x_i can be dropped from the aggregation procedure.
- If any input value is repeated, one can use just a copy of this value and add the corresponding weights, namely, if $x_i = x_j$ for some $i < j$, then

  $$Med_{\mathbf{w}}(\mathbf{x}) = Med_{\tilde{\mathbf{w}}}(x_1, \ldots, x_{j-1}, x_{j+1}, \ldots, x_n),$$

  where $\tilde{\mathbf{w}} = (w_1, \ldots, w_{i-1}, w_i + w_j, w_{i+1}, \ldots, w_{j-1}, w_{j+1}, \ldots, w_n)$.

As far as learning the weights of weighted medians from empirical data, Yager [Yag98a] presented a gradient-based local optimisation algorithm. Given that such a method does not guarantee the globally optimal solution, it is advisable to combine it with a generic global optimisation scheme, such as multistart local search or simulated annealing.

Based on the weighted median, Yager [Yag98a] also defined an ordinal OWA function using the following construction. We recall that $OWA_{\mathbf{w}}(\mathbf{x}) =< \mathbf{w}, \mathbf{x}_{\searrow} >$, i.e. the weighted mean of the vector $\mathbf{x}_{\searrow}$. By replacing the weighted mean with the weighted median, we obtain

Definition 6.27 (Ordinal OWA) The ordinal OWA function is

$$OOWA_{\mathbf{w}}(\mathbf{x}) = Med_{\mathbf{w}}(\mathbf{x}_{\searrow}).$$

Note 6.17 Since the components of the argument of the weighted median in Definition 6.27 are already ordered, calculation of the ordinal OWA is reduced to the formula

$$OOWA_{\mathbf{w}}(\mathbf{x}) = x_{(k)},$$

where k is the index obtained from the condition

$$\sum_{j=1}^{k-1} w_j < \frac{1}{2} \text{ and } \sum_{j=1}^{k} w_j \geqslant \frac{1}{2}.$$

A more general class of aggregation functions on an ordinal scale is that of weighted ordinal means, presented in [KMM07].

As mentioned previously, OWA functions can be considered as weighted sums of order statistics.

Definition 6.28 (Order Statistic) The k-th-order statistic is the function

$$kOS(\mathbf{x}) = x_{(k)},$$

i.e. its value is the k-th *smallest*[3] component of $\mathbf{x}$.

Note 6.18 The order statistics can be conveniently expressed as OWA functions with special weighting vectors. Let $\mathbf{w} = (0, 0, \ldots, 0, 1, 0 \ldots, 0)$, i.e. $w_i = 0$ for $i \neq n - k + 1$ and $w_{n-k+1} = 1$. Then $kOS(\mathbf{x}) = OWA_{\mathbf{w}}(\mathbf{x})$.

Note 6.19 Sorting is not required for computation of order statistics. It can be achieved by using a selection algorithm (e.g. Floyd and Rivest's algorithm), which has worst-case complexity $O(n)$. This is significantly less expensive when n is large.

6.9 OWMax and OWMin

As mentioned in Sect. 5.3, the Sugeno integral with respect to a symmetric capacity is known as the OWMax operator [DPT88, DP96]. Its definition is related to the weighted maximum (WMax).

Definition 6.29 (Weighted Maximum and Minimum (WMax and WMin)) Let $\mathbf{w}$ denote an ordinal weighting vector[4] $\mathbf{w} \in [0, 1]^n$ such that $\max_{i=1,\ldots,n} w_i = 1$. The weighted maximum (WMax) is given by

$$WMax(\mathbf{x}) = \max_{i=1,\ldots,n} \min\{w_i, x_i\}.$$

[3] Note that in OWA, weighted median, and ordinal OWA, $x_{(k)}$ denotes the k-th *largest* element of $\mathbf{x}$.

[4] Note that in this case, the weights need not sum to 1.

The weighted minimum (WMin) is given by

$$WMin(\mathbf{x}) = \min_{i=1,\dots,n} \max\{1 - w_i, x_i\}.$$

The ordered weighted versions are obtained by reordering the components of **x** as part of the calculation.

Definition 6.30 (Ordered Weighted Maximum and Minimum (OWMax and OWMin)) Let **w** denote an ordinal weight vector satisfying $1 = w_1 \geqslant w_2 \geqslant \dots \geqslant w_n$. The ordered weighted maximum (OWMax) is given by

$$OWMax(\mathbf{x}) = \max_{i=1,\dots,n} \min\{w_i, x_{(i)}\},$$

where the input vector is arranged in non-decreasing order, i.e. $x_{(1)} \leqslant x_{(2)} \leqslant \dots \leqslant x_{(n)}$. The ordered weighted minimum (OWMin) is defined with respect to a weighting vector $w_1 \leqslant w_2 \leqslant \dots \leqslant w_n = 1$ and is given by

$$OWMin(\mathbf{x}) = \min_{i=1,\dots,n} \max\{1 - w_i, x_{(i)}\}.$$

with the same non-decreasing permutation for the input vector.

Note 6.20 When expressed in the framework of the Sugeno integral, the weights w_i in the OWMax correspond with symmetric capacity weights $\mu(H_i)$.

The problem of learning symmetric capacities has been investigated in [GJ18, GJB19].

References

[Ahn06] B. Ahn, On the properties of OWA operator weights functions with constant level of orness. IEEE Trans. Fuzzy Syst. **14**, 511–515 (2006)

[Bel00] G. Beliakov, Shape preserving approximation using least squares splines. Approx. Theory Appl. **16**, 80–98 (2000)

[Bel01] G. Beliakov, Shape preserving splines in constructing WOWA operators: Comment on paper by V. Torra in Fuzzy Sets and Systems 113 (2000) 389–396. Fuzzy Sets Syst. **121**, 549–550 (2001)

[Bel03] G. Beliakov, How to build aggregation operators from data? Int. J. Intell. Syst. **18**, 903–923 (2003)

[Bel05] G. Beliakov, Learning weights in the generalized OWA operators. Fuzzy Optim. Decis. Making **4**, 119–130 (2005)

[Bel07] G. Beliakov, Construction of aggregation operators for auto-mated decision making via optimal interpolation and global optimization. J. Ind. Manag. Optim. **3**, 193–208 (2007)

[Bel17a] G. Beliakov, A method of introducing weights into OWA operators and other symmetric functions, in *Uncertainty Modeling. Dedicated to B. Kovalerchuk*, eds. by V. Kreinovich (Springer, Cham, 2017), pp. 37–52

[Bel17b] G. Beliakov, *Weighted OWA functions implementation in C++* (2017). https://github.com/gbeliako/wowa

[Bel18] G. Beliakov, Comparing apples and oranges: the weighted OWA function. Int. J. Intell. Syst. **33**, 1089–1108 (2018)

[BBC16] G. Beliakov, H. Bustince, T. Calvo, *A Practical Guide to Averaging Functions* (Springer, New York, Berlin, Heidelberg, 2016)

[BCF17] G. Beliakov, T. Calvo, P. Fuster, Implicit averaging functions. Inf. Sci. **417**, 96–112 (2017)

[BD16] G. Beliakov, J. Dujmovic, Extension of bivariate means to weighted means of several arguments by using binary trees. Inf. Sci. **331**, 137–147 (2016)

[BJ11] G. Beliakov, S. James, Induced ordered weighted averaging operators. In: Recent Developments in the Ordered Weighted Averaging Operators: Theory and Practice (Springer, Berlin Heidelberg, 2011), pp. 29–47

[BJL11] G. Beliakov, S. James, G. Li, Learning Choquet-integral-based metrics for semisupervised clustering. IEEE Trans. Fuzzy Syst. **19**(3), 562–574 (2011)

[BMV04] G. Beliakov, R. Mesiar, L. Valaskova, Fitting generated aggregation operators to empirical data. Int. J. Uncertainty Fuzziness Knowledge Based Syst. **12**, 219–236 (2004)

[BGW08] J. Bolton, P. Gader, J. Wilson, Discrete Choquet integral as a distance metric. IEEE Trans. Fuzzy Syst. **16**(4), 1107–1110 (2008)

[Cal+02] T. Calvo, A. Kolesárová, M. Komorníková, R. Mesiar, Aggregation operators: properties, classes and construction methods, in *Aggregation Operators. New Trends and Applications*, eds. by T. Calvo, G. Mayor, R. Mesiar (Physica-Verlag, Heidelberg, New York, 2002), pp. 3–104

[CMY04] T. Calvo, R. Mesiar, R. Yager, Quantitative weights and aggregation. IEEE Trans. Fuzzy Syst. **12**, 62–69 (2004)

[CMM97] M. Carbonell, M. Mas, G. Mayor, On a class of Monotonic Extended OWA Operators, in *6th IEEE International Conference on Fuzzy Systems*, vol. III, Barcelona, Spain (1997), pp. 1695–1700

[Chi+04] F. Chiclana, E. Herrera-Viedma, F. Herrera, S. Alonso, Induced ordered weighted geometric operators and their use in the aggregation of multiplicative preference relations. Int. J. Intell. Syst. **19**, 233–255 (2004)

[Chi+07] F. Chiclana, E. Herrera-Viedma, F. Herrera, S. Alonso, Some induced ordered weighted averaging operators and their use for solving group decision-making problems based on fuzzy preference relations. Eur. J. Oper. Res. **182**, 383–399 (2007)

[Cho54] G. Choquet, Theory of capacities. Annales de l'institut Fourier **5**, 131–295 (1954)

[DR95] V. Demyanov, A. Rubinov, *Constructive Nonsmooth Analysis* (Peter Lang, Frankfurt am Main, 1995)

[Den94] D. Denneberg, *Non-additive Measure and Integral* (Springer Science & Business Media, Dordrecht, 1994)

[DP96] D. Dubois, H. Prade, Semantics of quotient operators in fuzzy relational databases. Fuzzy Sets Syst. **78**(1), 89–93 (1996)

[DPT88] D. Dubois, H. Prade, C. Testemale, Weighted fuzzy pattern matching. Fuzzy Sets Syst. **28**, 313–331 (1988)

[DB17] J. Dujmovic, G. Beliakov, Idempotent weighted aggregation based on binary aggregation trees. Int. J. Intell. Syst. **32**, 31–50 (2017)

[EM14] A. Emrouznejad, M. Marra, Ordered weighted averaging operators 1988-2014: A citation-based literature survey. Int. J. Intell. Syst. **29**, 994–1014 (2014)

[FY98] D. Filev, R. Yager, On the issue of obtaining OWA operator weights. Fuzzy Sets Syst. **94**, 157–169 (1998)

[FM01] R. Fuller, P Majlender, An analytic approach for obtaining maximal entropy OWA operator weights. Fuzzy Sets Syst. **124**, 53–57 (2001)

[FM03] R. Fuller, P. Majlender, On obtaining minimal variability OWA operator weights. Fuzzy Sets Syst. **136**, 203–215 (2003)

[GJ18] M. Gagolewski, S. James, Fitting symmetric fuzzy measures for discrete Sugeno integration, in *Advances in Fuzzy Logic and Technology 2017*, eds. by J. Kacprzyk et al., vol. 642. *Advances in Intelligent Systems and Computing* (Springer, 2018), pp. 104–116

[GJB19] M. Gagolewski, S. James, G. Beliakov, Supervised learning to aggregate data with the Sugeno integral. IEEE Trans. Fuzzy Syst. **27**(4), 810–815 (2019)

[HHV03] F. Herrera, E. Herrera-Viedma, A study of the origin and uses of the Ordered Weighted Geometric operator in multi-criteria decision making. Int. J. Intell. Syst. **18**, 689–707 (2003)

[KSP14] A. Kishor, A. Singh, N. Pal, Orness measure of OWA operators: a new approach. IEEE Trans. Fuzzy Syst. **22**, 1039–1045 (2014)

[KMM07] A. Kolesárová, R. Mesiar, G. Mayor, Weighted ordinal means. Inf. Sci. **177**, 3822–3830 (2007)

[LJ14] J. Lin, Y. Jiang, Some hybrid weighted averaging operators and their application to decision making. Inf. Fusion **16**, 18–28 (2014)

[Liu05] X. Liu, On the properties of equidifferent RIM quantifier with generating function. Int. J. General Syst. **34**, 579–594 (2005)

[Liu07] X. Liu, The solution equivalence of minimax disparity and minimum variance problems for OWA operators. Int. J. Approx. Reason. **45**, 68–81 (2007)

[Lla15] B. Llamazares, Constructing Choquet integral-based operators that generalize weighted means and OWA operators. Inf. Fusion **23**, 131–138 (2015)

[Mer09] J. Merigo, On the use of the OWA operator in the weighted average and its application in decision making, in *Proc. WCE 2009 Conf.*, London, UK (2009), pp. 82–87

[Mer10] J. Merigo, Fuzzy generalized aggregation operators in a unified model between the probability, the weighted average and the OWA operator, in *Proc. Intl. Conf. on Fuzzy Systems* (IEEE, 2010), pp. 1–7

[Mer11] J. Merigó, A unified model between the weighted average and the induced OWA operator. Expert Syst. Appl. **38**, 11560–11572 (2011)

[MGL09] J. Merigó, A. Gil-Lafuente, The induced generalized OWA operator. Inf. Sci. **179**, 729–741 (2009)

[ME97] H. Mitchell, D. Estrakh, A modified OWA operator and its use in lossless DPCM image compression. Int. J. Uncertainty Fuzziness Knowledged Based Syst. **5**, 429–436 (1997)

[ME98] H. Mitchell, D. Estrakh, An OWA operator with fuzzy ranks. Int. J. Intell. Syst. **13**, 59–81 (1998)

[MS00] H. Mitchell, P. Schaefer, Multiple priorities in an induced ordered weighted averaging operator. Int. J. Intell. Syst. **15**, 317–327 (2000)

[Nar07] Y. Narukawa, *Distances Defined by Choquet Integral* (FUZZIEEE, London, UK, Jul. 2007), pp. 511–516

[NT05] Y. Narukawa, V. Torra, Fuzzy measure and probability distributions: distorted probabilities. IEEE Trans. Fuzzy Syst. **13**, 617–629 (2005)

[O'H88] M. O'Hagan, Aggregating template or rule antecedents in real-time expert systems with fuzzy set logic, in *22nd Annual IEEE Asilomar Conf. Signals, Systems, Computers*, Pacific Grove, CA (1988), pp. 681–689

[PY06] G. Pasi, R. Yager, Modeling the concept of majority opinion in group decision making. Inf. Sci. **176**, 390–414 (2006)

[Rub00] A. Rubinov, *Abstract Convexity and Global Optimization* (Kluwer, Dordrecht; Boston, 2000)

[Rud91] W. Rudin, *Functional Analysis* (McGraw-Hill, New York, 1991)

[Sch83] L. Schumaker, On shape preserving quadratic interpolation. SIAM J. Numer. Anal. **20**, 854–864 (1983)

[Sim49] E. Simpson, Measurement of diversity. Nature **163**, 688 (1949)

[Tor97] V. Torra, The weighted OWA operator. Int. J. Intell. Syst. **12**, 153–166 (1997)

[Tor98] V. Torra, On some relationships between WOWA operator and the Choquet integral, in *8th Int. Conf. on Information Processing and Management of Uncertainty*, Paris (1998), pp. 818–824

[Tor00] V. Torra, The WOWA operator and the interpolation function W*: Chen and Otto's interpolation revisited. Fuzzy Sets Syst. **113**, 389–396 (2000)

[Tor02] V. Torra, Learning weights for the quasi-weighted means. IEEE Trans. Fuzzy Syst. **10**, 653–666 (2002)

[Tor04] V. Torra, OWA operators in data modeling and reidentification. IEEE Trans. Fuzzy Syst. **12**, 652–660 (2004)

[Tor07] Y. Torra V. Narukawa, *Modeling Decisions. Information Fusion and Aggregation Operators* (Springer, Berlin, Heidelberg, 2007)

[TY05a] L. Troiano, R. Yager, A meaure of dispersion for OWA operators, in *Proc. of the 11th IFSA World Congress*, eds. by Y. Liu, G. Chen, M. Ying (Tsinghua University Press and Springer, Beijing, China, 2005), pp. 82–87

[TY05b] L. Troiano, R. Yager, Recursive and iterative OWA operators. Int. J. Uncertainty Fuzziness Knowledge Based Syst. **13**, 579–599 (2005)

[TY06] L. Troiano, R. Yager, On the relationship between the quantifier threshold and OWA operators, in *Modeling Decisions for Artificial Intelligence*, vol. LNAI 3885, eds. by V. Torra, Y. Narukawa, A. Valls, J. Domingo-Ferrer (Springer, Heidelberg, 2006), pp. 215–226

[WP05] Y. Wang, C. Parkan, A minimax disparity approach for obtaining OWA operator weights. Inf. Sci. **175**, 20–29 (2005)

[Xu05] Z. Xu, An overview of methods for determining OWA weights. Int. J. Intell. Syst. **20**, 843–865 (2005)

[XD02] Z. Xu, Q. Da, The ordered weighted geometric averaging operator. Int. J. Intell. Syst. **17**, 709–716 (2002)

[Yag88] R. Yager, On ordered weighted averaging aggregation operators in multicriteria decision making. IEEE Trans. Syst. Man Cybern. **18**, 183–190 (1988)

[Yag91] R. Yager, Connectives and quantifiers in fuzzy sets. Fuzzy Sets Syst. **40**, 39–76 (1991)

[Yag93] R. Yager, Families of OWA operators. Fuzzy Sets Syst. **59**, 125–148 (1993)

[Yag95] R. Yager, Measures of entropy and fuzziness related to aggregation operators. Inf. Sci. **82**, 147–166 (1995)

[Yag96] R. Yager, Quantifier guided aggregation using OWA operators. Int. J. Intell. Syst. **11**, 49–73 (1996)

[Yag98a] R. Yager, Fusion of ordinal information using weighted median aggregation. Int. J. Approx. Reason. **18**, 35–52 (1998)

[Yag98b] R. Yager, Including importances in OWA aggregations using fuzzy systems modeling. IEEE Trans. Fuzzy Syst. **6**, 286–294 (1998)

[Yag02a] R. Yager, The induced fuzzy integral aggregation operator. Int. J. Intell. Syst. **17**, 1049–1065 (2002)

[Yag02b] R. Yager, Using fuzzy methods to model nearest neighbor rules. IEEE Trans. Syst. Man Cybern. Part B Cybern. **32**, 512–525 (2002)

[Yag04a] R. Yager, Choquet aggregation using order inducing variables. Int. J. Uncertainty Fuzziness Knowledged Based Syst. **12**, 69–88 (2004)

[Yag04b] R. Yager, Generalized OWA aggregation operators. Fuzzy Optim. Decis. Making **3**, 93–107 (2004)

[Yag07a] R. Yager, Centered OWA operators. Soft Comput. **11**, 631–639 (2007)

[Yag07b] R. Yager, Using stress functions to obtain OWA operators. IEEE Trans. Fuzzy Syst. **15**, 1122–1129 (2007)

[Yag10] R. Yager, Norms induced from OWA operators. IEEE Trans. Fuzzy Syst. **18**(1), 57–66 (2010)

[YA14a] R. Yager, N. Alajlan, A generalized framework for mean aggregation: Toward the modeling of cognitive aspects. Inf. Fusion **17**, 65–73 (2014)

[YA14b] R. Yager, N. Alajlan, On characterizing features of OWA aggregation operators. Fuzzy Optim. Decis. Making **13**, 1–32 (2014)

[YA16] R. Yager, N. Alajlan, Some issues on the OWA aggregation with importance weighted arguments. Knowledge Based Syst. **100**, 89–96 (2016)

[YF94] R. Yager, D. Filev, *Essentials of Fuzzy Modelling and Control* (Wiley, New York, 1994)

[YF99] R. Yager, D.P. Filev, Induced ordered weighted averaging operators. IEEE Trans. Syst. Man Cybern. Part B Cybern. **20**, 141–150 (1999)

[YK97] R. Yager, J. Kacprzyk (eds.), *The Ordered Weighted Averaging Operators. Theory and Applications* (Kluwer, Boston, 1997)

[YKB11] R. Yager, J. Kacprzyk, G. Beliakov (eds.), *Recent Developments in the Ordered Weighted Averaging Operators: Theory and Practice*, vol. 265. *Studies in Fuzziness and Soft Computing* (Springer, Berlin, New York, 2011)

[YK03] R. Yager, V. Kreinovich, Universal approximation theorem for uninorm-based fuzzy systems modeling. Fuzzy Sets Syst. **140**, 331–339 (2003)

[ZSA08] M. Zarghami, F. Szidarovszky, R. Ardakanian, Sensitivity analysis of the OWA operator. IEEE Trans. Syst. Man Cybern. Part B **38**, 547–552 (2008)

Chapter 7
Sparse Capacities

Abstract A key challenge when working with capacities is in determining the exponentially increasing number of parameters required for their definition. The issue of tractability is not just with respect to the number of variables, but also in ensuring monotonicity or desirable properties to suit a given application such as submodularity and antibuoyancy. This chapter summarises several approaches that seek a trade-off between computability and the extent to which interaction between variables is still modelled. The resulting capacities are often sparse in one or another representation and can be interpreted in terms of simplifying assumptions or limitations on the interactions that take place. We present the key definitions and properties, as well as implications for capacity-based aggregation.

7.1 λ-Capacities

Additive capacities are often too simple for applications, whereas general capacities are sometimes too complicated. We now turn to a number of simplifying assumptions that can be used to define capacities with intermediate complexity.

The so-called λ-capacities (also called Sugeno measures and λ-fuzzy measures) were introduced by Sugeno [Sug74]. Such capacities are powerful enough to express interactions among variables, yet require much fewer than 2^n parameters to express them.

Definition 7.1 (λ-Capacity) Given a parameter $\lambda \in]-1, \infty[$, a λ-capacity is a capacity μ, which for all $\mathcal{A}, \mathcal{B} \subseteq \mathcal{N}$, $\mathcal{A} \cap \mathcal{B} = \emptyset$ satisfies

$$\mu(\mathcal{A} \cup \mathcal{B}) = \mu(\mathcal{A}) + \mu(\mathcal{B}) + \lambda\mu(\mathcal{A})\mu(\mathcal{B}). \tag{7.1}$$

Under these conditions, all the values $\mu(\mathcal{A})$ are immediately computed from n independent values $\mu(\{i\})$, $i = 1, \ldots, n$, by using the explicit formula

$$\mu(\bigcup_{i=1}^{m}\{i\}) = \frac{1}{\lambda}\left(\prod_{i=1}^{m}(1 + \lambda\mu(\{i\})) - 1\right), \quad \lambda \neq 0. \tag{7.2}$$

G. Beliakov et al., *Choquet Capacities and Fuzzy Integrals*, Theory and Applications of Computability, https://doi.org/10.1007/978-3-031-97070-2_7

If $\lambda = 0$, a λ-capacity becomes a probability measure. The coefficient λ is determined from the boundary condition $\mu(\mathcal{N}) = 1$, leading to

$$\lambda + 1 = \prod_{i=1}^{n}(1 + \lambda\mu(\{i\})), \tag{7.3}$$

which can be solved on $(-1, 0)$ or $(0, \infty)$ numerically (note that $\lambda = 0$ is always a solution). Thus, a λ-fuzzy measure is characterised by n independent values $\mu(\{i\}), i = 1, \dots, n$.

Example 7.1 For the singleton values $\mu(\{1\}) = 0.7$, $\mu(\{2\}) = 0.4$, $\mu(\{3\}) = 0.6$, solving numerically for λ gives -0.8845 (to 4 d.p.). This then results in the following capacity:

	1	
0.8523	0.9285	0.7877
0.7	0.4	0.6
	0	

If $\lambda \geqslant 0$, then clearly we have positive interaction effects and the resulting capacity is superadditive, while $\lambda \leqslant 0$ results in subadditive capacities.

Note 7.1 When $-1 < \lambda \leqslant 0$, a λ-capacity is a plausibility measure, and when $\lambda \geqslant 0$ it is a belief measure.

Note 7.2 The necessary and sufficient condition of antibuoyancy is

$$\frac{\mu(\{j\})}{1 + \lambda\mu(\{j\}} \leqslant \mu(\{i\}) \tag{7.4}$$

for all $i, j \in \mathcal{N}$. We can rewrite (7.4) using variables $a_i = \frac{1}{\mu(\{i\})} \geqslant n$. Then we obtain the conditions $a_j + \lambda \geqslant a_i$, which, by considering all pairs i, j, can be written as $|a_i - a_j| \leqslant \lambda$, where λ is determined from the nonlinear equation $1 + \lambda = \prod_i (1 + \frac{\lambda}{a_i})$.

A λ-capacity μ is related to a probability measure P through the relation

$$P(\mathcal{A}) = \frac{\log(1 + \lambda\mu(\mathcal{A}))}{1 + \lambda},$$

and, using $g(t) = ((1 + \lambda)^t - 1)/\lambda$ for $\lambda > -1, \lambda \neq 0$, and $g(t) = t$ for $\lambda = 0$,

$$g(P(\mathcal{A})) = \mu(\mathcal{A}).$$

Definition 7.2 (Dual λ-Capacity) For a given λ-capacity, the dual capacity satisfies $\lambda' = -\frac{\lambda}{\lambda+1}$ where λ' is said to be the conjugate of λ.

Example 7.2 For the λ-capacity obtained in Example 7.1, the dual λ-capacity has $\lambda' = -\frac{-0.8845}{-0.8845+1} = 7.658$ and is given by

	1	
0.4	0.6	0.3
0.2123	0.0715	0.1477
	0	

The Möbius transform of a λ-capacity μ for all $\emptyset \neq \mathcal{A} \subseteq \mathcal{N}$, $\lambda \neq 0$ is given by [Gra16],

$$M(\mathcal{A}) = \lambda^{|\mathcal{A}|-1} \prod_{i \in \mathcal{A}} \mu(\{i\}). \tag{7.5}$$

This, in conjunction with the Zeta transform, can be used to express the λ-capacity value for any subset in standard representation (noting that $\mu(\{i\}) = M(\{i\})$ for singletons).

$$\begin{aligned}\mu(\mathcal{A}) &= \sum_{\mathcal{B} \subseteq \mathcal{A}} \lambda^{|\mathcal{B}|-1} \prod_{i \in \mathcal{B}} M(\{i\}) \\ &= \sum_{i \in \mathcal{A}} M(\{i\}) + \lambda \sum_{\{i,j\} \subseteq \mathcal{A}} M(\{i\})M(\{j\}) + \cdots + \lambda^{|\mathcal{A}|} \prod_{i \in \mathcal{A}} M(\{i\}).\end{aligned}$$

Note 7.3 The set of all λ-capacities is not convex.

Note 7.4 Dirac measures (Example 1.5) are λ-capacities for all $\lambda \in]-1, \infty[$.

7.2 Decomposable Capacities and Distorted Probabilities

The λ-capacities described in the previous section allow for the construction of a capacity from singleton values alone, using λ to determine the value of larger subsets. They are a special case of decomposable capacities [GNW95].

Definition 7.3 (Decomposable Capacity) A decomposable capacity μ is a capacity, which for all $\mathcal{A}, \mathcal{B} \subseteq \mathcal{N}$, $\mathcal{A} \cap \mathcal{B} = \emptyset$ satisfies

$$\mu(\mathcal{A} \cup \mathcal{B}) = f(\mu(\mathcal{A}), \mu(\mathcal{B})) \tag{7.6}$$

for some function $f : [0, 1]^2 \to [0, 1]$ known as the decomposition function.

Note 7.5 It turns out that to get $\mu(\mathcal{N}) = 1$, f must necessarily be a t–conorm (see Chap. 1). In the case of λ-fuzzy measures, f is an Archimedean t–conorm with an additive generator $h(t) = \frac{\ln(1+\lambda t)}{\ln(1+\lambda)}$, $\lambda \neq 0$, which is a Sugeno-Weber t–conorm.

Note 7.6 Additive capacities are decomposable with respect to the Łukasiewicz t-conorm, $S_L(x, y) = \min(1, x + y)$. But not every S_L-decomposable capacity is a probability.

Example 7.3 The capacity below is S_L-decomposable and built from the singletons $\mu(\{1\}) = 0.3$, $\mu(\{2\}) = 0.6$, $\mu(\{3\}) = 0.5$. It is subadditive.

$$\begin{array}{ccc} & 1 & \\ 0.9 & 0.8 & 1 \\ 0.3 & 0.6 & 0.5 \\ & 0 & \end{array}$$

Note 7.7 Possibility measures are decomposable with respect to the maximum t–conorm $S_{\max}(x, y) = \max(x, y)$. Every $S_{\max}$-decomposable discrete capacity is a possibility measure.

Note 7.8 For a given t–conorm S and fixed n, the set of all S-decomposable capacities is not always convex.

Another construction of capacities involves a transformation of an additive probability by means of a function g.

Definition 7.4 (Distorted Probability Measure) A capacity μ is a distorted probability measure if there exists some non-decreasing function $g : [0, 1] \to [0, 1]$, $g(0) = 0$, $g(1) = 1$, and a probability measure P, such that for all $\mathcal{A} \subset \mathcal{N}$:

$$\mu(\mathcal{A}) = g(P(\mathcal{A})).$$

Such a construction clearly ensures the monotonicity and boundary conditions of the resulting capacity.

Example 7.4 Using the function $g(t) = t^2$ on the additive capacity below (left) results in the distorted probability measure (right), which is superadditive.

$$\begin{array}{ccc} & 1 & \\ 0.9 & 0.3 & 0.8 \\ 0.2 & 0.7 & 0.1 \\ & 0 & \end{array} \qquad \begin{array}{ccc} & 1 & \\ 0.81 & 0.09 & 0.64 \\ 0.04 & 0.49 & 0.01 \\ & 0 & \end{array}$$

The Choquet integral with respect to a distorted probability becomes a weighted OWA function (see Sect. 6.4). Distorted probabilities and their extension, m-dimensional distorted probabilities, have been studied in [NT05].

7.3 k-Additivity

One issue with capacities is their complexity. There are 2^n coefficients which represent importances and interaction of every possible subset of inputs. When a particular capacity is elicited, for example, from an expert in the field, it is unreasonable to expect that all coefficients could be specified. At most, the importances of individual inputs and a few input coalitions can be acquired in this way. On the other hand, capacities can be learned from empirical data. In this case, even if there are plenty of training data, the computational complexity of solving the respective linear, quadratic, and nonlinear programming problems becomes a bottleneck as the number of variables increases.

One way to reduce the complexity of aggregation functions based on capacities is to impose various linear constraints on their values. One type of constraint leads to k-additive capacities [Gra00, Gra97], which extend the additivity assumption on probability measures but limit it to small subsets of variables.

Definition 7.5 (k-Additive Capacity) A capacity μ is called k-additive ($1 \leqslant k \leqslant n$) if its Möbius transformation verifies

$$M(\mathcal{A}) = 0,$$

for any subset $\mathcal{A}$ with more than k elements, $|\mathcal{A}| > k$, and there exists a subset $\mathcal{B}$ with k elements such that $M(\mathcal{B}) \neq 0$.

Example 7.5 The following capacity is 2-additive, given in standard representation (left) and Möbius representation (right).

	1				0	
0.3	1	0.9		−0.2	0.1	−0.1
0.2	0.3	0.7		0.2	0.3	0.7
	0				0	

An alternative definition of k-additivity (which is also applicable to capacities on more general sets than $\mathcal{N}$) was given by Mesiar in [Mes99b, Mes99a]. It involves a weakly monotone[1] additive set function μ_k defined on subsets of $\mathcal{N}^k = \mathcal{N} \times \mathcal{N} \times \ldots \times \mathcal{N}$. A capacity μ is k-additive if $\mu(\mathcal{A}) = \mu_k(\mathcal{A}^k)$ for all $\mathcal{A} \subseteq \mathcal{N}$.

The next result due to Grabisch [Gra00, Gra97] establishes a fundamental property of k-additive capacities, which justifies their use in simplifying interactions between the criteria in multiple criteria decision-making. Recall that $I(\mathcal{A})$ and $J(\mathcal{A})$ are the Shapley and Banzhaf interaction indices for the subset $\mathcal{A}$, respectively (Definitions 3.7 and 3.9).

[1] Weakly monotone means $\forall \mathcal{A}, \mathcal{B} \subseteq \mathcal{N}$, $\mathcal{A} \subseteq \mathcal{B}$ implies $\mu_k(\mathcal{A}^k) \leqslant \mu_k(\mathcal{B}^k)$.

Proposition 7.1 *Let μ be a k-additive capacity, $1 \leqslant k \leqslant n$. Then*

- $I(\mathcal{A}) = 0$ *for every* $\mathcal{A} \subseteq \mathcal{N}$ *such that* $|\mathcal{A}| > k$.
- $I(\mathcal{A}) = J(\mathcal{A}) = M(\mathcal{A})$ *for every* $\mathcal{A} \subseteq \mathcal{N}$ *such that* $|\mathcal{A}| = k$.

Thus, k-additive capacities acquire an interesting interpretation. These are capacities that limit interaction among the criteria to groups of size at most k. For instance, for 2-additive capacities, there are pairwise interactions among the criteria but no interactions in groups of three or more. By limiting the class of capacities to those that are k-additive, one reduces their complexity, with the total number of linearly independent values being reduced from $2^n - 1$ to $\sum_{i=1}^{k} \binom{n}{i} - 1$.

The topic of k-additive aggregation functions was discussed in [Kol+18].

The particular case of 2-additive capacities has been well studied with a number of useful results. Here all $M(\mathcal{A}) = I(\mathcal{A}) = 0$ whenever $|\mathcal{A}| > 2$. In this case we have the following simplifications.

For 2-additive capacities the Choquet integral is expressed in terms of the interaction indices and Shapley value as [GMS00].

$$C_\mu(\mathbf{x}) = \sum_{I_{ij}>0} \min(x_i, x_j) I_{ij} + \sum_{I_{ij}<0} \max(x_i, x_j)|I_{ij}| + \sum_{i=1}^{n} x_i (\varphi(i) - \frac{1}{2}\sum_{i \neq j} |I_{ij}|), \tag{7.7}$$

subject to $\mu(\{i\}) = \varphi(i) - \frac{1}{2}\sum_{i \neq j} |I_{ij}| \geqslant 0$ for all $i = 1, \dots, n$.

The interaction indices need to satisfy a number of constraints [GMS00], p.429, namely,

$$\frac{1}{2}\left(\sum_{j \in \mathcal{N} \setminus \mathcal{A} \cup \{i\}} I_{ij} - \sum_{l \in \mathcal{A}} I_{il} \right) \leqslant \varphi(i), \tag{7.8}$$

for all $\mathcal{A} \subseteq \mathcal{N} \setminus \{i\}, i = 1, \dots, n$ and $\varphi(i)$ denoting the Shapley values. The constraints are satisfied if and only if μ is a 2-additive capacity.

With simplified notation $m_i = M(\{i\})$, $m_{ij} = M(\{i, j\})$ and based on [MGS20], we have a reduced number of constraints in Möbius representation.

Theorem 7.1 ([BBA25]) *A 2-additive set function is a fuzzy measure if and only if for all $i, j \in \mathcal{N}$:*

$$0 \leqslant m_i + \sum_{j \neq i | m_{ij} < 0} m_{ij} \leqslant 1, \tag{7.9}$$

$$0 \leqslant m_i \leqslant 1, \quad -1 \leqslant m_{ij} \leqslant 1, \tag{7.10}$$

and

$$\sum_i m_i + \sum_{j \neq i} m_{ij} = 1. \tag{7.11}$$

7.4 p-Symmetric Capacities

In this section the symmetry assumption is relaxed and generalised.

Definition 7.6 (Indifferent Elements [MGG02]) Two elements i and j of $\mathcal{N}$ are indifferent if

$$\forall \mathcal{A} \in \mathcal{N} \setminus \{i, j\} : \mu(\mathcal{A} \cup \{i\}) = \mu(\mathcal{A} \cup \{j\}).$$

That is, the fact that it is i or j in a coalition is irrelevant. This is generalised for subsets of higher cardinality.

Definition 7.7 (Subset of Indifference [MGG02]) Let μ be a capacity on $\mathcal{N}$. The subset $\mathcal{A} \subseteq \mathcal{N}$ is a subset of indifference with respect to μ if $\forall \mathcal{B}_1, \mathcal{B}_2 \subseteq \mathcal{A}$, $|\mathcal{B}_1| = |\mathcal{B}_2|$, we have $\mu(\mathcal{C} \cup \mathcal{B}_1) = \mu(\mathcal{C} \cup \mathcal{B}_2), \forall \mathcal{C} \subseteq \mathcal{N} \setminus \mathcal{A}$.

Any subset of a subset of indifference is also a subset of indifference, and any singleton subset is a subset of indifference.

Definition 7.8 (p-Symmetric Capacity [MGG02]) A capacity μ on $\mathcal{N}$ is said to be p-symmetric if the coarsest partition of $\mathcal{N}$ into subsets of indifference contains exactly p subsets $\mathcal{A}_1, \ldots, \mathcal{A}_p$, where $\mathcal{A}_i$ is a subset of indifference, $\mathcal{A}_i \cap \mathcal{A}_j = \emptyset$, $\cup_{i=1}^{p} \mathcal{A}_i = \mathcal{N}, i, j = 1, \ldots, p$, and a partition π is coarser than another partition π' if all subsets of π are the union of some subsets of π'. The partition $\{\mathcal{A}_1, \ldots, \mathcal{A}_p\}$ is called the basis of μ.

A 1-symmetric capacity is just the symmetric capacity. Let $\{\mathcal{A}_1, \ldots, \mathcal{A}_p\}$ be the basis of a p-symmetric capacity μ on $\mathcal{N}$. Then any subset $\mathcal{S} \subseteq \mathcal{N}$ can be identified with a p-dimensional vector $\mathbf{b}_{\mathcal{S}} = (b_1, \ldots, b_p)$, where $b_i = |\mathcal{S} \cap \mathcal{A}_i|$, $i = 1, \ldots, p$. That is, a p-symmetric capacity needs to define $\prod_{i=1}^{p}(|\mathcal{A}_i| + 1)$ coefficients [MGG02].

Example 7.6 The table below gives the capacity values for a 2-symmetric capacity on $\mathcal{N} = \{1, 2, 3, 4, 5\}$ with subsets of indifference $\mathcal{A}_1 = \{1, 2\}$ and $\mathcal{A}_2 = \{3, 4, 5\}$.

No. of elements	No. of elements from $\mathcal{A}_2$			
from $\mathcal{A}_1$	0	1	2	3
0	0	0	0	0
1	1/16	1/8	1/4	1/2
2	7/32	3/8	5/8	1

Each cell (i, j) of the table $(i = 0, \ldots, 2, j = 0, \ldots, 3)$ indicates the presence of i elements from $\mathcal{A}_1$ and j elements from $\mathcal{A}_2$. For example, with $\{1, 2, 5\}$ we have two elements from $\mathcal{A}_1$ and one from $\mathcal{A}_2$, and so reading from the table we have $\mu(\{1, 2, 5\}) = 3/8$. Then $\mu(\{3\}) = \mu(\{4\}) = \mu(\{5\}) = 0$ and so on. This particular capacity is an antibuoyant vertex following the construction method in [BGJ22b] (see Sects. 2.10 and 9.6 as well as Algorithm 9.7 and Fig. 9.2).

7.5 k-Tolerance and k-Intolerance

The concept of k-tolerant (and k-intolerant) capacities was proposed in [Mar04, Mar07].

Definition 7.9 (k-Tolerant Capacity) Let $k \in \{1, \ldots, n\}$. A capacity μ on $\mathcal{N}$ is said to be:

- k-tolerant if $\mu(\mathcal{A}) = 1$ for all $\mathcal{A} \subseteq \mathcal{N}$ such that $|\mathcal{A}| \geqslant k$ and there exists a subset $\mathcal{B} \subseteq \mathcal{N}$, with $|\mathcal{B}| = k - 1$, such that $\mu(\mathcal{B}) \neq 1$.
- k-intolerant if $\mu(\mathcal{A}) = 0$ for all $\mathcal{A} \subseteq \mathcal{N}$ such that $|\mathcal{A}| \leqslant n - k$ and there exists a subset $\mathcal{B} \subseteq \mathcal{N}$, with $|\mathcal{B}| = n - k + 1$, such that $\mu(B) \neq 0$.

The k-tolerant and k-intolerant capacities are dual to one another. To identify these two types of capacity, we only need to define $\sum_{i=1}^{k-1} \binom{n}{i}$ coefficients, i.e. $\mu(\mathcal{A}), 1 \leqslant |\mathcal{A}| \leqslant k - 1$ for k-intolerant capacities and $\mu(\mathcal{A}), n - k \leqslant |\mathcal{A}| \leqslant n - 1$ for k-tolerant capacities.

Example 7.7 The capacities on the left is 2-tolerant and its 2-intolerant dual is shown on the right, given in standard representation.

	1			1	
1	1	1	0	0.8	0.4
0.6	0.2	1	0	0	0
	0			0	

The Choquet integral with respect to a k-tolerant capacity is independent of the first $n - k$ smallest inputs and can be expressed as

$$C_\mu(\mathbf{x}) = \sum_{\mathcal{A} \subseteq \mathcal{N}, |\mathcal{A}| < k} \mu(\mathcal{A}) g_{\mathcal{A}}(\mathbf{x}) + x_{(n-k+1)}, \tag{7.12}$$

where, as given previously, $g_{\mathcal{A}}(\mathbf{x}) = \max(0, \min_{i \in \mathcal{A}} x_i - \max_{i \in \mathcal{N} \setminus \mathcal{A}} x_i)$.

7.6 k-Maxitivity and k-Minitivity

Inspired by the k-additive measures, the k-maxitive capacities were proposed in [CD98, Mes99b, Mes99a]. They are based on an alternative Möbius transform called the possibilistic Möbius transform. This can be seen as another framework for modelling k-order interaction in the coefficient space.

Definition 7.10 (Possibilistic Möbius Transform) For a capacity μ on $\mathcal{N}$, the possibilistic Möbius transform is a mapping $M_p : 2^{\mathcal{N}} \to [0, 1]$ defined by

$$M_p(\mathcal{A}) = \begin{cases} \mu(\mathcal{A}) & \text{if } \mu(\mathcal{A}) > \max\limits_{\mathcal{B} \subset \mathcal{A}} \mu(\mathcal{B}), \\ 0 & \text{otherwise.} \end{cases} \tag{7.13}$$

Such capacities will allocate 0 to subsets satisfying $\mu(\mathcal{A}) = \max\limits_{\mathcal{B} \subset \mathcal{A}} \mu(\mathcal{B})$, i.e. which applies to all subsets in the case of possibility measures (Definition 2.14).

Example 7.8 The following capacity is given in standard representation (left) and possiblistic Möbius representation (right). Note that it is only $\mu(\{1, 3\})$ and $\mu(\{2, 3\})$ that are set to 0.

	1			1	
0.8	0.6	0.7	0.8	0	0
0.3	0.7	0.6	0.3	0.7	0.6
	0			0	

The possibilistic Zeta transform of M_p is the mapping $Z_{M_p} : 2^{\mathcal{N}} \to [0, 1]$ defined by:

$$Z_{M_p}(\mathcal{A}) = \max_{\mathcal{B} \subseteq \mathcal{A}} M_p(\mathcal{B}). \tag{7.14}$$

Definition 7.11 (k-Maxitive Capacity) A capacity μ is called k-maxitive if its possibilistic Möbius transform satisfies $M_p(\mathcal{A}) = 0$ for any $\mathcal{A}$ such that $|\mathcal{A}| > k$ and there exists at least one subset $\mathcal{A}$ of $\mathcal{N}$ of exactly k elements such that $M_p(\mathcal{A}) \neq 0$.

Note 7.9 There are no specific constraints on M_p which ensure Z_{M_p} is monotone. It is clear from (7.14) that any non-negative values of M_p result in monotonicity of Z_{M_p} with respect to set inclusion. However, the conditions $M_p(\emptyset) = 0$ and $\max_{\mathcal{A} \subseteq \mathcal{N}} M_p(\mathcal{A}) = 1$ are needed to satisfy the normalisation conditions. In particular, for a k-maxitive capacity the latter implies $\mu(\mathcal{A}) = 1$ for at least one $\mathcal{A}$ with $|\mathcal{A}| = k$ (see Corollary 7.1 below).

Note 7.10 The choice of the value 0 in the case statement in (7.13) is not binding. In fact, given two set functions $M_p^1, M_p^2 : 2^{\mathcal{N}} \to [0, 1]$, one defined by (7.13) and the second by a modification of that equation where 0 is replaced by any value in the interval $[0, \max_{\mathcal{B} \subset \mathcal{A}} M_p^2(\mathcal{B})]$, both functions produce exactly the same function $Z_{M_p^1} = Z_{M_p^2}$. For this reason it is possible to perform technical work with the values of the possibilistic transform irrespective of the second case statement in (7.13), which results in a more convenient set of constraints for optimisation problems in later sections.

We know that the possibility and necessity measures are mutually dual. Recall that the possibility measure of any subset $\mathcal{A}$ can be obtained by

$$\mu(\mathcal{A}) = \bigvee_{i \in \mathcal{S}} \mu(\{i\}). \tag{7.15}$$

In contrast, the necessity measure of any subset $\mathcal{A}$ can be obtained by

$$\mu(\mathcal{A}) = \bigwedge_{\substack{\mathcal{B} \supseteq \mathcal{A} \\ |\mathcal{B}|=n-1}} \mu(\mathcal{B}). \tag{7.16}$$

That is, to identify a possibility or necessity measure, we need to define at most n coefficients.

One can determine that a k-maxitive capacity is totally defined by the capacity values of subsets whose cardinalities are equal to or less than k. If we have $\mu(\mathcal{A}), 1 \leqslant |\mathcal{A}| \leqslant k$ (suppose that there exists at least one subset $\mathcal{A}_0$ with $|\mathcal{A}_0| = k$ such that $M_p(\mathcal{A}_0) \neq 0$), we can obtain the capacity values of other subsets with higher cardinality and identify the unique k-maxitive capacity by one of the following three equivalent approaches:

$$\mu(\mathcal{A}) = \bigvee_{\mathcal{B} \subset \mathcal{A}} \mu(\mathcal{B}), |\mathcal{A}| \geqslant k+1, \tag{7.17}$$

$$\mu(\mathcal{A}) = \bigvee_{\substack{\mathcal{B} \subset \mathcal{A} \\ |\mathcal{B}| \leqslant k}} \mu(\mathcal{B}), |\mathcal{A}| \geqslant k+1, \tag{7.18}$$

$$\mu(\mathcal{A}) = \bigvee_{\substack{\mathcal{B} \subset \mathcal{A} \\ |\mathcal{B}|=|\mathcal{A}|-1}} \mu(\mathcal{B}), |\mathcal{A}| \geqslant k+1. \tag{7.19}$$

That is, to identify a k-maxitive capacity, we only require at most $\sum_{i=1}^{k} \binom{n}{i}$ coefficients.

Corollary 7.1 *If a capacity μ on $\mathcal{N}$ is k-maxitive, then there exists at least one subset $\mathcal{A}_0$ with $|\mathcal{A}_0| = k$ such that $\mu(\mathcal{A}_0) = 1$.*

Definition 7.12 (k-Minitive Capacity) A capacity μ on $\mathcal{N}$ is called a k-minitive capacity if its dual capacity μ^* is k-maxitive.

Let μ be a k-minitive capacity, μ^* its dual capacity, and M_p be the possibilistic Möbius transform of μ^*. Then for any $\mathcal{A} \subseteq \mathcal{N}$,

$$\mu(\mathcal{A}) = 1 - \mu^*(\mathcal{N}\backslash\mathcal{A}) = 1 - \bigvee_{\mathcal{B}\subseteq\mathcal{N}\backslash\mathcal{A}} M_p(\mathcal{B}) = \bigwedge_{\mathcal{B}\subseteq\mathcal{N}\backslash\mathcal{A}} \left(1 - M_p(\mathcal{B})\right). \quad (7.20)$$

Based on the above equation, we can derive the following notions concerning k-minitivity, which correspond with the respective notions of k-maxitivity.

Definition 7.13 (Necessity Möbius Transform) The necessity Möbius transform of a capacity μ on $\mathcal{N}$ is a mapping $M_n : 2^{\mathcal{N}} \rightarrow [0, 1]$ defined by

$$M_n(\mathcal{A}) = \begin{cases} \mu(\mathcal{A}) & \text{if } \mu(\mathcal{A}) < \min_{\mathcal{B}\supset\mathcal{A}} \mu(\mathcal{B}), \\ 1 & \text{otherwise.} \end{cases} \quad (7.21)$$

The necessity Zeta transform of M_n is the mapping $Z_{M_n} : 2^{\mathcal{N}} \rightarrow [0, 1]$ defined by:

$$Z_{M_n}(\mathcal{A}) = \min_{\mathcal{B}\subseteq\mathcal{A}} M_n(\mathcal{B}). \quad (7.22)$$

Theorem 7.2 *A capacity μ is called k-minitive if its necessity Möbius transform satisfies $M_n(\mathcal{A}) = 1$ for any $\mathcal{A}$ such that $|\mathcal{A}| < n - k$ and there exists at least one subset $\mathcal{A}_0$ of exactly $n - k$ elements such that $M_n(\mathcal{A}_0) \neq 1$.*

Example 7.9 The following capacities are 2-maxitive (left) and 2-minitive (right), respectively, given in standard representation. The 2-maxitive capacity can be easily identified from the value $\mu(\{1, 2\}) = 1$ (from which it follows that $M_p(\mathcal{N}) = 0$), while 2-minitivity is evident from $\mu(\{3\}) = 0$ (and hence $M_n(\emptyset) = 1$). The capacities shown are dual to one another.

	1			1	
1	0.5	0.8	0.5	0.6	0.7
0.3	0.4	0.5	0.2	0.5	0
	0			0	

In contrast to the case of k-maxitive capacities, a k-minitive capacity is totally determined by the capacity values of subsets whose cardinalities are equal to or greater than $n - k$. If we have all $\mu(\mathcal{A}), n - k \leqslant |\mathcal{A}| \leqslant n - 1$ (and suppose that there exists at least one subset $\mathcal{A}_0$ with $|\mathcal{A}_0| = n - k$ such that $M_n(\mathcal{A}_0) \neq 1$), we can obtain the capacity values of other subsets with lower cardinality and identify the unique k-minitive capacity by one of the following three equivalent approaches:

$$\mu(\mathcal{A}) = \bigwedge_{\mathcal{B}\supset\mathcal{A}} \mu(\mathcal{B}), |\mathcal{A}| \leqslant n - k - 1, \quad (7.23)$$

$$\mu(\mathcal{A}) = \bigwedge_{\substack{\mathcal{B}\supset\mathcal{A} \\ |\mathcal{B}|\geqslant n-k}} \mu(\mathcal{B}), |\mathcal{A}| \leqslant n - k - 1, \quad (7.24)$$

$$\mu(\mathcal{A}) = \bigwedge_{\substack{\mathcal{A}\subset\mathcal{B} \\ |\mathcal{B}|=|\mathcal{A}|+1}} \mu(\mathcal{B}), \, |\mathcal{A}| \leqslant n-k-1. \tag{7.25}$$

Similarly, we have the following corollary.

Corollary 7.2 *If a capacity μ on $\mathcal{N}$ is k-minitive, then there exists at least one subset $\mathcal{A}_0$ with $|\mathcal{A}_0| = n-k$ such that $\mu(\mathcal{A}_0) = 0$.*

We further note the following results:

- A capacity is 1-maxitive if and only if it is a possibility capacity. A capacity is 1-minitive if and only if it is a necessity capacity.
- A k-tolerant capacity is a k-maxitive capacity. A k-intolerant capacity is a k-minitive capacity.

Interestingly enough, any Sugeno integral can be rewritten using possibilistic Möbius values (which can be ordinal) as follows [Gra+09]:

$$S_v(\mathbf{x}) = \bigvee_{\mathcal{A}\subseteq\mathcal{N}} \left(\bigwedge_{i\in\mathcal{A}} x_i \wedge M_p(\mathcal{A}) \right). \tag{7.26}$$

The computation of $S_v(\mathbf{x})$ with Eq. (7.26) involves 2^n sets; however, this simplifies when we have sets such that $M_p(\mathcal{A}) = 0$. For example, this is precisely the case for k-maxitive capacities, which can be seen as ordinal counterparts of k-additive capacities (see [Mes99a]).

The Sugeno integral can hence be reformulated as follows for k-maxitive capacities:

$$\begin{aligned}
S_v(\mathbf{x}) &= \bigvee_{\mathcal{A}\subseteq\mathcal{N}} \left(\bigwedge_{i\in\mathcal{A}} x_i \wedge M_p(\mathcal{A}) \right) \\
&= \left(\bigvee_{\substack{\mathcal{A}\subseteq\mathcal{N} \\ |\mathcal{A}|\leqslant k}} \left(\bigwedge_{i\in\mathcal{A}} x_i \wedge M_p(\mathcal{A}) \right) \right) \vee \left(\bigvee_{\substack{\mathcal{A}\subseteq\mathcal{N} \\ |\mathcal{A}|> k}} \left(\bigwedge_{i\in\mathcal{A}} x_i \wedge M_p(\mathcal{A}) \right) \right) \\
&= \left(\bigvee_{\substack{\mathcal{A}\subseteq\mathcal{N} \\ |\mathcal{A}|\leqslant k}} \left(\bigwedge_{i\in\mathcal{A}} x_i \wedge M_p(\mathcal{A}) \right) \right) \vee \left(\bigvee_{\substack{\mathcal{A}\subseteq\mathcal{N} \\ |\mathcal{A}|> k}} \left(\bigwedge_{i\in\mathcal{A}} x_i \wedge \mu(\emptyset) \right) \right) \\
&= \bigvee_{\substack{\mathcal{A}\subseteq\mathcal{N} \\ |\mathcal{A}|\leqslant k}} \left(\bigwedge_{i\in\mathcal{A}} x_i \wedge M_p(\mathcal{A}) \right).
\end{aligned}$$

This compact formulation of the Sugeno integral involves only $\binom{n}{k}$ coefficients, significantly simplifying the calculation.

7.6.1 k-Maxitive and k-Minitive Aggregation Functions

In a similar way that k-additive capacities lead to k-additive aggregation functions, k-maxitive and minitive capacities lead to k-maxitive (minitive) aggregation functions. We follow the definition from [MK18].

Definition 7.14 (**k-Maxitive Aggregation Function**) Let $k \in \{1, \ldots, n\}$. An aggregation function $F : [0, 1]^n \to [0, 1]$ is called k-maxitive if for any $\mathbf{x}_1, \ldots, \mathbf{x}_{k+1} \in [0, 1]^n$ there is a proper subset $\mathcal{I}$ of $\{1, \ldots, k+1\}$ such that

$$F\left(\bigvee_{i=1}^{k+1} \mathbf{x}_i\right) = F\left(\bigvee_{i \in \mathcal{I}} \mathbf{x}_i\right). \tag{7.27}$$

Based on the monotonicity property of aggregation functions, the k-maxitivity of an n-ary aggregation function F can be defined equivalently by requiring the equality [MK18],

$$F\left(\bigvee_{i=1}^{k+1} \mathbf{x}_i\right) = \bigvee_{i=1}^{k+1} F\left(\bigvee_{j \neq i} \mathbf{x}_i\right) \tag{7.28}$$

for any $\mathbf{x}_1, \ldots, \mathbf{x}_{k+1} \in [0, 1]^n$, where $i, j \in \{1, \ldots, k+1\}$.

An aggregation function is called proper k-maxitive if it is k-maxitive but not $(k-1)$-maxitive [MK18]. For any proper k-maxitive aggregation function $F : [0, 1]^n \to [0, 1]$, the set function $\mu : 2^{\mathcal{N}} \to [0, 1]$ given by $\mu(\mathcal{A}) = F(1_{\mathcal{A}})$, where $1_{\mathcal{A}}(i) = 1$ if $i \in \mathcal{A}$, and 0 otherwise, is the characteristic function of the set $\mathcal{A}$ and is a k-maxitive capacity [MK18]. Hence maxitive aggregation functions can be seen as particular monotone extensions of maxitive capacities [MK18].

It should be pointed out that if F is a k-maxitive aggregation function, then $\mu(\mathcal{A}) = F(1_{\mathcal{A}})$ is not necessarily k-maxitive and can be a $(k-1)$ or smaller-order maxitive capacity. For convenience, we will refer to the k-maxitive capacity given in Definition 7.14 as a *proper* k-maxitive capacity. We further give the definition of a *general* k-maxitive capacity as follows: a capacity μ is said to be *generally* k-maxitive if its possibilistic Möbius transform satisfies $M_p(\mathcal{A}) = 0$ for any $\mathcal{A}$ such that $|\mathcal{A}| > k$. That is, a capacity μ is a *proper* k-maxitive capacity if and only if μ is a *general* k-maxitive but not a *general* $k-1$-maxitive capacity. Following from this, we can say that if F is a k-maxitive aggregation function, then $\mu(\mathcal{A}) = F(1_{\mathcal{A}})$ is a *general* k-maxitive capacity.

It can be checked that for any $k + 1$ nonempty different subsets of $\mathcal{N}$, denoted $\mathcal{A}_1, \ldots, \mathcal{A}_{k+1}$, their union's cardinality is equal to or larger than $k + 1$,

$$\left| \bigcup_{i=1}^{k+1} \mathcal{A}_i \right| \geqslant k + 1. \tag{7.29}$$

Based on the above, we can see that Eqs. (7.27) and (7.28) just extend Eqs. (7.17) and (7.19).

Inspired by the extension process of k-maxitive aggregation functions, we have a definition of k-minitive aggregation functions which is correspondingly a monotone extension of the k-minitive capacity.

Definition 7.15 (k-Minitive Aggregation Function) Let $k \in \{1, \ldots, n\}$. An aggregation function $G : [0, 1]^n \to [0, 1]$ is called k-minitive if for any $\mathbf{x}_1, \ldots, \mathbf{x}_{k+1} \in [0, 1]^n$ there is a proper subset $\mathcal{I}$ of $\{1, \ldots, k + 1\}$ such that

$$G\left(\bigwedge_{i=1}^{k+1} \mathbf{x}_i\right) = G\left(\bigwedge_{i\in\mathcal{I}} \mathbf{x}_i\right). \tag{7.30}$$

Similarly, based on the monotonicity property of aggregation functions, the k-minitivity of an n-ary aggregation function G can be defined equivalently by requiring the equality,

$$G\left(\bigwedge_{i=1}^{k+1} \mathbf{x}_i\right) = \bigwedge_{i=1}^{k+1} G\left(\bigwedge_{j\neq i} \mathbf{x}_i\right) \tag{7.31}$$

for any $\mathbf{x}_1, \ldots, \mathbf{x}_{k+1} \in [0, 1]^n$, $j \in \{1, \ldots, k + 1\}$.

An aggregation function is called proper k-minitive if it is k-minitive but not $(k - 1)$-minitive. Similar to Eq. (7.29), we have for any $k + 1$ different nonempty proper subsets of $\mathcal{N}$, $\mathcal{A}_1, \ldots, \mathcal{A}_{k+1}$,

$$\left| \bigcap_{i=1}^{k+1} \mathcal{A}_i \right| \leqslant n - k - 1. \tag{7.32}$$

Hence, one can see that minitive aggregation functions are monotone extensions of minitive capacities, and Eqs. (7.30) and (7.31) just extend Eqs. (7.23) and (7.25). Further, as before, we refer to the k-minitive capacity given in Definition 7.12 as a *proper* k-minitive capacity and give the definition of *general* k-minitive capacities as follows: a capacity μ is said to be *generally* k-minitive if its necessity Möbius transform satisfies $M_n(\mathcal{A}) = 1$ for any $\mathcal{A}$ such that $|\mathcal{A}| < n - k$. That is, a capacity μ is a *proper* k-minitive capacity if and only if μ is a *general* k-minitive but not a *general* $k-1$-minitive capacity. Now, we get that if G is a k-minitive aggregation function, then $\mu(\mathcal{A}) = G(1_{\mathcal{A}})$ is a *general* k-minitive capacity.

The following theorem pertaining to the Sugeno integral (Sect. 5.3) has been proposed in [MK18].

Theorem 7.3 *The Sugeno integral w.r.t. a general k-maxitive capacity is a k-maxitive aggregation function.*

We also have the following theorem regarding k-minitive capacities.

Theorem 7.4 *The Sugeno integral w.r.t. a general k-minitive capacity is a k-minitive aggregation function.*

As might be expected, it also holds that a k-maxitive aggregation function is the dual function of a k-minitive aggregation function, that is,

$$G(\mathbf{x}) = 1 - F(1 - \mathbf{x}).$$

Theorem 7.5 *If an aggregation function is a k-maxitive (resp.k-minitive) aggregation function, then its dual function is a k-minitive (resp.k-maxitive) aggregation function.*

The following theorem for identifying a k-maxitive aggregation function is given in [MK18].

Theorem 7.6 *An aggregation function* $F : [0,1]^n \to [0,1]$ *is k-maxitive if and only if* $F(\mathbf{x}) = \bigvee_{\mathcal{B}\subseteq\mathcal{N},|\mathcal{B}|=k} F(\mathbf{x}_{\mathcal{B}})$, $\forall \mathbf{x} \in [0,1]^n$, *where* $\mathbf{x}_{\mathcal{B}} = ((x_{\mathcal{B}})_1, \ldots, (x_{\mathcal{B}})_n)$ *with* $(x_{\mathcal{B}})_i = x_i$ *if* $i \in \mathcal{B}$ *and* $(x_{\mathcal{B}})_i = 0$ *if* $i \notin B$.

By Theorems 7.5 and 7.6, we can obtain the following theorem concerning k-minitive aggregation functions.

Theorem 7.7 *An aggregation function* $G : [0,1]^n \to [0,1]$ *is k-minitive if and only if* $G(\mathbf{x}) = \bigwedge_{B\subseteq N,|\mathcal{B}|=n-k} G(\mathbf{x}_{\bar{\mathcal{B}}})$, $\forall \mathbf{x} \in [0,1]^n$, *where* $\mathbf{x}_{\bar{\mathcal{B}}} = ((x_{\bar{\mathcal{B}}})_1, \ldots, (x_{\bar{\mathcal{B}}})_n)$ *with* $(x_{\bar{\mathcal{B}}})_i = 1$ *if* $i \in \mathcal{B}$ *and* $(x_{\bar{\mathcal{B}}})_i = x_i$ *if* $i \notin \mathcal{B}$.

7.7 k-Interactivity

Recall that nonadditivity (Definition 3.10) also allows an equivalent representation of a capacity such that indices $n_\mu(\mathcal{A})$ become 0 whenever the interaction behaviour within the subset $\mathcal{A}$ is additive. A simplification based on this representation can hence be given as follows.

Definition 7.16 (k-Nonadditive Capacity) Let $k \in \{1, \ldots, n\}$. A capacity μ on $\mathcal{N}$ is said to be k-nonadditive if its nonadditivity satisfies $n_\mu(\mathcal{A}) = 0$ for all $\mathcal{A} \subseteq \mathcal{N}$ such that $|\mathcal{A}| > k$ and there exists at least one subset $\mathcal{A}$ of k elements such that $n_\mu(\mathcal{A}) \neq 0$.

It can be shown that a 1-nonadditive capacity is an additive capacity. To identify a k-nonadditive capacity μ on $\mathcal{N}$, we also only need to define $\sum_{i=1}^{k}\binom{n}{i}$ nonadditivity indices.

Another approach to reduce the complexity of capacity identification was proposed in [BW19] as a way of reducing not only the number of capacity values but also the number of monotonicity constraints. It works by fixing the values of the capacity for all subsets of cardinality greater than k in some appropriate way. The approach, which maximises the partial entropy of the capacity (calculated over subsets of cardinality greater than k), results in the following definition.

Definition 7.17 (k-Interactive Capacity) A capacity μ on $\mathcal{N}$ is called k-interactive if for some chosen $K \in [0, 1]$ and $1 \leqslant k \leqslant n$

$$\mu(\mathcal{A}) = K + \frac{|\mathcal{A}| - k - 1}{n - k - 1}(1 - K), \text{ for all } \mathcal{A}, |\mathcal{A}| > k.$$

In particular, the values $\mu(\mathcal{B}) = K$ for all $\mathcal{B}, |\mathcal{B}| = k + 1$, are fixed at K; see Fig. 7.1. In the marginal contribution representation we have that all $\Delta_i(\mathcal{A}) = \frac{1-K}{n-k-1}$ whenever $|\mathcal{A}| > k$. These conditions simplify various expressions, in particular the Choquet integral and the orness measure, and significantly reduce the number of variables and constraints when constructing capacities from data. The k-tolerant capacities arise as special cases.

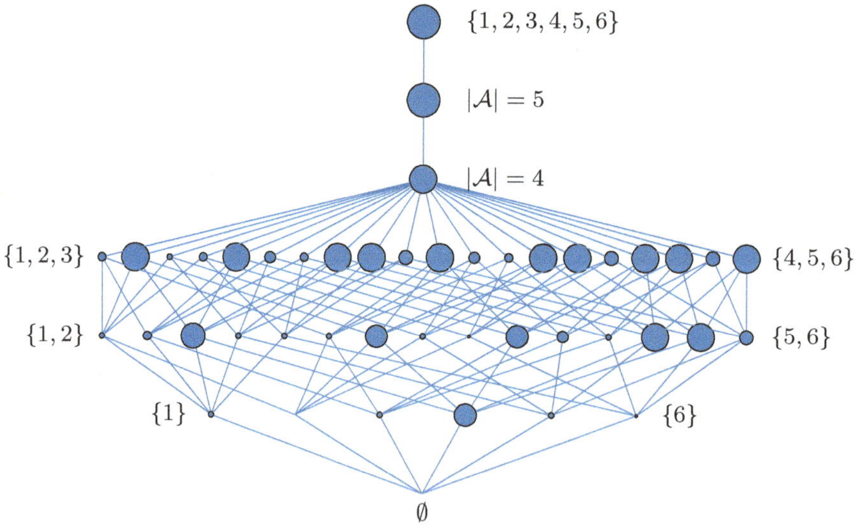

Fig. 7.1 A 3-interactive capacity for $n = 6$ and the implied partial order. The sizes of the circles reflect the values of the respective capacities. Since all $\mu(\mathcal{A})$ are the same for every fixed cardinality $|\mathcal{A}| = 4, 5, 6$, they are treated as equivalent and hence represented by single nodes

Note that the interactions in subsets larger than k still take place, but are predefined by interactions in smaller subsets and the values of k, K. Also note that the particular formula in Definition 7.17 is obtained by applying the maximum entropy principle [BW19], which maximises the average contribution of the $n-k-1$ smallest inputs.

It turns out that k-interactive capacities have an interesting representation in co-Möbius representation. Namely, the Möbius values of their *duals* are all zero for subsets of cardinalities ranging from 2 to $n-k$, provided that $n-k \geqslant 2$. Therefore, they can also be represented compactly in Möbius representation through their duals. Using the relation to co-Möbius representation, $\check{M}(\mathcal{A}) = (-1)^{|\mathcal{A}|+1} M^*(\mathcal{A})$

Theorem 7.8 *k-Interactive capacities have co-Möbius values $\check{M}(\mathcal{A}) = 0, |\mathcal{A}| = 2, \ldots, n-k$.*

The duals of k-interactive capacities instead have zero Möbius values for $|\mathcal{A}| = 2, \ldots, n-k$. A generalisation to k-lower and k-upper interactive capacities that replace the fixed subset behaviour with additivity was proposed in [BGJ22a].

Definition 7.18 A capacity is k lower interactive for $k \in \{1, 2, \ldots, n-2\}$ when for all $\mathcal{A} \subseteq \mathcal{N}$ and $1 < |\mathcal{A}| \leqslant n-k$, it holds that $M(\mathcal{A}) = 0$. The boundary cases of $k = 0$ and $k = n-1$ correspond with additive and general capacities, respectively. A capacity is k-upper interactive if it is dual to a k-lower interactive capacity.

A k-lower interactive capacity takes a weighted mean of the highest $n-k$ inputs and allows for interaction in the lowest k inputs (see Fig. 7.2). If all weights assigned to singletons are equal, we obtain the k-interactive capacities as originally defined.

Another property of the k-interactive capacities can be found by expressing $d_\mu(\mathcal{A})$ for $|\mathcal{A}| = k+1, \ldots, n$ by applying formula (3.18). For $|\mathcal{A}| = n$ we have $d_\mu(\mathcal{N}) = n\Delta - \sum_i \mu(\{i\})$, and for $\mathcal{A} = \mathcal{N} \setminus \{i\}$ we get

$$d_\mu(\mathcal{A}) = (n-1)\Delta - \sum_{j \in \mathcal{A}} \mu(\{j\}) = d_\mu(\mathcal{N}) - \Delta + \mu(\{i\}),$$

and so on. Thus, we obtain that the derivative $\Delta_i d_\mu(\mathcal{A}) = \Delta - \mu(\{i\})$ for all $|\mathcal{A}| = k+1, \ldots, n$. That is, the nonmodularity values for subsets larger than k depend only on singletons. Furthermore, the second derivative $\Delta_{ij} d_\mu(\mathcal{A}) = 0$ for $|\mathcal{A}| > k+1$.

Note 7.11 The duals of k-interactive capacities are not k-interactive (unlike k-additive capacities).

7.7.1 The k-Interactive Choquet Integral

The Choquet integral with respect to a k-interactive capacity can be written as

$$\mathcal{C}_\mu(\mathbf{x}) = \frac{1-K}{n-k-1}\sum_{i=1}^{n-k-1} x_{(i)} + Kx_{(n-k)} + \sum_{\mathcal{A}\subseteq\mathcal{N}, |\mathcal{A}|\leqslant k} \mu(\mathcal{A})g_{\mathcal{A}}(\mathbf{x}). \quad (7.33)$$

We see that the contribution of the $n-k-1$ smallest inputs is averaged with the arithmetic mean, while the interactions are explicitly accounted for in the case of the remaining inputs. Importantly, k-interactive capacities significantly reduce both the number of parameters and the monotonicity constraints and make it feasible to fit capacities to data for larger n by solving an optimisation problem detailed in the next chapter.

7.8 Capacities Resulting in Hierarchical Aggregation

The relationship between the Choquet integral and hierarchical aggregation frameworks has been of interest to researchers for some time, with Sugeno, Fujimoto, and Murofushi establishing the conditions under which a Choquet integral can be represented as a Choquet integral of smaller Choquet integrals defined over its subsets [SFM95] and Mesiar and Vivona defining the two-step integral in [MV99].

In [BGJ22a, Bel+24] it was observed that some sparse capacities in Möbius representation can be interpreted in the framework of aggregation hierarchies, where interaction only occurs within small subsets.

For example, the k-lower interactive capacities (Definition 7.17 in the previous section) have zero Möbius values for all subsets with cardinality $1 < |\mathcal{A}| \leqslant n-k$. Rather than applying the Choquet integral with respect to a capacity defined on $\mathcal{N}$, we could instead perform aggregation in two steps, taking a weighted arithmetic mean of the lowest $n-k$ inputs and a Choquet integral of the largest k inputs (with respect to a capacity defined over this smaller set), and then take a weighted average of these in the second step (see Fig. 7.2).

The hierarchical framework is particularly useful when the interacting subsets can be grouped to form a partition or covering with minimal overlap. The capacity represented in Fig. 7.3b only has nonzero Möbius values for subsets of $\mathcal{A}_1 = \{1, 2, 3\}$ and $\mathcal{A}_2 = \{4, 5\}$. This is a special type of 3-additive capacity, however with far fewer parameters, resulting in aggregation equivalent to taking Choquet integrals of the non-interacting subsets and then aggregating them with a weighted arithmetic mean (Fig. 7.3a).

Being able to represent and interpret capacities in this framework is especially useful when n is high, allowing us to work comfortably with capacities defined over 100s of inputs where it would otherwise be infeasible when it comes to practical implementation. As well as requiring fewer parameters, the number of constraints required to ensure monotonicity is also significantly reduced. Regression experiments performed on synthetic datasets in [Bel+24] showed that, for $n = 30$, partitioning the input set into random groups of 5 (requiring 186 variables) slightly outperformed 2-additive capacities (465 variables) in accuracy.

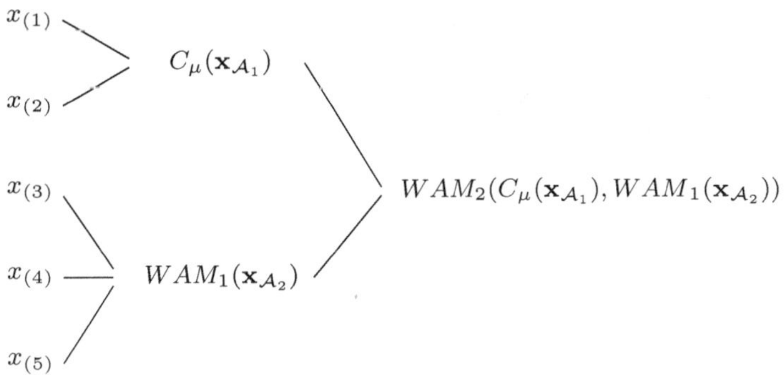

(a) Hierarchical aggregation architecture with ordered inputs

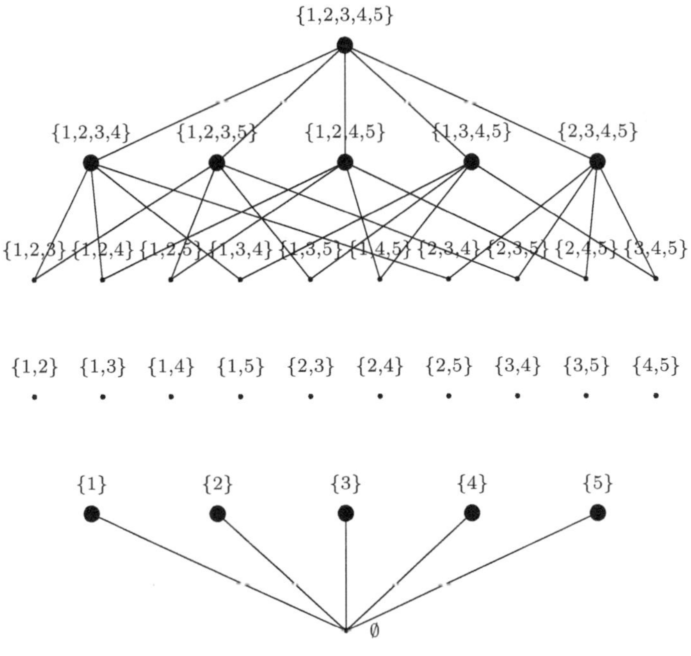

(b) Diagram showing which capacities in Möbius representation are non-zero

Fig. 7.2 Hierarchical aggregation where $x_{(1)}$ and $x_{(2)}$ are the largest inputs. The architecture in (**a**) results in aggregation equivalent to a Choquet integral with respect to the k-lower interactive capacity depicted in (**b**)

As well as being able to understand sparse capacities in terms of these hierarchical aggregation structures, the overall capacity can also be interpreted according to the value and interaction indices presented in Chap. 3. This framework hence makes great headway when it comes to addressing two of the big challenges when it comes to capacities: interpretability and tractability.

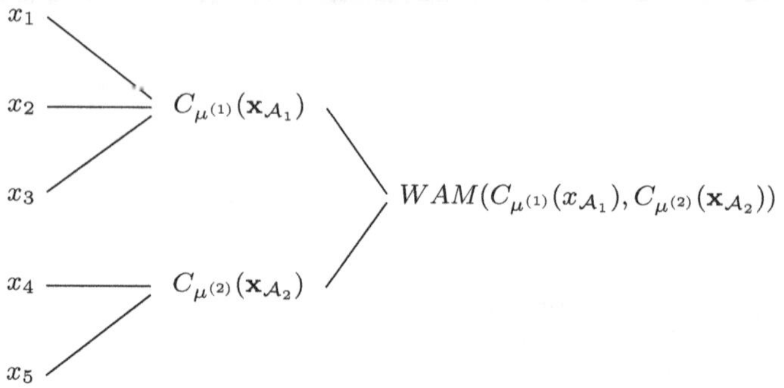

(a) Hierarchical aggregation architecture for a partition of the inputs

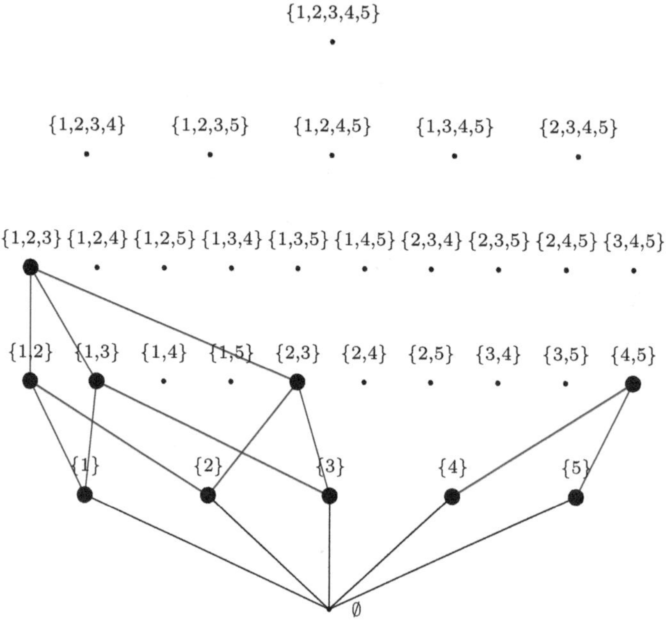

(b) Interaction diagram showing non-zero Möbius values

Fig. 7.3 Architecture and corresponding interaction diagram

7.8.1 *Determining Möbius Values of Sparse Capacities Incrementally*

In this section we provide a little algorithm to determine nonzero Möbius values from some basic assumptions about the sets of interaction (like the ones shown in Fig. 7.3b), and help the restaurant manager solve the price calculation problem formulated in Sect. 1.4.

We recall the motivating problem: We are given a set of values of singletons (menu items) $M(\{i\}) = \mu(\{i\})$, as well as some values of subsets of items $\mu(\mathcal{A}_k), k = 1, \dots, K$, the discounted tourist menu prices. We assume no interaction except in subsets $\mathcal{A}_k$, and hence the Möbius values are set to 0 elsewhere.

We need to determine the nonzero Möbius values $M(\mathcal{A}_k)$ and hence be able to calculate easily the prices of any combination of items by using the Choquet integral (in Möbius representation).

We proceed by iterating the Möbius and Zeta transforms, starting from the subsets of smallest cardinality (singletons) and moving up by increasing cardinality until we reach the largest $|\mathcal{A}_k|$. For all larger subsets the Möbius values are 0 and are not required.

The following formulas result from applying Möbius and Zeta transforms in the mentioned order.

$$M(\mathcal{A}) = \mu(\mathcal{A}) + \sum_{\mathcal{B} \subset \mathcal{A}} (-1)^{|\mathcal{A} \setminus \mathcal{B}|} \left(\sum_{\mathcal{C} \subseteq \mathcal{B}} M(\mathcal{C}) \right). \tag{7.34}$$

All the values on the right are known (and $\mathcal{C}$ runs through subsets of nonzero values $\mathcal{A}_k \subseteq \mathcal{B}$), so we are able to determine the unknown value on the left. Proceeding incrementally we end up with all required $M(\mathcal{A}_k)$.

Now the Choquet integral is calculated by

$$C(\mathbf{x}) = \sum_{i=1}^{n} \mu(i) x_i + \sum_{k=1}^{K} M(\mathcal{A}_k) \min_{i \in \mathcal{A}_k} x_i.$$

Thus, the process of eliciting a sparse capacity relies on the domain experts providing the values of singletons and some subsets where they think the value is not the sum of singletons (or calculated otherwise through the already known values of smaller subsets).

References

[BGJ22a] G. Beliakov, M. Gagolewski, S. James, Hierarchical data fusion processes involving the Möbius representation of capacities. Fuzzy Sets Syst. **433**, 1–21 (2022)

[BGJ22b] G. Beliakov, M. Gagolewski, S. James, Reduction of variables and constraints in fitting antibuoyant fuzzy measures to data using linear programming. Fuzzy Sets Syst. **451**, 266–284 (2022)

[BW19] G. Beliakov, J.-Z. Wu, Learning fuzzy measures from data: simplifications and optimisation strategies. Inf. Sci. **494**, 100–113 (2019)

[BBA25] G. Beliakov, C. Beliakov Amaya, Efficient methods for verifying monotonicity of 2-additive fuzzy measures. Comput. Appl. Math. **44**, 25 (2025)

[Bel+24] G. Beliakov, S. James, J.-Z. Wu, M. Gagolewski, A Tractable Approach to Fitting the Choquet Integral for Explainable Prediction and Analysis, in *2024 IEEE International Conference on Fuzzy Systems (FUZZ-IEEE)* (2024), pp. 1–8

[CD98] T. Calvo, B. De Baets, Aggregation operators defined by k-order additive/maxitive fuzzy measures. Int. J. Uncertainty Fuzziness Knowledge Based Syst. **6**(06), 533–550 (1998)

[Gra97] M. Grabisch, k-Order additive discrete fuzzy measures and their representation. Fuzzy Sets Syst. **92**, 167–189 (1997)

[Gra00] M. Grabisch, The interaction and Möobius representation of fuzzy measures on finite spaces, k-additive measures: A survey, in *Fuzzy Measures and Integrals. Theory and Applications*, eds. by M. Grabisch, T. Murofushi, M. Sugeno (Physica-Verlag, Heidelberg, 2000), pp. 70–93

[Gra16] M. Grabisch, *Set Functions, Games and Capacities in Decision Making* (Springer, Berlin, New York, 2016)

[Gra+09] M. Grabisch, J.-L. Marichal, R. Mesiar, E. Pap, *Aggregation Functions. Encyclopedia of Mathematics and Its Foundations* (Cambridge University Press, Cambridge, 2009)

[GMS00] M. Grabisch, T. Murofushi, M. Sugeno (eds.), *Fuzzy Measures and Integrals. Theory and Applications* (Physica-Verlag, Heidelberg, 2000)

[GNW95] M. Grabisch, H. Nguyen, E. Walker, *Fundamentals of Uncertainty Calculi, with Applications to Fuzzy Inference* (Kluwer, Dordrecht, 1995)

[Kol+18] A. Kolesárová, J. Li, R. Mesiar, k-additive aggregation functions and their characterisation. Eur. J. Oper. Res. **265**, 985–992 (2018)

[Mar04] J.-L. Marichal, Tolerant or intolerant character of interacting criteria in aggregation by the Choquet integral. Eur. J. Oper. Res. **155**, 771–791 (2004)

[Mar07] J.-L. Marichal, k-intolerant capacities and Choquet integrals. Eur. J. Oper. Res. **177**, 1453–1468 (2007)

[Mes99a] R. Mesiar, Generalizations of k-order additive discrete fuzzy measures. Fuzzy Sets Syst. **102**, 423–428 (1999)

[MV99] R. Mesiar, D. Vivona, Two-step integral with respect to fuzzy measure. Tatra Mount. Math. Publ. **16**, 359–368 (1999)

[Mes99b] R. Mesiar, k-order additive fuzzy measures. Int. J. Uncertainty Fuzziness Knowledge Based Syst. **7**(06), 561–568 (1999)

[MK18] R. Mesiar, A. Kolesärovä, k-maxitive aggregation functions. Fuzzy Sets Syst. **346**, 127–137 (2018)

[MGS20] P. Miranda, P. Garcï a-Segador, Combinatorial structure of the polytope of 2-additive measures. IEEE Trans. Fuzzy Syst. **28**, 2864–2874 (2020)

[MGG02] P. Miranda, M. Grabisch, P. Gil, p-symmetric fuzzy measures. Int. J. Uncertainty Fuzziness Knowledge Based Syst. **10**(supp01), 105–123 (2002)

[NT05] Y. Narukawa, V. Torra, Fuzzy measure and probability distributions: distorted probabilities. IEEE Trans. Fuzzy Syst. **13**, 617–629 (2005)

[Sug74] M. Sugeno, Theory of Fuzzy Integrals and Applications. PhD thesis. Tokyo Inst. of Technology, 1974

[SFM95] M. Sugeno, K. Fujimoto, T. Murofushi, A hierarchical decomposition of Choquet integral model. Int. J. Uncertainty Fuzziness Knowledge Based Syst. **3**(1), 1–15 (1995)

Chapter 8
Learning Capacities

Abstract While a decision expert may be able to manually specify the capacity values for a small input set, most real-world applications will require software and data-driven approaches. In this chapter, we focus on optimisation-based methods for learning capacities from data. The capacity data-fitting problem can be approached with respect to different metrics, resulting in either convex or non-convex optimisation problems. We will show that the sparse capacities of the previous chapter result in simplified problems that can help us address tractability issues when working with a large number of variables. Specialised software packages used to implement these methods are also presented.

8.1 Learning General Capacities

Elicitation of capacities is a complicated process because of the large number of parameters and constraints that need to be satisfied even for quite small values of n. We remind that there are $2^n - 2$ non-negative parameters to be fitted (assuming that the empty and full set values are fixed) and $n2^{n-1}$ monotonicity constraints to be enforced. The assumption of additivity greatly simplifies the learning process and reduces it to finding the weights of individual criteria, which for small n can be elicited from domain experts. However, additivity negates any interaction between the criteria, which is often a great oversimplification. The whole point of using capacities is precisely that dependencies and interaction between inputs can be incorporated into our model. This, of course, comes at the cost of computational complexity and problems when perceiving and interpreting capacity values. For example, eliciting capacity values directly from domain experts is nearly impossible, as people cannot provide so many numerical values satisfying even more constraints.

For this reason we approach the learning problem from the data-fitting perspective. We assume that we have access to data representing input-output pairs, like that shown in Table 8.1. Such data can come from different sources. Firstly, the data can be observational and constitute recorded stimuli (inputs) along with the decisions made by people in response (outputs). The goal of the fitting process here is to build

G. Beliakov et al., *Choquet Capacities and Fuzzy Integrals*, Theory and Applications of Computability, https://doi.org/10.1007/978-3-031-97070-2_8

Table 8.1 A dataset D with inputs $\mathbf{x}$ and target values y

x_1	x_2	x_3	...	x_n	y
x_1^1	x_2^1	x_3^1	...	x_n^1	y_1
x_1^2	x_2^2	...		x_n^2	y_2
$\vdots$	$\vdots$	$\ddots$		$\vdots$	$\vdots$
x_1^J	x_2^J	...	...	x_n^J	y_J

a model capable of emulating human decision-making. Alternatively, the data may relate to measurable features of objects, items, or individuals, and we aim to build a model that predicts the value of some *target* variable from the others.

Secondly, the data can constitute landmark points—the desired outputs corresponding to prototypical inputs. For example, a number of prototypical case studies can be presented to an expert in a field (or a group of experts), and their corresponding evaluations or decisions recorded. Here we assume that the dataset is rather small, and therefore additional information is also sought, like the relative importance of criteria or their interaction.

As a variation, or as part of an iterative elicitation process, prototypical inputs and outputs could be generated automatically (based on some model) and then be given to experts for their approval or rejection. This can be done explicitly, e.g. through questionnaires, or also implicitly, by recording people's decision patterns. Examples of the latter scenario are when customers are presented with a range of options (online advertisements, links in Internet search engines, suggested goods, movies, or hotels, etc.).

The data itself can be numerical (or converted into numerical values), but also ordinal, or given as pairwise preferences or class labels. All this is very typical to many machine learning problems and applications, and we treat capacity learning from that perspective. Hence, we consider a range of machine learning or data-fitting regression problems, where the aggregation model is based on capacities.

The decision variables of the learning problem are hence the values $\mu(\mathcal{A})$ for all $\mathcal{A} \subset \mathcal{N}$, and the constraints required are those that ensure monotonicity of the capacity, in addition to other suitable constraints associated with k-additivity, k-maxitivity, estimated values of importance, interaction, or other indices.

The key features of the capacity learning problem are as follows: (a) the large number of parameters representing input interactions, (b) the large number of monotonicity and other constraints, (c) semantical interpretability of the learned parameters, and (d) preservation of monotonicity in the input-output space. The two latter points in particular distinguish capacity-based models from many others—neural networks, nearest neighbour rules, and statistical multiple regression to name a few.

At first we look at fitting Choquet integral-based models in the case of numerical data and then extend our methods to ordinal and preference data. The large number of parameters and constraints are illustrated in Fig. 8.1.

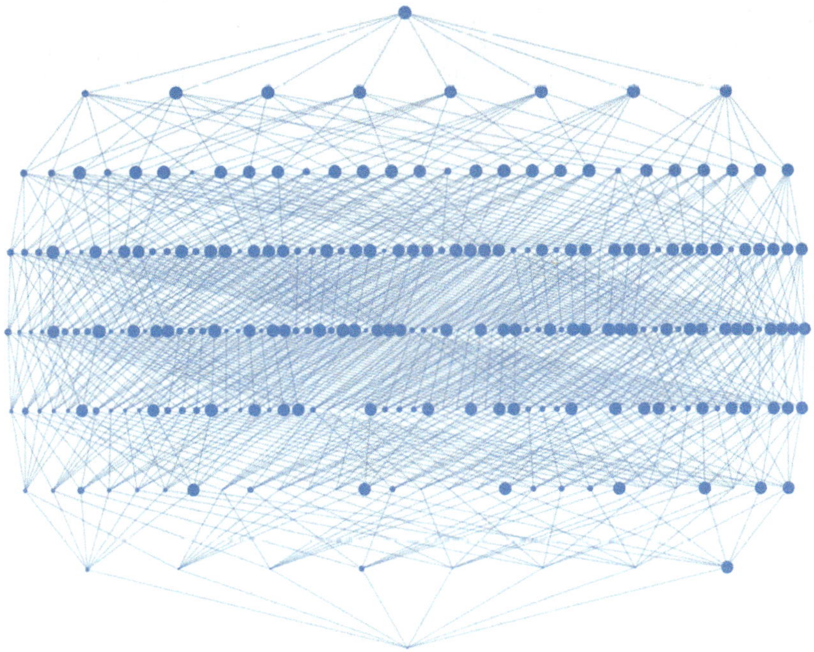

Fig. 8.1 A graphical illustration of a capacity with $n = 8$ in which the number of parameters (vertices) and constraints (edges in the graph) can be appreciated

8.1.1 Fitting in the Least Squares Sense

Among the first approaches to learning capacities based on the discrete Choquet integral is the heuristic least squares method presented in [Gra95, GKM08]. The Choquet integral fitting model is formulated as follows.

Assume there is a dataset D composed of J samples described by n attributes $x_i^j, i = 1, \dots, n, j = 1, \dots, J$ and the target values $y_j, j = 1, \dots, J$, organised into a table (see Table 8.1). The goal is to determine a capacity μ such that the Choquet integral $C_\mu(\mathbf{x}^j)$ matches the target value y_j for all the samples $j = 1, \dots, J$. This is achieved by minimising the l_2 norm of the residuals

$$\text{Minimise } ||\mathbf{r}|| = ||C_\mu(\mathbf{x}) - \mathbf{y}||_2,$$

where $C_\mu(\mathbf{x})$ in this instance indicates the vector of J predicted values and $\mathbf{y}$ is the vector of observed values. This will be subject to monotonicity constraints, which we write as

$$\mu(\mathcal{A}) \geqslant \mu(\mathcal{A} \setminus \{i\}), \forall i \in \mathcal{A} \text{ for all } A \subseteq \mathcal{N},$$

and boundary conditions $\mu(\emptyset) = 0$, $\mu(\mathcal{N}) = 1$.

Because the Choquet integral is a *linear* function of μ (i.e. in terms of the basis functions $g_{\mathcal{A}}(\mathbf{x})$ given in Eq. (5.5)), the minimisation problem is a convex quadratic programming problem, which has a unique minimum (possibly several minimisers). A key difference when compared to a typical linear regression problem is that there are multiple constraints here, which can be accommodated by the general methods of quadratic programming [BV04, GW03, Fle00, NW99].

However, even for a relatively small $n > 5$, the number of inequality constraints is too big for efficient handling by general quadratic programming methods. The heuristic least mean squares (HLMS) method by Grabisch [Gra95] is an example of a heuristic approach taking the specifics of capacities into account. This algorithm starts from the equilibrium point (the additive and symmetric capacity corresponding to the arithmetic mean function) and then iteratively adjusts capacity values by presenting it one input-output pair at a time and reducing the fitting error while enforcing monotonicity on the corresponding maximal chain and the neighbouring nodes. This way most of the capacity values at each iteration are unaffected, and numerical efficiency is achieved. While the HLMS process is not guaranteed to converge to the globally optimal solution, numerical experiments in [Gra95] and subsequent studies [GKM08] show that the optimum is only marginally different to those obtained by Lemke's standard quadratic optimisation algorithm.

The HLMS method is implemented in the *Kappalab* software package [KG05, GKM15] written in the R programming language [R D25].

8.1.2 Fitting as a Linear Programming Problem

We now look at data-fitting in the sense of least absolute deviations (LAD). Here the following objective is minimised:

$$\text{Minimise } ||\mathbf{r}||_1 = ||C_\mu(\mathbf{x}) - \mathbf{y}||_1 = \sum_{j=1}^{J} |C_\mu(\mathbf{x}^j) - y_j|,$$

subject to the monotonicity, boundary, and other constraints. There are two reasons to consider this formulation. Firstly, fitting in the LAD sense is less sensitive to potential outliers in the data, i.e. grossly erroneous or highly unusual values. Secondly, this problem is converted into a linear programming problem, and linear programming (LP) methods are very efficient in dealing with large numbers of equality and inequality constraints. Furthermore, the capacity monotonicity constraints are very sparse, and sparse matrix algebra is efficiently implemented in LP software.

Let us now formulate the fitting problem as an LP. Using the standard expression for the residuals $r_j = C_\mu(\mathbf{x}^j) - y_j$ and splitting them into positive and negative parts $r_j = r_j^+ - r_j^-$, we get $|r_j| = r_j^+ + r_j^-, r_j^+, r_j^- \geqslant 0$. Taking the following set of non-negative decision variables, $r_j^+, r_j^-, j = 1, \dots, J$, $\mu(\mathcal{A})$, $\mathcal{A} \subseteq \mathcal{N}$, $\mathcal{A} \neq \emptyset$, we have an equivalent linear programming problem:

$$\text{minimise} \quad \sum_{j=1}^{J} r_j^+ + r_j^-, \tag{8.1}$$

$$\text{s.t.} \quad r_j^+ - r_j^- = \sum_{\mathcal{A} \subseteq \mathcal{N}} \mu(\mathcal{A}) g_{\mathcal{A}}(\mathbf{x}^j) - y_j, \; j = 1, \dots J$$

$$\mu(\mathcal{A}) \geqslant \mu(\mathcal{A} \setminus \{i\}), \forall i \in \mathcal{A} \text{ for all } \mathcal{A} \subseteq \mathcal{N}$$

$$\mu(\mathcal{N}) = 1,$$

where the decision variables μ are organised into a vector with the relevant binary- or cardinality-based numbering system.

This method is implemented in the FMTools software [VBL14, Bel07] available for R [Bel24] and Python and as a C/C++ package [Bel07].

8.1.3 Other Constraints on Capacities

There are several other types of constraints that can be specified in the least squares or linear programming formulations. The first type is related to predefined (or desired) Shapley values or different interaction indices. Using the fact that the Shapley (or Banzhaf) values and interaction, nonadditivity, nonmodularity, and other indices are linear combinations of capacity values, their desired values, or even relations between them, are written as linear equality or inequality constraints.

Consider the following additional requirements on the capacity:

- The desired Shapley value for the criterion i is in $[0.1, 0.2]$.
- The criterion i is more important than j.
- The interaction index of criteria i, j is in the interval $[0.2, 0.3]$.
- The nonmodularity index of the set $\mathcal{S}$ is larger than that of $\mathcal{T}$.

These requirements are written in the form of linear constraints as follows:

- $0.1 \leqslant \varphi(i) \leqslant 0.2$.
- $\varphi(i) \geqslant \varphi(j)$.
- $0.2 \leqslant I_{ij} \leqslant 0.3$.
- $d_\mu(\mathcal{S}) \geqslant d_\mu(\mathcal{T})$.

The desired orness value of the Choquet integral can also be incorporated as a linear constraint.

Another requirement one may wish to impose is superadditivity (or subadditivity) of the capacity. This also involves a set of linear constraints to be added along with the other inequalities; however, these involve a significantly higher number of constraints. The sub-/supermodularity conditions given in Sect. 2.7 are simplified by removing many redundant constraints in [CJ89b] (see also [BJL11]). For submodularity we have, in Möbius representation,

$$\sum_{\mathcal{C} \subseteq \mathcal{B} \subseteq \mathcal{A},} M(\mathcal{B}) \leqslant 0, \quad \forall \mathcal{A} \text{ and } \forall \mathcal{C} \subseteq \mathcal{A} \subseteq \mathcal{N} \text{ such that } |\mathcal{C}| = 2. \tag{8.2}$$

Supermodularity is dealt with by using duality. Here we only require combinations of every subset $\mathcal{A}$, $|\mathcal{A}| \geqslant 2$ with each of its subset pairs, $|\mathcal{C}| = 2$. There are $\frac{n!}{k!(n-k)!}$ subsets $\mathcal{A}$ of size k, and for each such subset there are $\frac{k(k-1)}{2}$ pairs. This gives a total of

$$\sum_{k=2}^{n} \frac{n!k(k-1)}{2 \cdot k!(n-k)!} = \frac{1}{8} 2^n n(n-1)$$

constraints.

Note 8.1 The constraints on the indices can be specified in the standard, Möbius, or another representation of the capacity, and in fact the optimisation problem can be set in the standard or another representation based on convenience. We have specified the conversion matrices in Chap. 4, so that the conversion is achieved by matrix multiplication. Of course, for numerical efficiency purposes it is worth stating the constraints in terms of the optimisation problem variables directly.

Another type of constraint accommodates the relative ordering of the inputs. While the LAD criterion ensures that the predicted outputs are close to the observed, there is no provision for order preservation. An additional set of requirements can be added to ensure that if two observed outputs satisfy $y_j \leqslant y_k$, then the predicted values should also preserve that ordering $C_\mu(\mathbf{x}^j) \leqslant C_\mu(\mathbf{x}^k)$. Such conditions, however, could be specified for observations that would contradict the monotonicity of the Choquet integral, i.e. when $\mathbf{x}^j > \mathbf{x}^k$ (e.g. the correct ordering can be broken by the errors in the abscissas $\mathbf{x}$ or observed values y), in which case it need not be enforced. Again, the order preservation conditions are specified as additional inequality constraints, or as extra terms in the objective function.

8.2 Nonadditive Ordinal Regression

The nonadditive robust ordinal regression (NAROR) model has been developed in [AGM10, Cor+13, Ang+16, CGI16] whereby capacities are constructed with respect to preferences over the dataset as well as over the Shapley importance and interaction indices.

8.2.1 Problem Formulation

In summary, the NAROR model has the following three steps:

Step 1 Elicit preference information characterising all compatible capacities. The preferences are usually expressed by two types of linear constraints. The first type reflects preferences on some, not necessary all, decision alternatives. For example:

- The alternative **a** is at least as good as **b**: $\mathbf{a} \succsim \mathbf{b} \Leftrightarrow C_\mu(\mathbf{a}) \geqslant C_\mu(\mathbf{b})$.
- **a** is preferred to **b** at least as much as **c** is preferred to **d**: $(\mathbf{a}, \mathbf{b}) \succsim (\mathbf{c}, \mathbf{d}) \Leftrightarrow C_\mu(\mathbf{a}) - C_\mu(\mathbf{b}) \geqslant C_\mu(\mathbf{c}) - C_\mu(\mathbf{d})$.

The second type reflects the preferences in the decision criteria space, mainly the pairwise comparison of the importance of some decision criteria as well as the interactions between two criteria. Examples of this type include:

- Criterion i is at least as important as criterion j: $i \succsim j \Leftrightarrow \varphi(i) \geqslant \varphi(j)$.
- The difference of importance between criteria i and j is at least as big as the difference of importance between criteria k and l: $(i, j) \succsim (k, l) \Leftrightarrow \varphi(i) - \varphi(j) \geqslant \varphi(k) - \varphi(l)$.
- The sign of interaction of pairs of criteria is positive or negative: $[i, j] \geqslant (\leqslant)0 \Leftrightarrow I_{ij} \geqslant (\leqslant)0$.
- The interaction intensity between criteria i and j is at least as strong as the interaction intensity between criteria k and l: $|[i, j]| \succsim |[k, l]| \Leftrightarrow |I_{ij}| \geqslant |I_{kl}|$.
- The difference of interaction intensity between criteria i and j and interaction intensity between criteria k and l is at least as strong as the difference of interaction intensity between criteria r and s and interaction intensity between criteria t and w: $(|[i, j]|, |[k, l]|) \succsim (|[r, s]|, |[t, w]|) \Leftrightarrow |I_{ij}| - |I_{kl}| \geqslant |I_{rs}| - |I_{tw}|$.

The constraints involving the absolute value can be translated into normal linear constraints if the signs of interactions are given. Further, in practice we may encounter the indifference "$\sim$" and preference "$\succ$" relationships which will correspond to "=" and "$>$" in the above constraints. These preference constraints, combined with the boundary and monotonicity constraints, constitute the feasible domain of all compatible capacities.

In addition, we can also use the nonadditivity indices in the same spirit as the Shapley values above.

Step 2 Check for preference inconsistency and adjust if needed. The constraints in Step 1 can be collected into the constraint set E^{AC}. As mentioned above, there are three types of constraints: those with respect to equality, weak inequality, and strict inequality. By introducing an auxiliary variable ε, strict inequalities can be converted to weak inequalities. For example, we can write the constraints as:

- $C_\mu(\mathbf{a}) = C_\mu(\mathbf{b})$ if $\mathbf{a} \sim \mathbf{b}$.
- $C_\mu(\mathbf{a}) \geqslant C_\mu(\mathbf{b})$ if $\mathbf{a} \succsim \mathbf{b}$.
- $C_\mu(\mathbf{a}) \geqslant C_\mu(\mathbf{b}) + \varepsilon$ if $\mathbf{a} \succ \mathbf{b}$.

For convenience, we will only use weak inequalities, so that E^{AC} is represented as follows:

$$E^{AC} \Leftarrow \begin{cases} C_\mu(\mathbf{a}) \geqslant C_\mu(\mathbf{b}) + \varepsilon \\ C_\mu(\mathbf{a}) - C_\mu(\mathbf{b}) \geqslant C_\mu(\mathbf{c}) - C_\mu(\mathbf{d}) + \varepsilon, \\ \varphi(i) \geqslant \varphi(j) + \varepsilon, \\ \varphi(i) - \varphi(j) \geqslant \varphi(k) - \varphi(l) + \varepsilon, \\ I_{ij} \geqslant \varepsilon \quad (I_{ij} \leqslant -\varepsilon), \\ |I_{ij}| \geqslant |I_{kl}| + \varepsilon, \\ |I_{ij}| - |I_{kl}| \geqslant |I_{rs}| - |I_{tw}| + \varepsilon, \\ \mu(\emptyset) = 0, \, \mu(\mathcal{N}) = 1, \\ \mu(\mathcal{A}) \leqslant \mu(\mathcal{B}), \forall \mathcal{A}, \mathcal{B} \subseteq \mathcal{N}, \mathcal{A} \subseteq \mathcal{B}, \end{cases}$$

where ε is an auxiliary variable.

If E^{AC} is feasible and $\varepsilon^* = \max \varepsilon > 0$ subject to E^{AC}, then there exists at least one compatible capacity. Otherwise, one should check for inconsistency and adjust the constraints by using some appropriate techniques, among which the most common is the 0-1 linear programming-based method [AGM10, Ang+16].

Step 3 Exploit the preferences on all the decision alternatives, by obtaining sets of necessary and possible preferences from the following two types of linear program.

The *necessary* preference of alternative pair **x**, **y** (over which preference was not expressed in Step 1) can be confirmed by the following linear programming problem if it has a nonpositive optimal value:

$$\begin{aligned} \text{maximise} \quad & \varepsilon \\ \text{s.t.} \quad & C_\mu(\mathbf{y}) \geqslant C_\mu(\mathbf{x}) + \varepsilon, \\ & E^{AC}. \end{aligned} \tag{8.3}$$

If the programming problem (8.3) has an optimal nonpositive value with $\varepsilon \leqslant 0$, then $C_\mu(\mathbf{x}) \geqslant C_\mu(\mathbf{y})$ for all compatible capacities.

In a similar fashion, the *possible* preference of an alternative pair **x**, **y** (not given in Step 1) can be confirmed by the following linear programming program if it has a positive optimal value:

$$\begin{aligned} \text{maximise} \quad & \varepsilon \\ \text{s.t.} \quad & C_\mu(\mathbf{x}) \geqslant C_\mu(\mathbf{y}), \\ & E^{AC}. \end{aligned} \tag{8.4}$$

If the program (8.4) has an optimal positive value with $\varepsilon > 0$, then $C_\mu(\mathbf{x}) \geqslant C_\mu(\mathbf{y})$ for at least one compatible capacity.

We highlight that in contrast to usual regression problems, in NAROR the actual values of the Choquet integral to be fitted are not provided. This implies that the NAROR method does not identify one particular capacity that fits best the decision-maker's preferences, but rather a subset of feasible capacities. Our goal here is to formulate an additional optimisation criterion to determine in some sense the "best" feasible capacity.

The decision-maker's preferences in NAROR are expressed in the form of the constraints E^{AC}, which can be converted into a set of linear constraints by representing the modulus function with the help of two auxiliary non-negative variables d_r^+, d_r^-, so that the rth variable $a_r = d_r^+ - d_r^-$ and $|a_r| = d_r^+ + d_r^-$. Hence, we assume that all the constraints in E^{AC} have been converted into an equivalent set of linear constraints which we also denote by E^{AC}.

The set of constraints E^{AC} provides a feasible set D of all capacities consistent with the decision-maker's preferences, which can be empty in the case of inconsistent preferences. We discuss the latter case in a subsequent section, and at the moment focus on nonempty D. Different capacities from D can produce different orderings of the available alternatives, and it is a significant challenge to determine some particular capacity from D which is optimal in some sense.

Moreover, typically the set of constraints E^{AC} is rather small compared to the total number of parameters which identify a capacity even for moderate n, because the decision-makers provide only a few preferences or alternative rankings.

The principle of maximum entropy can be used as an additional criterion to select the optimal capacity consistent with E^{AC}. The rationale is that maximum entropy capacities tend to take into account as many criteria as possible and weight them rather uniformly. In the absence of the decision-maker's preferences, the principle of maximum entropy results in the additive symmetric capacity, for which the Choquet integral is the arithmetic mean function. When the preferences are specified, we expect the optimal compatible capacity to be the one from D closest to the additive symmetric capacity.

Thus the maximum entropy capacity identification problem is formulated as follows:

$$\begin{aligned} \text{maximise} \quad & H(\mu) \\ \text{s.t.} \quad & \mathbf{A}\mu \preceq \mathbf{b} \\ & \mathbf{C}\mu = \mathbf{d}, \end{aligned} \tag{8.5}$$

where the set of constraints E^{AC} is written as systems of inequality and equality constraints.

Recall the expression for $H(\mu)$ as given in Definition 3.24:

$$H(\mu) = \sum_{i \in \mathcal{N}} \sum_{\mathcal{A} \subseteq \mathcal{N} \setminus \{i\}} \frac{(n - |\mathcal{A}| - 1)!|\mathcal{A}|!}{n!} h(\mu(\mathcal{A} \cup \{i\}) - \mu(\mathcal{A})), \tag{8.6}$$

with $h(t) = -t \log t$, if $t > 0$ and $h(0) = 0$.

We notice that this does not involve functions $h(\mu(\mathcal{A}))$ but rather h of the differences between the values of μ. Therefore, to convert problem (8.5) into the standard entropy maximisation problem, we use the marginal contribution representation of capacities (4.22) in which the decision variables are $\Delta_i(\mathcal{A}) = \mu(\mathcal{A} \cup \{i\}) - \mu(\mathcal{A})$:

$$\begin{aligned} \text{maximise } & H(\Delta) \\ \text{s.t. } & \mathbf{A}\Delta \preceq \mathbf{b} \\ & \mathbf{C}\Delta = \mathbf{d}, \end{aligned} \tag{8.7}$$

where the constraints are rewritten in terms of the variables Δ (we shall use the same letters to denote the matrices and the right-hand sides).

The set of equality constraints stems from Eq. (4.23), whereas the inequality constraints involve simple non-negativity of the variables (monotonicity of the capacity) and the constraints in E^{AC} excluding monotonicity. The Choquet integral which appears in E^{AC} can be written in terms of the marginal contributions as

$$C_\mu(\mathbf{x}) = \sum_{i=1}^{n} x_{\pi(i)} \Delta_{\pi(i)} \mu(H_{\pi(i+1)}), \tag{8.8}$$

where $x_{\pi(.)}$ is a non-decreasing permutation induced by $x_i, i = 1, \ldots, n$, i.e. $x_{\pi(1)} \leqslant \ldots \leqslant x_{\pi(n)}$, and $H_{\pi(i)} = \{\pi(i), \ldots, \pi(n)\}$, $H_{\pi(n+1)} = \emptyset$, by convention.

Once the objective and the constraints have been set, the numerical solution to problem (8.7) can be found by methods of convex optimisation, found in the package CVX [GB17], which is available as a Python, Matlab, or R library.

8.2.2 *Sparse Matrices and Large n: Maximising Min-Entropy*

While maximisation of the Shannon entropy (8.7) is achievable through convex optimisation, for sufficiently large n (of the order of 20) it becomes numerically expensive to do so because of the sheer number of non-negativity and equality constraints. In this section we consider a different type of entropy, the Renyi entropy H_∞, also called the min-entropy. Therefore the objective is to maximise the Renyi H_∞ entropy from Definition 3.25,

$$\max H_\infty(\Delta) = -\log(\max(\Delta)),$$

which translates (using monotonicity of the log function) into minimising

$$\min \max(\Delta),$$

subject to the same set of linear constraints as before.

This optimisation problem can be solved by methods of linear programming, namely, by minimising

$$\begin{aligned}
&\text{minimise } t \\
&\quad \text{s.t. } t \geqslant \Delta_i, \text{ for all } i \\
&\quad \mathbf{A}\Delta \preceq \mathbf{b} \\
&\quad \mathbf{C}\Delta = \mathbf{d}.
\end{aligned} \tag{8.9}$$

The numerical advantages include the significant numerical efficiency of linear programming and the widespread ability to handle sparse matrices of constraints using LP software. The automatic incorporation of non negativity of the variables (which reflects monotonicity of the capacity) is also a significant bonus. Indeed, the matrices of constraints are very sparse: for example, there are at most n nonzero entries per row in the matrix $\mathbf{C}$ even without k-interactivity, which also helps in simplifying the problem.

One can also switch back to the standard capacity representation in this case and write the problem as

$$\begin{aligned}
&\text{minimise } t \\
&\quad \text{s.t. } t \geqslant \mu(\mathcal{A} \cup \{i\}) - \mu(\mathcal{A}), \text{ for all } \mathcal{A} \subset \mathcal{N} \text{ and } i \notin \mathcal{A} \\
&\quad \mu(\mathcal{A} \cup \{i\}) - \mu(\mathcal{A}) \geqslant 0 \\
&\quad \mathbf{A}\mu \preceq \mathbf{b} \\
&\quad \mathbf{C}\mu = \mathbf{d}.
\end{aligned} \tag{8.10}$$

This significantly reduces the number of variables and constraints and allows one to take full advantage of the sparsity of the matrices of constraints.

8.2.3 Preference Inconsistency

If the preference set E^{AC} is infeasible and $\varepsilon^* \leqslant 0$, where $\varepsilon^* = \max \varepsilon$ subject to E^{AC}, then we need to check and adjust the preferences accordingly. Different to the 0-1 linear programming-based method, we give a multiple goal linear programming-based consistency check method, which can help decision-makers to remove or adjust the inconsistent constraints according to its deviation degree, reduce the redundant constraints to a certain extent, and even provide some suggestions for adjusting the inconsistent and redundant constraints.

First, positive and negative deviation variables are introduced, d_r^+ and d_r^-, where $r = 1, ..., p$, with p the total number of preference constraints in E^{AC}, which are then associated with each preference constraint, for example, the constraints $C_\mu(\mathbf{a}) \geqslant C_\mu(\mathbf{b}) + \varepsilon$ become $C_\mu(\mathbf{a}) - C_\mu(\mathbf{b}) - d_r^+ + d_r^- = 0$. We denote the goal linear preference constraints and the boundary and monotonicity constraints of capacity by E_G^{AC}.

The following multiple goal linear programming problem is then constructed and solved:

$$\begin{aligned} &\text{minimise} \sum_{r=1}^{p} d_r^+ + d_r^- \\ &\text{s.t.} \quad E_G^{AC}, \end{aligned} \tag{8.11}$$

where $d_r^+, d_r^- \geqslant 0$ and $d_r^+ \times d_r^- = 0$. We can find one subset of consistent constraints with both optimal deviation variables equal to zero, $d_r^{+*} = d_r^{-*} = 0$, and another subset of inconsistent (not necessarily contradictory—they may also be redundant) constraints with optimal positive or negative deviation not zero, $d_r^{+*} = 0$ or $d_r^{-*} = 0$ and $d_r^{+*} \neq d_r^{-*}$.

Here, one can see that $\max\{d_r^{+*}, d_r^{-*}\}$ serves as an inconsistency degree of each preference constraint. More specifically, the value d_r^{+*} can be interpreted as the contradiction degree and d_r^{-*} can be regarded as the redundancy degree of the rth preference constraint, where the original preference constraints in E^{AC} are "equal or larger than", $\geqslant$, type inequalities. On the other hand, if the original inequality is a $\leqslant$ type inequality, then d_r^{+*} and d_r^{-*} are regarded as the redundancy degree and contradiction degree, respectively. We refer to the preference constraints with redundancy degree (resp. contradiction degree) nonzero as the redundant (resp. contradictory) constraints.

The infeasibility of E^{AC} is caused by the existence of contradictory constraints. There are two strategies to deal with these. One strategy is to iteratively remove the contradictory constraints from E^{AC} whose corresponding goal constraint is the largest until there are no contradictory constraints left. Another strategy is to adjust all the contradictory constraints in E^{AC} by minus d_r^{+*} (resp. plus d_r^{-*}) on the left-hand side if they are $\geqslant$ (resp. $\leqslant$) type inequalities. As for the redundant constraints, it is better that they be kept rather than removed because these constraints are helpful for keeping the diversity of feasible capacities. Of course, the above adjustment strategies should be carried out with the close cooperation with the field analysts and decision-makers.

8.3 Learning k-Additive Capacities

Most of the existing works on capacity identification are related to acquiring the k-additive capacities, especially the most popular family, the 2-additive capacities [GKM08, WPS18, WB18]. The monotonicity and boundary conditions of ordinary capacities in terms of Möbius representation can be found in Eqs. (1.10) and (4.1). In the case of learning k-additive capacities, the variables, i.e. the Möbius representation or Shapley simultaneous interaction indices, with cardinality greater than k will be omitted.

For the decision variables we use $M(\mathcal{A})$ such that $|\mathcal{A}| \leqslant k$ in the cardinality ordering indexing system (p. 88). This is a much reduced set of variables ($\sum_{i=1}^{k} \binom{n}{i} - 1$ compared to $2^n - 2$). In this framework, monotonicity of the capacity based on Eq. (4.1), together with k-additivity, is expressed:

$$\sum_{\mathcal{B} \subseteq \mathcal{A}, i \in \mathcal{B}, |\mathcal{B}| \leqslant k} M(\mathcal{B}) \geqslant 0, \quad \text{for all } \mathcal{A} \subseteq \mathcal{N} \text{ and all } i \in \mathcal{A}.$$

The (non-redundant) set of non-negativity constraints $\mu(\{i\}) \geqslant 0, i = 1, \ldots, n$, is a special case of the previous formula when $\mathcal{A}$ is a singleton, which simply becomes

$$\sum_{\mathcal{B}=\{i\}} M(\mathcal{B}) = M(\{i\}) \geqslant 0, \; i = 1, \ldots, n.$$

Finally, the condition $\mu(\mathcal{N}) = 1$ is just $\sum_{\mathcal{B} \subseteq \mathcal{N}, |\mathcal{B}| \leqslant k} M(\mathcal{B}) = 1$.

Summarising, we have the simplified problem:

$$\begin{aligned}
\text{minimise} \quad & \sum_{j=1}^{J} \left| \sum_{\mathcal{A}, |\mathcal{A}| \leqslant k} h_{\mathcal{A}}(\mathbf{x}^j) M(\mathcal{A}) - y_j \right| \qquad (8.12) \\
\text{s.t.} \quad & \sum_{\mathcal{B} \subseteq \mathcal{A}, i \in \mathcal{B}, |\mathcal{B}| \leqslant k} M(\mathcal{B}) \geqslant 0 \\
& \qquad \text{for all } \mathcal{A} \subseteq \mathcal{N}, \; |\mathcal{A}| > 1, \; \text{and all } i \in \mathcal{A} \\
& M(\{i\}) \geqslant 0, \; i = 1, \ldots, n \\
& \sum_{\mathcal{B} \subseteq \mathcal{N}, |\mathcal{B}| \leqslant k} M(\mathcal{B}) = 1,
\end{aligned}$$

where $h_{\mathcal{A}}(\mathbf{x}) = \min_{i \in \mathcal{A}} x_i$. Note that only the specified $M(\mathcal{B})$ are non-negative; the others are unrestricted.

The case of 2-additive capacities is frequently used, as it allows further reduction of the number of constraints. The monotonicity and boundary conditions can be written [BBA25], denoting singletons and pairs by $m_i = M(\{i\}), m_{ij} = M(\{i, j\})$ as follows. A 2-additive set function is a capacity if and only if for all $i, j \in N$:

$$0 \leqslant a_i = m_i + \sum_{j \neq i | m_{ij} < 0} m_{ij} \leqslant 1, \tag{8.13}$$

$$0 \leqslant m_i \leqslant 1, \quad -1 \leqslant m_{ij} \leqslant 1. \tag{8.14}$$

and

$$\sum_i m_i + \sum_{j \neq i} m_{ij} = 1. \tag{8.15}$$

In the Shapley index representation, using $I_i = I(\{i\}) = \varphi(i)$ and $I_{\mathcal{A}} = I(\mathcal{A})$ for simplicity, we have [Wu+15]:

$$I_i - \frac{1}{2} \sum_{j \in \mathcal{N} \setminus \mathcal{A}} I_{ij} + \frac{1}{2} \sum_{j \in \mathcal{A} \setminus i} I_{ij} \geqslant 0, \ \forall \mathcal{A} \subseteq \mathcal{N}, \forall i \in \mathcal{A} \tag{8.16}$$

$$I_\emptyset = \frac{1}{2} \sum_{i \in \mathcal{N}} I_i - \frac{1}{6} \sum_{\{i,j\} \in \mathcal{N}} I_{ij}, \quad \sum_{i=1}^{n} I_i = 1$$

$$I_{\mathcal{A}} = 0, \ \forall \mathcal{A} \subseteq \mathcal{N}, |\mathcal{A}| > 2.$$

The decision-maker's explicit preferences regarding the multiple decision criteria [GKM08, Wu+14, MGL11, KM07, MR00, Rou96] can be provided in the form of comparisons, e.g. criterion i is more important than criterion j, the interaction between criteria i and j is greater than that between criteria k and l, and so on. Generally, this explicit preference information about the inputs constitutes a feasible region in the set of capacities, and some additional selection principle should be applied to identify the most satisfactory one(s) [Wu+14]. Most of the additional selection principles depend on a learning set [GKM08], which can be given in terms of the desired overall evaluations of all the alternatives or by a weak partial order.

In addition to the linear programming formulation (8.12), there are least squares-based approaches [GMS00], maximum split approaches [Mar00, MR00], TOMASO (tool for ordinal multi-attribute sorting and ordering) approaches [MR05], the nonadditive robust ordinal regression model [Ang+16, AGM10, GMS14], the maximum log-likelihood principle-based optimisation model [Teh+12], and so on. The major obstacle for this type of identification method in practical application is that the construction of the learning set is indeed a very time-consuming process [GKM08].

Other selection principles that can be incorporated into the learning process include the maximum entropy principle [KM07, KMR05, Mar02, WZ10], the compromise principle [Wu+14], the least squares and absolute deviation principle [Wu+15], and the interaction-oriented principle [WPS18].

8.4 Learning k-Maxitive and k-Minitive Capacities

While much of the attention on learning simplified capacities has focused on k-additive capacities, there is also the potential to reduce the complexity of the fitting problem via other k-order assumptions. Recall that k-maxitive capacities allow a reduction in the number of defining values required since for all subsets with cardinality $|\mathcal{A}| > k$ it holds that $\mu(\mathcal{A}) = \max_{\mathcal{B} \subset \mathcal{A}} \mu(\mathcal{B})$. Here we will look at some recently proposed methods for their identification.

8.4.1 *Fitting in the Euclidean Norm*

We now consider the fitting problem

$$\text{Minimise } ||\mathbf{r}|| = ||C_\mu(\mathbf{x}) - y||,$$

in the l_2 or l_1 norm, subject to k-maxitivity of the capacity.

A new heuristic algorithm for learning k-maxitive capacities from data called *k-HLMS* is presented in [MGB17].[1] This algorithm uses the Euclidean norm of the residuals and a reduced set of variables $\mu(\mathcal{A})$, $|\mathcal{A}| \leqslant k$. The values of $\mu(\mathcal{A})$ for larger subsets are computed as

$$\mu(\mathcal{A}) = \max_{\mathcal{B} \subset \mathcal{A}} \mu(\mathcal{B}), \ |\mathcal{A}| > k.$$

The value $\mu(\mathcal{N})$ is set to 1.

The variables are initialised at the equilibrium state, $|\mathcal{A}|/n$, and then iteratively updated based on the learning rate parameter and the difference between the computed and predicted values of the Choquet integral for a particular sample. Monotonicity of the capacity is enforced at every iteration. To avoid bias related to data presentation order, the sample order is randomised at every iteration.

This heuristic algorithm *k-HLMS* does not ensure that for at least one $\mathcal{A}$ with cardinality k, $\mu(\mathcal{A}) = 1$, which is necessary for k-maxitive capacities. So strictly speaking, it does not always produce a k-maxitive measure (but in many cases it does, as confirmed in experiments). However, since the main aim of using k-maxitivity here is to simplify the capacity, the substitute constraint $\mu(\mathcal{N}) = 1$

[1] Available at https://github.com/murillojavieriv/khlms

results in a quasi- k-maxitive capacity, which would be suitable enough for most applications. As with the majority of heuristic algorithms, there is no guarantee that the result is actually the best (k-maxitive) capacity that fits the data, because the algorithm can become stuck in a local but not global optimum. This can be confirmed numerically by noticing that different runs of the *k-HLMS* algorithm with the same input data results in different (although similar) output solutions. Results of some experiments show that this might only happen when the number of data is relatively small (under 100 samples); for larger datasets the effect is less pronounced.

8.4.2 Mixed Integer Programming Formulation

Let us now formulate the fitting problem for k-maxitive capacities as a standard optimisation problem with linear constraints. We favour the proposed formulation for two reasons: (a) as mentioned previously, minimisation of the sum of absolute residuals is less sensitive to outliers in the data compared to the least squares fit approach, and (b) the resulting optimisation program is more efficient to solve by standard linear and mixed integer programming algorithms than the corresponding quadratic optimisation problem (when minimising the sum of squared residuals) [Bel09]. As becomes apparent, strict adherence to k-maxitivity comes at a higher computational cost. We hence show how that cost can be reduced by using a simple relaxation technique and the notion of k-tolerance from [Mar04, Mar07].

To reduce the number of variables, we shall work in the possibilistic Möbius representation of capacities (Definition 7.10), i.e. our set of (non-negative) parameters to fit consists of all $M_p(\mathcal{A})$, $\mathcal{A} \subseteq \mathcal{N}$, $|\mathcal{A}| \leqslant k$. These parameters are organised into a vector of decision variables (which we also denote M_p) in cardinality ordering.

Now, we need to express the Choquet integral in terms of M_p. We shall use the representation of the input vector using basis functions $g_{\mathcal{A}}(\mathbf{x})$ (Eq. (5.6)). Then with Eq. (7.14), we have

$$C_\mu(\mathbf{x}) = \sum_{\mathcal{A}\subseteq\mathcal{N}} \max_{\mathcal{B}\subseteq\mathcal{A}} M_p(\mathcal{B}) g_{\mathcal{A}}(\mathbf{x}) = \sum_{\mathcal{A}\subseteq\mathcal{N}} z(\mathcal{A}) g_{\mathcal{A}}(\mathbf{x}), \tag{8.17}$$

where the variables $z(\mathcal{A}) = \max_{\mathcal{B}\subseteq\mathcal{A}} M_p(\mathcal{B})$ satisfy the constraints,

$$z(\mathcal{A}) \geqslant M_p(\mathcal{B}), \forall \mathcal{B} \subseteq \mathcal{A} \text{ for all } \mathcal{A} \subseteq \mathcal{N} \text{ and } |\mathcal{B}| \leqslant k.$$

In addition, it is required that if $|\mathcal{A}| > k$ then $z(\mathcal{A}) = z(\mathcal{B})$ for some $\mathcal{B} \subset \mathcal{A}$ with $|\mathcal{B}| = k$ to comply with k-maxitivity. This last condition presents a challenge because the actual subset $\mathcal{B}$ is unknown.

To eliminate some redundant constraints we transform them into an equivalent system

$$z(\mathcal{A}) \geqslant z(\mathcal{A} \setminus \{i\}), \forall i \in \mathcal{A},$$

$$\text{for all } \mathcal{A} \subseteq \mathcal{N}, |\mathcal{A}| > k, \exists i \in \mathcal{A} \text{ such that } z(\mathcal{A}) = z(\mathcal{A} \setminus \{i\}),$$

$$\text{and } z(\mathcal{A}) \geqslant M_p(\mathcal{A}) \text{ for } |\mathcal{A}| \leqslant k.$$

We also require $\max_{\mathcal{A} \subseteq \mathcal{N}} M_p(\mathcal{A}) = 1$.

Let us now formulate the fitting problem. We look for the values M_p that minimise the least absolute deviation criterion,

$$\text{Minimise } \sum_{j=1}^{J} |C_\mu(\mathbf{x}^j) - y_j|,$$

subject to μ being a capacity. We also identify the decision variables $z(\mathcal{A}) = \mu(\mathcal{A})$ and eliminate the variables $M_p(\mathcal{A})$ from the problem.

Using the standard expression for the residuals $r_j = C_\mu(\mathbf{x}^j) - y_j$ and splitting them into positive and negative parts $r_j = r_j^+ - r_j^-$, we get $|r_j| = r_j^+ + r_j^-$, $r_j^+, r_j^- \geqslant 0$. Taking the following set of non-negative decision variables, $r_j^+, r_j^-, j = 1, \ldots, J, \mu(\mathcal{A}), \mathcal{A} \subseteq N, \mathcal{A} \neq \emptyset$, and enforcing k-maxitivity by adding 0-1 integer variables $c(\mathcal{A}, i)$ for all $\mathcal{A} \subseteq \mathcal{N}$, $|\mathcal{A}| > k$ and $i \in \mathcal{A}$, which determine whether the constraint $z(\mathcal{A}) = z(\mathcal{A} \setminus \{i\})$, we transform linear programming problem (8.1) into a mixed integer problem (MIP)

$$\text{minimise} \quad \sum_{j=1}^{J} r_j^+ + r_j^-, \tag{8.18}$$

$$\begin{aligned}
\text{s.t.} \quad & r_j^+ - r_j^- = \sum_{A} \mu(\mathcal{A}) g_A(\mathbf{x}^j) - y_j, \ j = 1, \ldots, J \\
& \mu(\mathcal{A}) \geqslant \mu(\mathcal{A} \setminus \{i\}), \forall i \in \mathcal{A} \text{ for all } \mathcal{A} \subseteq \mathcal{N} \\
& \mu(\mathcal{A}) - \mu(\mathcal{A} \setminus \{i\}) \leqslant c(\mathcal{A}, i) M, \quad \forall i \in \mathcal{A} \text{ and } |\mathcal{A}| > k \\
& \sum_{i \in A} c(\mathcal{A}, i) \leqslant |\mathcal{A}| - 1, \ \text{for all } \mathcal{A} \subseteq \mathcal{N}, |\mathcal{A}| > k, \\
& \mu(\mathcal{N}) = 1, \quad c(\mathcal{A}, i) \in \{0, 1\},
\end{aligned}$$

for some positive number M. So for every $\mathcal{A}$, $|\mathcal{A}| > k$, we have $\mu(\mathcal{A}) = \mu(\mathcal{A} \setminus \{i\})$ whenever $c(\mathcal{A}, i) = 0$, and a void constraint if $c(\mathcal{A}, i) = 1$. At least one $c(\mathcal{A}, i)$ must be 0 for every $\mathcal{A}$, $|\mathcal{A}| > k$, which is precisely the condition of k-maxitivity.

The MIP problem given by (8.18) is certainly feasible, i.e. all but one variable μ set to zero is a feasible point, with the remaining variable equal to one. Standard MIP solvers based on the simplex algorithm using a branch-and-bound technique will deliver an optimal solution.

A relaxation of the MIP problem was presented in [BW21], proposing the following algorithm. This way we solve two sparse linear programming problems,

---

Algorithm 8.1: k-Maxitive capacity MIP relaxation algorithm

---

1. Solve the LP problem (8.18) with $c(\mathcal{A}, i) \in [0, 1]$ using the Simplex method. Denote the optimal solution by r^*, μ^*, c^*.
2. Fix k-maxitivity constraints and solve the second LP
   a. For each $\mathcal{A}$ such that $|\mathcal{A}| > k$ find $i^* = \arg\min_j c^*(\mathcal{A}, j)$ and fix the constraint $c(\mathcal{A}, i^*) = 0$.
   b. Solve problem (8.18) with all $c(\mathcal{A}, i) \in [0, 1]$ and the equality constraints from step 2a using the Simplex method.
3. Return the optimal k-maxitive capacity μ.

---

which is significantly cheaper than solving the MIP. The price is a suboptimal solution compared to that of MIP; however, the optimality gaps between the values of the objective at the relaxed solution f_{rel} (problem at Step 1), at our suboptimal solution f_{sub} (problem at Step 2), and at the MIP solution f_{MIP} are less than 10%.

The identification of k-minitive capacities is done by using duality.

Note 8.2 Recall that in the case of k-additive capacities, while there is reduction in the number of variables, the number of monotonicity constraints is not drastically reduced (see [BBC16, Bel07, Bel09]). The use of k-maxitive capacities only marginally simplifies the problem. Even though the number of variables μ is reduced, the number of monotonicity constraints is not (note that all the terms $\mu(\mathcal{A})g_A(\mathbf{x})$ are still used, and hence, monotonicity conditions need to be enforced for all $\mu(\mathcal{A})$). Furthermore, incorporating k-maxitivity actually made the fitting problem change from a linear program to a mixed integer program, a more challenging problem—not to mention that if fitting the data in the least squares sense, we would have an even more complex mixed integer quadratic problem. Therefore we do not recommend enforcing k-maxitivity purely for simplification purposes, but rather when the semantics of the problem warrants it.

8.4.3 An Illustrative Example

Suppose a buyer adopts five criteria to evaluate vehicles:

1. The selling price
2. The space and comfort
3. The safety and convenience
4. The engine and transmission system
5. The maintenance cost for 5 years

Denoting the criteria set by $\mathcal{N} = \{1, 2, 3, 4, 5\}$, the decision-maker believes that:

(a) The capacity μ on $\mathcal{N}$ should be 2-minitive. This means that the decision-maker is mainly concerned with capacity values and interaction effects for subsets of three or more decision criteria, paying less attention to measures of singletons and pairs. That is, the decision-maker provides a relatively macroscopic view of these five criteria.

(b) The comparative importance of some of the subsets with $|\mathcal{A}| = 4$ is expressed as follows: $\mu(\{1,3,4,5\}) > \mu(\{2,3,4,5\})$ with threshold 0.1 (i.e. "the space and comfort" (criterion 2) has relatively more positive influence than "selling price" (criterion 1) when combined with the remaining criteria); $\mu(\{1,2,4,5\}) > \mu(\{1,3,4,5\})$, $\mu(\{1,2,3,5\}) > \mu(\{1,2,4,5\})$, and $\mu(\{1,2,3,4\}) > \mu(\{1,2,3,5\})$, all with threshold 0.05.

(c) The comparative importance between some of the subsets with three and four criteria is expressed: $\mu(\{2,3,4,5\}) > \mu(\{2,3,5\})$ with threshold 0.1 (this means that a minimum increment to importance is set when criterion 4 is added); $\mu(\{1,2,3,5\}) > \mu(\{1,3,5\})$, $\mu(\{1,3,4,5\}) > \mu(\{1,3,4\})$, and $\mu(\{1,2,3,4\}) > \mu(\{1,2,4\})$, all with threshold 0.1.

(d) Some comparisons for groups of three are given as follows: $\mu(\{1,2,5\}) > \mu(\{3,4,5\})$ with threshold 0.05, $\mu(\{1,4,5\}) > \mu(\{1,2,5\})$ with threshold 0.2, and then $\mu(\{1,2,3\}) > \mu(\{1,4,5\})$, $\mu(\{2,4,5\}) > \mu(\{1,2,3\})$, and $\mu(\{2,3,4\}) > \mu(\{2,4,5\})$, all with threshold 0.1.

(e) Comparison of pairs: $\mu(\{1,2\}) > \mu(\{3,4\})$, $\mu(\{2,3\}) > \mu(\{4,5\})$, and $\mu(\{1,3\}) > \mu(\{2,5\})$ with thresholds 0.15, 0.2, and 0.1, respectively.

(f) Minimum increments between pairs and triples: $\mu(\{2,3,5\}) > \mu(\{3,5\})$, $\mu(\{1,2,4\}) > \mu(\{2,4\})$, $\mu(\{1,3,5\}) > \mu(\{1,5\})$, and $\mu(\{1,3,4\}) > \mu(\{1,4\})$, all with threshold 0.1.

(g) The objective of the model is to maximise the orness index of the capacity, i.e. subject to the above preferences, the capacity that maximises

$$orness(\mu) = \sum_{\mathcal{A} \subset \mathcal{N}} \frac{(n - |\mathcal{A}|)!|\mathcal{A}|!}{n!(n-1)} \mu(\mathcal{A})$$

should be chosen.

Based on the decision-maker's specifications, we can construct the following identification model:

$$\begin{aligned} \text{maximise } z &= orness(\mu) \\ &= 0.05\mu(\{1\}) + \cdots + 0.025\mu(\{1,2\}) + \cdots + 0.025\mu(\{1,2,3\}) + \\ &\quad \cdots + 0.05\mu(\{1,2,3,4\}) + \cdots + 0.05\mu(\{2,3,4,5\}) \end{aligned}$$

subject to:

boundary conditions:

$$\mu(\emptyset) = 0$$
$$\mu(\mathcal{N}) = 1$$

monotonicity conditions:

$$\mu(\{1\}) \geqslant 0$$
$$\ldots$$
$$\mu(\mathcal{N}) - \mu(\{1, 2, 3, 4\}) \geqslant 0$$

constraints for a 2-minitive capacity:

$$\mu(\{1, 2, 3\}) - \mu(\{1, 2\}) \leqslant c(\{1, 2\}, 3)$$
$$\mu(\{1, 2, 4\}) - \mu(\{1, 2\}) \leqslant c(\{1, 2\}, 4)$$
$$\mu(\{1, 2, 5\}) - \mu(\{1, 2\}) \leqslant c(\{1, 2\}, 5)$$
$$c(\{1, 2\}, 3) + c(\{1, 2\}, 4) + c(\{1, 2\}, 5) \leqslant 2$$
$$\ldots$$

constraints stemming from (b) $\sim$ (f):

$$\mu(\{1, 3, 4, 5\}) - \mu(\{2, 3, 4, 5\}) \geqslant 0.1$$
$$\ldots$$
$$\mu(\{1, 3, 4\}) - \mu(\{1, 4\}) \geqslant 0.1.$$

Solving the above linear programming problem, we obtain an optimal capacity μ with the objective value $orness(\mu) = 0.53375$ and the capacity values as shown in Table 8.2. It can be verified that the capacity μ is 2-minitive.

Table 8.2 The 2-minitive capacity μ on criteria set $\{1, 2, 3, 4, 5\}$

$\mathcal{A}$	$\mu(\mathcal{A})$	$\mathcal{A}$	$\mu(\mathcal{A})$	$\mathcal{A}$	$\mu(\mathcal{A})$	$\mathcal{A}$	$\mu(\mathcal{A})$
$\emptyset$	0.00	$\{1, 4\}$	0.45	$\{1, 2, 3\}$	0.55	$\{2, 4, 5\}$	0.65
$\{1\}$	0.25	$\{1, 5\}$	0.25	$\{1, 2, 4\}$	0.90	$\{3, 4, 5\}$	0.00
$\{2\}$	0.25	$\{2, 3\}$	0.55	$\{1, 2, 5\}$	0.25	$\{1, 2, 3, 4\}$	1.00
$\{3\}$	0.00	$\{2, 4\}$	0.65	$\{1, 3, 4\}$	0.75	$\{1, 2, 3, 5\}$	0.95
$\{4\}$	0.00	$\{2, 5\}$	0.25	$\{1, 3, 5\}$	0.85	$\{1, 2, 4, 5\}$	0.90
$\{5\}$	0.00	$\{3, 4\}$	0.00	$\{1, 4, 5\}$	0.45	$\{1, 3, 4, 5\}$	0.85
$\{1, 2\}$	0.25	$\{3, 5\}$	0.00	$\{2, 3, 4\}$	0.75	$\{2, 3, 4, 5\}$	0.75
$\{1, 3\}$	0.55	$\{4, 5\}$	0.00	$\{2, 3, 5\}$	0.65	$\{1, 2, 3, 4, 5\}$	1.00

Table 8.3 The 2-maxitive capacity μ^* on criteria set $\{1, 2, 3, 4, 5\}$

$\mathcal{A}$	$\mu^*(\mathcal{A})$	$\mathcal{A}$	$\mu^*(\mathcal{A})$	$\mathcal{A}$	$\mu^*(\mathcal{A})$	$\mathcal{A}$	$\mu^*(\mathcal{A})$
$\emptyset$	0.00	$\{1,4\}$	0.35	$\{1,2,3\}$	1.00	$\{2,4,5\}$	0.45
$\{1\}$	0.25	$\{1,5\}$	0.25	$\{1,2,4\}$	1.00	$\{3,4,5\}$	0.75
$\{2\}$	0.15	$\{2,3\}$	0.55	$\{1,2,5\}$	1.00	$\{1,2,3,4\}$	1.00
$\{3\}$	0.10	$\{2,4\}$	0.15	$\{1,3,4\}$	0.75	$\{1,2,3,5\}$	1.00
$\{4\}$	0.05	$\{2,5\}$	0.25	$\{1,3,5\}$	0.35	$\{1,2,4,5\}$	1.00
$\{5\}$	0.00	$\{3,4\}$	0.75	$\{1,4,5\}$	0.45	$\{1,3,4,5\}$	0.75
$\{1,2\}$	1.00	$\{3,5\}$	0.10	$\{2,3,4\}$	0.75	$\{2,3,4,5\}$	0.75
$\{1,3\}$	0.35	$\{4,5\}$	0.45	$\{2,3,5\}$	0.55	$\{1,2,3,4,5\}$	1.00

From this example, one can see that a k-minitive capacity provides a top-down perspective: the decision-maker needs only focus on the larger subsets and the relationships of higher-order criteria coalitions; then according to the recursive equations (7.23)–(7.25), the lower-order coalition values are obtained.

For comparison, we can model the above example from the dual capacity perspective, i.e. to obtain a 2-maxitive capacity from preferences specified over the lower-order capacity values.

The following items (a') $\sim$ (g') represent the dual preferences corresponding to items (a) $\sim$ (g), respectively:

(a') The capacity should be 2-maxitive.
(b') The comparative importance for singletons is provided as follows: $\mu^*(\{1\}) > \mu^*(\{2\})$ with threshold 0.1; $\mu^*(\{2\}) > \mu^*(\{3\})$ with threshold 0.05; ...

...

(g') The objective of the model is to minimise the orness index of the capacity.

Solving the mixed integer programming problem in this case, we obtain μ^*, the optimal objective $orness(\mu^*) = 0.46625$ (note that $orness(\mu)+orness(\mu^*) = 1$), and the capacity values shown in Table 8.3. The 2-maxitive capacity μ^* can be recognised as the dual of the capacity μ.

8.5 Learning k-Tolerant and k-Intolerant Capacities

Recall from Definition 7.9 that k-tolerant and k-intolerant capacities are dual to one another. A Choquet integral with respect to a k-tolerant capacity is independent of the first $n-k$ smallest inputs. For example, if a student's work were to be evaluated based on n assignments, such a Choquet integral would ignore the marks from the lowest two assignments, and the student could miss at most two assignments with no consequences for their final mark. A k-tolerant capacity is also k-maxitive, satisfying $\mu(\mathcal{A}) = 1$ whenever $|\mathcal{A}| \geqslant k$.

In this situation, the Choquet integral is expressed

$$C_\mu(\mathbf{x}) = \sum_{\mathcal{A}\subseteq\mathcal{N},|\mathcal{A}|<k} \mu(\mathcal{A})g_A(\mathbf{x}) + \sum_{\mathcal{A}\subseteq\mathcal{N},|\mathcal{A}|\geqslant k} g_A(\mathbf{x}),$$

whereby not only the number of variables but also the set of monotonicity constraints is reduced. When fitting a k-tolerant capacity to empirical data, we convert the mixed integer program (8.18) to a linear programming problem with a much reduced set of constraints,

$$\text{minimise} \quad \sum_{j=1}^{J} r_j^+ + r_j^- \tag{8.19}$$

$$\text{s.t.} \quad r_j^+ - r_j^- = \sum_{\mathcal{A}} \mu(\mathcal{A})g_A(\mathbf{x}^j) - y_j, \; j = 1, \ldots, J$$

$$\mu(\mathcal{A}) \geqslant \mu(\mathcal{A}\setminus\{i\}), \forall i \in \mathcal{A} \text{ for all } \mathcal{A} \subseteq \mathcal{N}, |\mathcal{A}| \leqslant k$$

$$\mu(\mathcal{A}) = 1, \text{ for all } \mathcal{A}, \; |\mathcal{A}| \geqslant k.$$

8.6 Learning k-Interactive Capacities

A limitation of many of the methods presented above is that the solutions are computationally feasible only for small $n \leqslant 10$ due to the excessive number of monotonicity constraints. One approach to addressing this challenge is the use of k-interactive capacities, which can be learned with respect to a significantly reduced number of constraints.

Let us now formulate the fitting problem for k-interactive capacities. We shall use the l_1 norm as it allows us to translate the problem into a linear programming problem. As earlier, using the standard expression for the residuals $r_j = C_\mu(\mathbf{x}^j) - y_j$ and splitting them into positive and negative parts $r_j = r_j^+ - r_j^-$, and then taking the following set of non-negative decision variables, $r_j^+, r_j^-, j = 1, \ldots, J, \mu(\mathcal{A})$, $\mathcal{A} \subseteq \mathcal{N}, \mathcal{A} \neq \emptyset$, we have an equivalent linear programming problem to (8.1), where the decision variables μ are organised into a vector with the relevant cardinality-based indexing system.

Now, since we have

$$C_\mu(\mathbf{x}) = \frac{1-K}{n-k-1}\sum_{i=1}^{n-k-1} x_{(i)} + Kx_{(n-k)} + \sum_{\mathcal{A}\subseteq\mathcal{N},|\mathcal{A}|\leqslant k} \mu(\mathcal{A})g_A(\mathbf{x}), \tag{8.20}$$

for fixed k and K, we have the problem

$$\text{minimise} \quad \sum_{j=1}^{J} r_j^+ + r_j^- \tag{8.21}$$

$$\text{s.t.} \quad r_j^+ - r_j^- = \sum_{\mathcal{A}\subset\mathcal{N}, |\mathcal{A}|\leqslant k} \mu(\mathcal{A}) g_A(\mathbf{x}^j)$$

$$+ \frac{1-K}{n-k-1} \sum_{i=1}^{n-k-1} x_{(i)}^j + K x_{(n-k)}^j - y_j, \quad j = 1, \ldots, J$$

$$\mu(\mathcal{A}) \geqslant \mu(\mathcal{A} \setminus \{i\}), \forall i \in \mathcal{A} \text{ for all } \mathcal{A} \subset \mathcal{N}, 0 < |\mathcal{A}| \leqslant k$$

$$\mu(\mathcal{A}) \leqslant K, |\mathcal{A}| = k.$$

Notice the drastic reduction in the number of variables and constraints, because monotonicity is implied for subsets of cardinality greater than k. We now discuss various ways to set the values $\mu(\mathcal{A})$ for these subsets, which may give a different expression to (8.20).

8.6.1 Maximum Entropy Approach

Let us fix the maximal value K of a capacity at any subset of cardinality k, that is, $\mu(\mathcal{A}) \leqslant K$ for all $\mathcal{A}$, $|\mathcal{A}| = k$. Our first step is to define the values of $\mu(\mathcal{B})$, $|\mathcal{B}| > k$ in a suitable way. We fix the values $\mu(\mathcal{B}) = K$ for all $\mathcal{B}$, $|\mathcal{B}| = k+1$ and then define the values at the larger sets by optimising some criterion.

One choice is to fix the values $\mu(\mathcal{B}) = 1$ when $|\mathcal{B}| > k$, which results in k-tolerant capacities. However, these are somewhat restrictive as the $n-k$ smallest components are not used in the aggregation process.

Here we choose the values $\mu(\mathcal{B})$ that will maximise the entropy of the capacity, given by (3.36), which we express here in a different form,

$$H(\mu) = \sum_{\mathcal{A}\subset\mathcal{N}} \sum_{i\notin\mathcal{A}} \frac{(n-|\mathcal{A}|-1)!|\mathcal{A}|!}{n!} h\left(\mu(\mathcal{A}\cup\{i\}) - \mu(\mathcal{A})\right). \tag{8.22}$$

where $h(x) = -x\ln(x)$ if $x > 0$ and 0 if $x = 0$.

Since the values of μ for subsets smaller than k are determined by the data, and their contribution to the total entropy is therefore governed by the data, we optimise only the contribution of the larger subsets, which is calculated independently, as we can split the sum into

$$H(\mu) = \sum_{\mathcal{A}\subset\mathcal{N},|\mathcal{A}|>k} \sum_{i\notin\mathcal{A}} \gamma_a h\left(\mu(\mathcal{A}\cup\{i\}) - \mu(\mathcal{A})\right) + \sum_{\mathcal{A}\subset\mathcal{N},|\mathcal{A}|\leqslant k} \sum_{i\notin\mathcal{A}} \gamma_a h\left(\mu(\mathcal{A}\cup\{i\}) - \mu(\mathcal{A})\right), \tag{8.23}$$

where $\gamma_a = \frac{(n-|\mathcal{A}|-1)!|\mathcal{A}|!}{n!}$. Since maximum entropy capacities are additive, our rationale is that by optimising partial entropy for subsets of cardinality larger than k, we obtain a nearly additive measure with respect to those subsets. Discussion of maximum entropy (or minimum variance) capacities has been presented in [Koj07].

Let us represent this as an optimisation problem

$$\text{maximise} \quad H_k(\mu) = \sum_{\mathcal{A}\subset\mathcal{N},|\mathcal{A}|>k} \sum_{i\notin\mathcal{A}} \gamma_a h\left(\mu(\mathcal{A}\cup\{i\}) - \mu(\mathcal{A})\right) \tag{8.24}$$
$$\text{s.t.} \quad \mu(\mathcal{A}) = K \text{ for all } \mathcal{A}, |\mathcal{A}| = k+1.$$

The solution to this problem is given explicitly as

$$\mu(\mathcal{A}) = K + \frac{|\mathcal{A}| - k - 1}{n - k - 1}(1 - K), \quad \text{for all } \mathcal{A} : |\mathcal{A}| > k, \tag{8.25}$$

as shown in the following proposition [BW19].

Proposition 8.1 *The maximum of the partial entropy $H_k(\mu)$ is reached at μ given by* (8.25).

8.6.2 Minimising the Nonadditivity Index

Another way to specify the values of $\mu(\mathcal{B})$ for $|\mathcal{B}| > k$ presented in [WB18] is to use the total amount of nonadditivity instead of the entropy. The nonadditivity index of a subset $\mathcal{A}$ measures how far $\mathcal{A}$ and its subsets are from being additive, that is, the extent to which the additivity condition is violated.

By summing the weighted nonadditivity indices over all subsets, we obtain an overall measure of nonadditivity for μ. Since we are interested in minimising the combined nonadditivity over subsets of higher cardinality $\mathcal{C} = \{\mathcal{A} \subseteq \mathcal{N} \mid |\mathcal{A}| > k\}$, we use the partial nonadditivity given by

$$Total(n_\mu(\mathcal{C})) = \sum_{\mathcal{A}\in\mathcal{C}} |n_\mu(\mathcal{A})| = \sum_{\mathcal{A}\in\mathcal{C}} \left| \mu(\mathcal{A}) - \frac{1}{2^{|\mathcal{A}|-1} - 1} \sum_{\mathcal{B}\subset\mathcal{A},\mathcal{B}\in\mathcal{C}} \mu(\mathcal{B}) \right|. \tag{8.26}$$

Note that we use the modulus of $n_\mu(\mathcal{A})$, as values range over $[-1, 1]$, with 0 corresponding with additivity. We arrive at the following problem:

$$\begin{aligned} \text{minimise} \quad & Total(n_\mu(\mathcal{C})) \\ \text{s.t.} \quad & \mu(\mathcal{A}) = K \text{ for all } \mathcal{A}, |\mathcal{A}| = k+1 \\ & \mathcal{C} = \{\mathcal{A} \subseteq \mathcal{N} | |\mathcal{A}| > k\} \\ & \text{monotonicity of the capacity.} \end{aligned} \tag{8.27}$$

Here we can also use the permutation symmetry of $n_\mu(\mathcal{A})$ and express the objective as (a constant multiple of)

$$Total(n_\mu(C)) = \sum_{a=k+1}^{n} \left| \mu_a - \frac{1}{2} \sum_{i=k+1}^{a-1} \binom{a}{i} \left(\sum_{b=k+1}^{n-1} \binom{a}{b} \mu_b \right) \right| . \tag{8.28}$$

Here $\mu_a = \mu(\mathcal{A})$ for convenience, which is the same for all $\mathcal{A}$ for a fixed a. Now problem (8.27) can be set up as a linear programming problem with respect to the variables $\mu_a, a = k+1, \ldots, n$:

$$\begin{aligned} \text{minimise} \quad & \sum_{a=k+1}^{n} r_a^+ + r_a^- \\ \text{s.t.} \quad & r_a^+ - r_a^- = \mu_a - \frac{1}{2} \sum_{i=k+1}^{|\mathcal{A}|-1} \binom{|\mathcal{A}|}{i} \sum_{|\mathcal{B}|=k+1}^{n-1} \binom{|\mathcal{A}|}{b} \mu_b, \\ & \qquad |\mathcal{A}| = k+1, \ldots, n \\ & \mu_{k+1} = K, \ \mu_n = 1 \\ & \mu_a \leqslant \mu_{a+1}, \ |\mathcal{A}| = k+1, \ldots, n-1. \end{aligned} \tag{8.29}$$

While we cannot derive an explicit solution here, any linear programming software would be capable of providing the optimal values of μ_a for fixed k, K and n.

Similarly to optimising the nonadditivity index over the sets of cardinality greater than k, we can set up linear programming problems for optimising the Shapley and Banzhaf indices, as they too are linear combinations of the capacity values. Thus while we would have a different set of coefficients in the linear expression in (8.28) under the modulus sign, the linear programming formulation based on (8.29) would be similar.

8.6.3 Maximising Orness or Andness

Another criterion to optimise is the orness value (see Definition 5.2) of the Choquet integral [Mar04], calculated from the capacity μ by

$$orness(C_\mu) = \frac{1}{n-1} \sum_{\mathcal{A} \subseteq \mathcal{N}} \frac{(n-|\mathcal{A}|)!|\mathcal{A}|!}{n!} \mu(\mathcal{A}).$$

Again, by using the symmetry and cancelling the coefficients, we can set up the problem (using notation $\mu_a = \mu(\mathcal{A})$ for $|\mathcal{A}| > k$)

$$\begin{aligned} \text{minimise/maximise} \quad & \sum_{a=k+1}^{n} \mu_a \\ \text{s.t.} \quad & \mu_{k+1} = K, \; \mu_n = 1, \\ & \text{monotonicity of capacity.} \end{aligned} \tag{8.30}$$

The solution to this problem will be the k-tolerant capacity if we maximise orness (all $\mu_a = 1$), and $\mu_a = K, |\mathcal{A}| = k+1, \ldots, n-1$, and $\mu_n = 1$ if we minimise orness (and maximise andness).

8.6.4 *Learning the Value of K*

One additional parameter in the k-interactivity approach is the value of K. It could be assigned a specific value, like k/n; however, this would not guarantee optimality with respect to the data-fitting criterion. Another approach is to solve a bi-level optimisation problem, where at the inner level we optimise μ for a fixed $K \in [0, 1]$, and at the outer level we optimise K:

$$\min_K \min_\mu \sum_{j=1}^{J} r_j^+ + r_j^-,$$

where the inner problem is the same as (8.21).

Because K is a single parameter of the outer problem, the quasi-Newton method, Brent method, or the Golden section method will deliver the optimal value efficiently. The objective function is itself a solution to another LP problem (8.21). However, that inner problem is an LP with the same structure, since only the values of the right-hand side in the constraints are affected by K. Therefore an efficient algorithm that sets up the inner problem only once and then only adjusts its constraints each time K is changed can be designed.

8.7 Learning Convex Capacities

Convex or concave capacities may be required to help facilitate formulation of optimisation problems where the Choquet integral is used as an objective (see

Chap. 10) or to meet some context-specific requirements. Many of the conditions here are also summarised in Table. 4.1.

8.7.1 Sub- and Supermodularity Constraints

Recall from Definition 2.9 that a capacity is supermodular if

$$\mu(\mathcal{A} \cup \mathcal{B}) + \mu(\mathcal{A} \cap \mathcal{B}) \geqslant \mu(\mathcal{A}) + \mu(\mathcal{B}). \tag{8.31}$$

Linear constraints can be formulated in multiple representations, equivalent to ensuring, for all pairs $\{i, j\}$ and sets $\mathcal{A} \subseteq \mathcal{N} \setminus \{i, j\}$, that

$$\mu(\mathcal{A} \cup \{i, j\}) + \mu(\mathcal{A}) - \mu(\mathcal{A} \cup \{i\}) - \mu(\mathcal{A} \cup \{j\}) \geqslant 0. \tag{8.32}$$

It is not necessary to formulate constraints for all subsets $\mathcal{A}, \mathcal{B}$ since many of these combinations will be redundant.

The condition in Eq. (8.32) is precisely that of the second discrete derivatives Δ_{ij} in Definition 3.1 being greater than 0 or, in terms of the first derivatives and marginal contribution representation, that

$$\Delta_i \mu(\mathcal{A} \cup \{j\}) - \Delta_i \mu(\mathcal{A}) \geqslant 0. \tag{8.33}$$

In Möbius representation, Eq. (8.32) simplifies to [CJ89a]:

$$\sum_{\mathcal{B} \subseteq \mathcal{A} | i, j \in \mathcal{B}} M(\mathcal{B}) \geqslant 0, \tag{8.34}$$

for all $\mathcal{A} \subseteq \mathcal{N}$ and pairs $\{i, j\} \subseteq \mathcal{A}$.

For submodular capacities, the inequalities in Eqs. (8.32), (8.33), and (8.34) are reversed.

8.7.2 Buoyant and Antibuoyant Capacities

Buoyancy and antibuoyancy are stronger conditions. For buoyancy, which results in a subset of submodular capacities, we require for all pairs $\{i, j\}$ and all $\mathcal{A} \subseteq \mathcal{N} \setminus \{i, j\}$ that

$$\mu(\mathcal{A} \cup \{i, j\}) - 2\mu(\mathcal{A} \cup \{j\}) + \mu(\mathcal{A}) \leqslant 0. \tag{8.35}$$

Hence, the number of constraints required is equivalent to that of submodularity (with the latter being redundant).

In Möbius representation, the condition can be expressed as

$$\sum_{\mathcal{B}\subseteq\mathcal{A}} M(\mathcal{B}\cup\{i,j\}) + \sum_{\mathcal{B}\subseteq\mathcal{A}} [M(\mathcal{B}\cup\{j\}) - M(\mathcal{B}\cup\{i\})] \geqslant 0, \tag{8.36}$$

for all $\{i, j\} \subseteq \mathcal{N}$ and all $\mathcal{A} \subseteq \mathcal{N} \setminus \{i, j\}$.

In marginal contribution representation, over the same sets, we have

$$\Delta_i(\mathcal{A}\cup\{j\}) - \Delta_j(\mathcal{A}) \leqslant 0. \tag{8.37}$$

For the special case of 2-additive capacities, the buoyancy requirements can be expressed, for each i, j and $\mathcal{A} \subseteq \mathcal{N} \setminus \{i, j\}$, as

$$M(\{i,j\}) - M(\{i\}) + M(\{j\}) - \sum_{k\in\mathcal{A}} (M(\{i,k\}) - M(\{j,k\})) \leqslant 0, \tag{8.38}$$

with the inequalities being reversed in the case of antibuoyancy.

8.7.3 Buoyancy on Average

Given the high number of constraints required for supermodularity and antibuoyancy (even in the case of 2-additive capacities), simpler conditions resulting in special sub-types of capacity may be preferred. For example, belief measures (Definition 2.15), which are a special case of supermodular capacities, require only that $M(\mathcal{A}) \geqslant 0$ for all $\mathcal{A} \subseteq \mathcal{N}$.

The following property was proposed in [BJW25] using Shapley values and interaction indices, referred to as *buoyancy on average*:

$$I_{ij} + \varphi_j \leqslant \varphi_i, \forall i, j \in \mathcal{N}. \tag{8.39}$$

It can also be written as

$$I_{ij} \leqslant -|\varphi_i - \varphi_j|, \forall i, j \in \mathcal{N}, i < j, \tag{8.40}$$

noting that, for submodular capacities, $-1 \leqslant I_{ij} \leqslant 0$.

The dual, antibuoyancy on average, is hence expressed in terms of the reverse inequality, i.e.

$$I_{ij} \geqslant |\varphi_i - \varphi_j|, \forall i, j \in \mathcal{N}, i < j. \tag{8.41}$$

This requires considerably fewer constraints when compared with submodularity and buoyancy and hence can be used in cases where buoyant or antibuoyant behaviour is desirable but not necessarily a strict requirement. It was shown in

[BJW25] that when the Choquet integral is used as an objective function, the location of the optima will often be similar for certain capacities satisfying the buoyancy on average property to those that are buoyant.

For the 2-additive case, we have

$$M(\{i,j\}) \leqslant M(\{i\}) - M(\{j\}) + \frac{1}{2} \sum_{k \in \mathcal{N} \setminus \{i,j\}} (M(\{i,k\}) - M(\{j,k\})). \qquad (8.42)$$

In the case of standard buoyancy, the capacity monotonicity constraints become redundant; however, this is not necessarily the case for capacities only satisfying (8.39), and hence, some additional constraints are required. The following n inequalities can also be included:

$$\mu(\{i\}) = M(\{i\}) = \varphi_i - \frac{1}{2} \sum_{j \neq i} |I_{ij}| \geqslant 0. \qquad (8.43)$$

Buoyancy can also be enforced on the cardinality indices (Sect. 3.7), i.e. so that

$$\vartheta_i \leqslant \vartheta_{i+1}, i = 1, \ldots, n-1, \qquad (8.44)$$

or on the weights corresponding with a particular position, e.g. so that the effective weight applied to the largest input is always larger than that applied to the second largest [BJW25].

8.8 Further Reduction of the Number of Variables

Another approach to reduce the complexity of the fitting problem is based on the observation that it may not be possible for all the values of μ to be determined from the data given. Recall that the Choquet integral is a piecewise linear monotone continuous function on $[0, 1]^n$ with the idempotency property $C_\mu(t, t, \ldots, t) = t$ for all $t \in [0, 1]$. There are exactly $n!$ linear segments in the function, the same as the number of permutations of the components in $\mathbf{x}$. Thus, at least $n!$ data points are needed in order for each linear segment to have one datum. Of course, the linear segments are not independent, and ultimately only $2^n - 2$ capacity values are to be determined. Each datum affects $n - 1$ values, which further relaxes the requirements on J. It has been shown that at least $n!/[(n/2)!]^2$ (for even n) and $n!/[((n-1)/2)!((n+1)/2)!]$ (for odd n) data are needed [GN94]. However, for sufficiently large n, J could be substantially smaller than those quantities.

One method proposed is based on using only decision variables corresponding to the values $\mu(\mathcal{A})$ that can be determined from the dataset. The rest can be calculated after the fitting process, while ensuring satisfaction of the monotonicity constraints. This way we reduce the number of variables of the fitting problem.

Consider the inclusion relation $\subseteq$ over the set $2^{\mathcal{N}} = \mathcal{P}(\mathcal{N})$. It is a finite bounded lattice with the least element $\emptyset$ and the greatest element $\mathcal{N}$. There are $n!$ maximal chains in that lattice, the same as the number of sorting orders of the input vectors, as there is a one-to-one correspondence between the maximal chains and permutations of the inputs; see, e.g. [Koj07].

Now, each datum from the dataset D corresponds to one or more (in the case of equal components of $\mathbf{x}$) maximal chains. Let us construct the set $\mathcal{C} \subseteq \mathcal{P}(\mathcal{N})$ of those values of μ that belong to any of the maximal chains that correspond to the dataset D, with $|\mathcal{C}| \leqslant |\mathcal{P}(\mathcal{N})| = 2^n$. Then only the values $\mu(\mathcal{A})$, $\mathcal{A} \in \mathcal{C}$, can be directly identified from the dataset.

The rest of the values of μ still need to be consistent with those constructed from the data due to the monotonicity constraints; however, this problem will be solved separately. Therefore, we now focus on the identification of capacity values on $\mathcal{C}$. The rationale is that $\mathcal{C}$ is much smaller than $\mathcal{P}(\mathcal{N})$ (for a large enough n), and also, the number of monotonicity constraints on $\mathcal{C}$ is much smaller than those on $\mathcal{P}(\mathcal{N})$.

While the reduction of the number of variables to identify is straightforward to implement, the monotonicity constraints require the following considerations. Let us build, algorithmically, all the monotonicity constraints on the set $\mathcal{C}$. Firstly, on each maximal chain MC_j we have the constraints $\mu(\mathcal{A}) \leqslant \mu(\mathcal{B})$, $\mathcal{A}, \mathcal{B} \in MC_j$; however, we can note that there are only $n-1$ non-redundant constraints, when $|\mathcal{B}| = |\mathcal{A}| + 1$. Although we refer to these as trivial constraints, they are not the only restraints required. Consider "neighbouring" maximal chains when $\mathcal{A} \in MC_j$, $\mathcal{B} \in MC_l$, $\mathcal{B} \subset \mathcal{A}$, but $\mathcal{B} \notin MC_j$. We must include the non-trivial constraint $\mu(\mathcal{A}) \geqslant \mu(\mathcal{B})$, although it may be redundant. To distill only the non-redundant non-trivial constraints, we proceed algorithmically.

Fitting the capacity is now done by solving a modification of LP (8.1):

$$\begin{aligned} \text{minimise} \quad & \sum_{j=1}^{J} r_j^+ + r_j^- \qquad (8.45)\\ \text{s.t.} \quad & r_j^+ - r_j^- = \sum_{\mathcal{A}\in\mathcal{C}} \mu(\mathcal{A}) g_A(\mathbf{x}^j) - y_j, \; j = 1, \ldots, J \\ & \text{monotonicity constraints in } Cons \\ & \mu(\mathcal{N}) = 1. \end{aligned}$$

The maximal chains are constructed algorithmically from the outputs of the sort operations for every $\mathbf{x}^j$.

To calculate $\mu(\mathcal{A})$ for $\mathcal{A} \notin \mathcal{C}$, we formulate the maximal entropy problem

$$\text{Maximise } H = \sum_{i=1}^{n} \sum_{\mathcal{A}\in\mathcal{C}, i\notin\mathcal{A}} \frac{(n - |\mathcal{A}| - 1)!|\mathcal{A}|!}{n!} h\left(\mu(\mathcal{A} \cup \{i\}) - \mu(\mathcal{A})\right), \qquad (8.46)$$

Algorithm 8.2: Enumeration of monotonicity constraints

Input: the set of maximal chains $MCS = \{MC\}$ already processed, new maximal chain $MCnew$

Output: Set of monotonicity constraints $Cons$

Process the elements of $MCnew$

1. Trivial constraints
   For i from 1 to $n-1$ do
   Add (to $Cons$) constraint $\mu(MC_i) \leqslant \mu(MC_{i+1})$
2. Non-trivial constraints
   a. $Used = \emptyset$
      For i from 2 to n do
      For all $MC \in MCS$ do
      For m from i-1 to 1 do
      If $MCnew_i \supset MC_m$ and $MC_m \notin Used$
      add constraint $\mu(MCnew_i) \geqslant \mu(MC_m)$ and break;
      else continue;
   b. $Used = \emptyset$
      For i from $n-1$ to 2 do
      For all $MC \in MCS$ do
      For m from $i+1$ to n do
      If $MCnew_i \subset MC_m$ and $MC_m \notin Used$
      add constraint $\mu(MCnew_i) \leqslant \mu(MC_m)$ and break;
      else continue;
3. Add $MCnew$ to MCS;
4. Return $Cons$.

with $h(x) = -x \ln x$ and subject to monotonicity constraints. To find the set of non-redundant constraints, we now use Algorithm 8.2 again, but initialise it not with $MCS = \emptyset$ as we do for problem (8.45), but from the already computed set MCS, to which we add the remaining maximal chains.

In order not to process all possible remaining maximal chains, of which there are $n! - J$, we only take one maximal chain for every $\mathcal{A} \notin \mathcal{C}$, by identifying the corresponding maximal chain from the datum $\mathbf{x}$ with $x_i = 1$ if $i \in \mathcal{A}$ and 0 otherwise. Thus we limit the number of new maximal chains by $|\mathcal{P}(\mathcal{N}) - \mathcal{C}|$.

Let us examine the computational cost of our approach compared to using all the values of the capacity in the fitting process (8.1). Firstly, we have reduced the number of variables and constraints in problem (8.45). Although the total number of variables to identify in both problems (8.45) and (8.46) is the same ($2^n - 2$) and the total number of monotonicity constraints is also the same as in (8.1), we have split them into two consecutive smaller problems, and consequently (given superlinear time complexity of each problem) solving two smaller problems is more efficient than solving one larger problem.

However, problem (8.46) is not an LP, but a nonlinear convex constrained optimisation problem, and our aim was for a more significant reduction in complexity. Therefore, next, we simplify (8.46) to avoid solving such a problem altogether. To this end we establish the upper and lower bounds on the variables $\mu(\mathcal{A}), \mathcal{A} \notin \mathcal{C}$. This is easily accomplished by setting $UB(\mathcal{A}) = \min_{\mathcal{B}\in\mathcal{C},\mathcal{B}\supset\mathcal{A}} \mu(\mathcal{B})$ and $LB(\mathcal{A}) = \max_{\mathcal{B}\in\mathcal{C},\mathcal{B}\subset\mathcal{A}} \mu(\mathcal{B})$. Now we can proceed in the direction of maxitive capacities and set $\mu(\mathcal{A}) = LB(\mathcal{A})$.

Alternatively, the suitable values of each $\mu(\mathcal{A}), \mathcal{A} \notin \mathcal{C}$ can be set to $\frac{UB(\mathcal{A})+LB(\mathcal{A})}{2}$, which will be consistent with the monotonicity constraints as for $\mathcal{A} \subset \mathcal{B}, \mathcal{A}, \mathcal{B} \notin \mathcal{C}$, $UB(\mathcal{A}) \leqslant UB(\mathcal{B})$ and $LB(\mathcal{A}) \leqslant UB(\mathcal{B})$ by construction. To further increase the spread, we can take the weighted mean $\frac{aUB(\mathcal{A})+(n-a)LB(\mathcal{A})}{n}$ with $a = |\mathcal{A}|$, which would ensure that for $\mathcal{A} \subset \mathcal{B}, \mathcal{A}, \mathcal{B} \notin \mathcal{C}$ with both $\mathcal{A}, \mathcal{B}$ having equal bounds, $\mu(\mathcal{A}) < \mu(\mathcal{B})$. Satisfaction of the monotonicity constraints is guaranteed in all three cases.

To efficiently determine the upper and lower bounds, we proceed by calculating all the values of the bounds on $\mu(\mathcal{A})$ for each cardinality $a = 1, 2, \ldots, n$ (for the lower bound) and in the reverse order for the upper bound in turn. Consider calculation of the lower bounds. For $a = 1$ we trivially have $LB(\mathcal{A}) = 0$ for each singleton. For $a = 2$ we first take $LB(\mathcal{A}) = \mu(\mathcal{A})$ for $\mathcal{A} \in \mathcal{C}$ and then $LB(\mathcal{A}) = \max_{\mathcal{B}, b=a-1} LB(\mathcal{B})$ for the remaining $\mathcal{A} \notin \mathcal{C}$. We proceed in this way until $a = n$. The number of operations we perform is the same as the total number of monotonicity constraints $n2^{n-1}$. However, in contrast to setting up a mathematical programming problem with a higher number of constraints and then solving it, here we perform only the elementary operations (taking maximum), which in addition can also be easily parallelised (for shared, distributed memory systems or GPU). This is where the main gain in complexity reduction comes from.

A combination of both approaches—k-interactivity and reducing the number of variables to those directly identifiable from the data—leads to even further reduction of both the variables and the constraints. For example, for $n = 20$, $k = 4$ we have 6196 variables (vs 1,048,574 when $k = 20$) and 100,720 constraints (vs 10 million). When we add the second method with $J = 1000$, on average in our experiments we recorded 3700 variables with 5500 constraints (the numbers vary depending on the input data distribution; we generated uniformly distributed data as a worst-case scenario in terms of the reduction of variables).

8.9 Learning in Marginal Contribution Representation

We now set up the learning problem in the marginal contribution representation presented in Sect. 4.8. We have the Choquet integral in terms of marginal contributions as

$$C_\mu(\mathbf{x}) = \sum_{i=1}^{n} x_{\pi(i)} \Delta_{\pi(i)} \mu(H_{\pi(i+1)}), \tag{8.47}$$

where $x_{\pi(.)}$ is a non-decreasing permutation induced by x_i, $i = 1, \ldots, n$, and $H_{\pi(i)} = \{\pi(i), \ldots, \pi(n)\}$, $H_{\pi(n+1)} = \emptyset$, by convention.

Further, if all the elements of $\mathbf{x}$ are all different, we can have the following expression for the Choquet integral.

Theorem 8.1 *If* $\forall x_i, x_j \in \mathbf{x}$, $x_i \neq x_j$, *then*

$$\begin{aligned} C_\mu(\mathbf{x}) &= \sum_{(i,\mathcal{A})} y_{i,\mathcal{A}} \Delta_i \mu(\mathcal{A}) x_i \\ &= \sum_{(i,\mathcal{A})} y_{i,\mathcal{A}} \Delta_i \mu(\mathcal{A}) \overline{x}_{\{i\}\cup\mathcal{A}}, \end{aligned} \tag{8.48}$$

where $y_{i,\mathcal{A}} = 1 - \text{sign}(\max_{j\in\mathcal{N}\setminus\mathcal{A}} x_j - \min_{j\in\{i\}\cup\mathcal{A}} x_j)$, sign *is the sign function,* $\overline{x}_\mathcal{A} = \min_{i\in\mathcal{A}}(x_i)$.

In practice, if there are equal evaluations of the criteria, we can introduce small random perturbations, e.g. within $10^{-6\sim-8}$, added to the equal inputs. The small perturbation does not significantly affect the overall evaluation and at the same time ensures that Eq. (8.48) works. This formulation is implemented in the FMTools package [Bel07, Bel24].

It is then possible to formulate the capacity learning problem in terms of the marginal contributions as a linear programming problem. There is no need to enforce monotonicity as, by default, in linear programming the variables are non-negative. Additional constraints on the desired orness value or interaction indices can also be added.

Note 8.3 If we adopt the marginal contributions as the variables for representing the capacity, the number of (non-negative) variables is $n \times 2^{(n-1)}$, and the number of (possibly redundant) constraints is $n!$ as shown in Theorem 4.9. This contrasts with the much smaller number of variables and constraints in standard or Möbius representation. The rationale for using the marginal contribution representation is the efficient treatment of sub- and supermodularity constraints.

Theorem 8.2 ([WB20]) *For a capacity* μ, *if its marginal contributions on every maximal chain increase (resp. decrease) w.r.t. the cardinality of the subset, i.e.* $\Delta_{\pi(i)}\mu(H_{\pi(i-1)}) \geqslant$ *(resp.* $\leqslant$ *)* $\Delta_{\pi(j)}\mu(H_{\pi(j-1)})$, $\forall \pi(i) \geqslant \pi(j)$, *then* μ *is supermodular (resp. submodular).*

Thus specification of the sub- and supermodularity constraints also involves a very sparse matrix, easily constructed at the same time as the data-fitting conditions (the equality constraints involving the residuals in (8.21)) on each maximal chain, which favourably compares to the standard or Möbius representations; c.f. [BJL11].

Another use of the marginal representation is an explicit representation of the entropy (3.36), which can be written as

$$H(\mu) = \sum_{i=1}^{n} \sum_{\mathcal{A} \subseteq \mathcal{N} \setminus \{i\}} \frac{(n - |\mathcal{A}| - 1)!|\mathcal{A}|!}{n!} h\left(\Delta_i(\mathcal{A})\right). \tag{8.49}$$

This way one can formulate the entropy maximisation problem as a standard convex optimisation problem with linear inequality and equality constraints, and then apply standard solution machinery [BV04].

It is straightforward to specify the respective conditions for submodular (concave) capacities by using duality. The increasingness on maximal chains changes to decreasingness, and conjunctive behaviour changes to disjunctive. Submodular capacities correspond to convex Choquet integrals, which are used, for instance, to define metrics [BJL11].

While the condition in Theorem 8.2 is sufficient but not necessary, it provides a valuable tool for constructing capacities in marginal representation from data. The reason is that checking supermodularity using conditions as stated in [BJL11] is computationally more expensive, whereas the increasingness on the maximal chains involves a smaller set of simple inequality constraints. Because fitting the Choquet integral to data also involves constraints on the maximal chains, the additional monotonicity constraints in the marginal contributions representation can be added with very little computational effort.

8.10 (Max, +) Transform and Capacity Fitting

The following transform was presented in [Tor22, Tor23a, Tor23b].

Definition 8.1 ((Max, +) Transform) Let μ be a capacity on $\mathcal{N}$. Then, we define the (max,+) transform as the set function $\tau_\mu : \mathcal{P}(\mathcal{N}) \to [0, 1]$ such that:

$$\tau_\mu(B) = \mu(\mathcal{B}) - \max_{\mathcal{A} \subset \mathcal{B}} \mu(\mathcal{A}). \tag{8.50}$$

It can be seen from this definition that τ_μ is always positive and that for an arbitrary normalised capacity μ and any set $\mathcal{B} \subset \mathcal{N}$, the (max, +) transform τ_μ is at most one. When μ is an additive measure, $\tau_\mu(\mathcal{B}) = \min_{i \in B} \mu(\{i\})$.

The inverse transform is obtained by recursion,

$$\mu(\mathcal{B}) = \tau_\mu(\mathcal{B}) + \max_{\mathcal{A} \subset \mathcal{B}} \mu(\mathcal{A}). \tag{8.51}$$

One can see here that the monotonicity condition is automatically satisfied, but in contrast to the possibilistic transform, the values $\tau_\mu(\mathcal{A})$ express the marginal contribution to the total value of the set $\mathcal{A}$. Some of those contributions can be explicitly set to 0, similar to the Möbius values. The transformations between Möbius, possibilistic Möbius transforms, and (max, +) transform are specified in [Tor22].

Compared with the k-additive capacities, which in spite of the reduction in the number of parameters still require the same number of linear constraints to satisfy monotonicity, the values τ_μ require no additional monotonicity constraints, which has advantages when learning capacities in that particular representation.

By replacing the operation $+$ with any co-norm operation $\oplus$ (while keeping max unchanged), and $-$ with the corresponding norm operation $\ominus$, we obtain the (max, $\oplus$) transform, which exhibits similar properties to the (max, +) transform [Tor22].

8.11 Learning Capacities for Aggregation with the Sugeno Integral

While the Sugeno integral has been proposed as a suitable aggregation tool in various applications related to fuzzy sets and decision-making, the great majority of works focusing on learning capacities have been based on Choquet integral models. Without learning and optimisation frameworks such as those detailed in the above sections, the applicability of other fuzzy integrals such as the Sugeno integral becomes limited to small problems and examples. However, the problem of identifying capacity values from data when the aggregation is to be performed with the Sugeno integral presents additional challenges on top of the increasing number of variables and monotonicity constraints. Namely, the Sugeno integral cannot be expressed as a linear combination of the capacity values and inputs, and hence, the methods based on linear and quadratic objectives are not applicable.

As with (8.1), we assume we have access to a dataset in the format of Table 8.1 with observed $\mathbf{x}^j$ and y_j. In practice, the Sugeno integral should be suitable for data aggregation over ordinal sets, e.g. linguistic labels like “high”, “very high”, “low”, etc.; however, for the purpose of fitting we assume that the data can be transformed to a numeric interval.

The objective becomes

$$\text{Minimise}||\mathbf{r}|| = ||S_\mu(\mathbf{x}) - \mathbf{y}||$$

where $||\cdot||$ is some norm, i.e. for least absolute deviation we have

$$||\mathbf{r}|| = ||S_\mu(\mathbf{x}) - \mathbf{y}||_1 = \sum_{j=1}^{J} |S_\mu(\mathbf{x}^j) - y_j|,$$

subject to the monotonicity and boundary constraints on the capacity.

Algorithmic approaches to the task of Sugeno integral fitting include those based on neural networks [YK96] and genetic algorithms [AKH10], as well as those which search for compatible, rather than “best-fitting”, measures [PRS09]. More recently, branch-and-bound type approaches minimising the squared error and least absolute deviations have been developed in [GJ18, GJB19] and a bi-level approach

Table 8.4 Three vectors aggregated using the Sugeno integral with respect to an unknown capacity. Underlined values indicate the inputs that y_j falls between

j	$x^j_{(1)}$	$x^j_{(2)}$	$x^j_{(3)}$	$x^j_{(4)}$	$x^j_{(5)}$	y_j
1	0.21	0.30	<u>0.31</u>	<u>0.70</u>	0.73	0.33
2	0.60	<u>0.78</u>	<u>0.90</u>	0.91	1.00	0.78
3	0.33	<u>0.42</u>	<u>0.67</u>	0.71	0.86	0.63

for minimising the maximum error has been proposed in [BGJ20a]. It was noted in [GJ18] that the solution to the least squares minimisation of error becomes particularly difficult because there can be multiple local minima.

If the data can be fit exactly, we can infer the values of $\mu(\mathcal{A})$ in a relatively straightforward manner, by exploiting the fact that the Sugeno integral always returns either one of the inputs or one of the capacity values as the output.

Consider the example data presented in Table 8.4 with $n = 5$ and J consisting of three instances. The vectors have been arranged into non-decreasing order.

Recall that the Sugeno integral can be expressed as

$$S_\mu(\mathbf{x}) = \max_{i=1,\dots,n} \min\{x_{(i)}, \mu(H_i)\}, \tag{8.52}$$

where $\mathbf{x}_\nearrow = (x_{(1)}, x_{(2)}, \dots, x_{(n)})$ is a non-decreasing permutation of the input $\mathbf{x}$ and $H_i = \{(i), \dots, (n)\}$.

For $j = 1$ in Table 8.4, the observed output falls between the third and fourth inputs, and so we can deduce that $\mu(H_4) = 0.33$. Since $\mu(H_4) \leqslant \mu(H_3)$, we have:

$$\mu(H_3) \wedge x^1_{(3)} = 0.31, \qquad \mu(H_4) \wedge x^1_{(4)} = 0.33,$$

and therefore the output is indeed equal to 0.33.

We have slightly different reasoning for $j = 2$, because y_2 is *equal* to the second input. As $\mu(H_3) \leqslant \mu(H_2)$, we necessarily have that $\mu(H_2) \geqslant 0.78$ as well as $\mu(H_3) \leqslant 0.78$. This gives:

$$\mu(H_2) \wedge x^2_{(2)} = 0.78, \qquad \mu(H_3) \wedge x^2_{(3)} \leqslant 0.78,$$

and so the output is 0.78.

In the case of $j = 3$, the input falls between the second and third inputs, so (similar to the case of $j = 1$) we will have $\mu(H_3) = 0.63$.

In general, we can say that if $x_{(a-1)}$ is the largest input such that $x_{(a-1)} < y^j$, then $\mu(H_a) \geqslant y^j$, and if $x_{(b)}$ is the smallest input such that $x_{(b)} > y_j$, then $\mu(H_b) \leqslant y_j$. If y_j falls between two values, then this results in $H_a = H_b$ and $\mu(H_a) = y_j$, while if it coincides with one of the values, then the measure of the H_i corresponding with that value will need to be greater than the observed y and H_{i+1} will be lower.

Such inequalities, in conjunction with the monotonicity constraints on the capacity, can be used to identify capacities in the case of zero error, e.g. a constraint

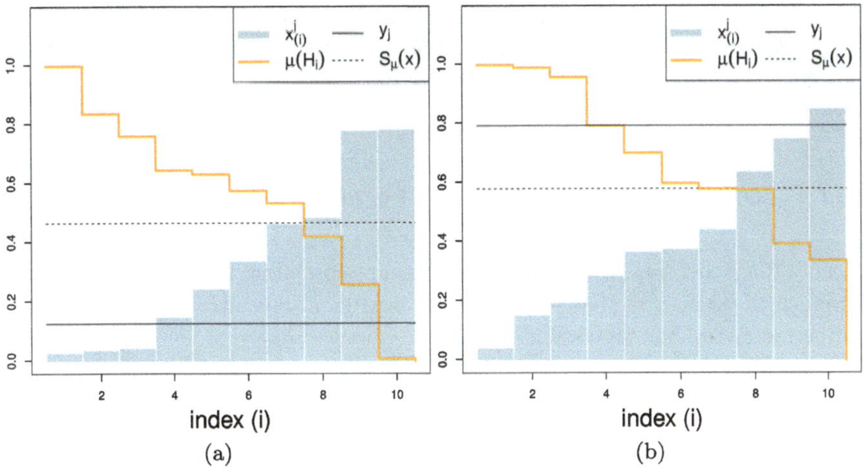

Fig. 8.2 Example input vectors $\mathbf{x}$ along with the actual $\mu(H_i)$ values and the output $S_\mu(\mathbf{x})$ compared to the observed (noisy) output y. The relevant H_i cannot be inferred from the observed value

satisfaction problem with respect to the following constraints:

$$\mu(H_{a_j}) \geqslant y_j, \; j = 1, \ldots, J \tag{8.53}$$
$$\mu(H_{b_j}) \leqslant y_j, \; j = 1, \ldots, J$$
$$\mu(\mathcal{A}) \geqslant \mu(\mathcal{A} \setminus \{i\}), \forall i \in \mathcal{A} \text{ for all } \mathcal{A} \subseteq \mathcal{N}$$
$$\mu(\mathcal{N}) = 1,$$

where a_j is the largest index such that $x_{(a_j-1)} < y^j$ and b_j is the smallest index such that $x_{(b_j)} > y_j$ for each j and H_{a_j} is the relevant set $\{(a_j), (a_j+1), \ldots, (n)\}$ and similarly for H_{b_j}.

However, where the aim is to minimise the l_1 or l_2 error when the data contains noise or erroneous inputs (which would usually be the case), such inferences about the position of the output with respect to the inputs can cause problems.

Figure 8.2 helps illustrate this. In (a), the actual capacity values produce an output $S_\mu(\mathbf{x})$ that is equal to $x_{(7)}$; however, the observed y lies between $x_{(3)}$ and $x_{(4)}$ (so we would infer that $\mu(H_4)$ is well below its actual value in this case). Similarly in (b), the observed value lies between $x_{(9)}$ and $x_{(10)}$, while the actual output should lie between $x_{(7)}$ and $x_{(8)}$.

The constraints in (8.53) are used as the basis for a bi-level fitting approach minimising the maximum error in [BGJ20a], whereby additional parameters are incorporated to model an auxiliary residual and a value which expands to coincide with the max error (see Sect. 8.11.5 below).

As further methods for fitting the Sugeno integral are developed, addressing its ordinal capabilities—in particular taking into account that distances between values may not be defined—will be a key consideration.

8.11.1 Ordinal Learning Problem Formulation

When fitting the Sugeno integral, the learning problem becomes much more complicated because of the non-convex multiextremal objective [GJ18, BGJ20a]. Furthermore, ordinal regression [AGM10, Cor+13, Ang+16, CGI16] implies a different objective: it is not required to fit the target outputs but rather preserve their relative ordering.

We assume that the data is provided in the form of linguistic variable labels, which are mapped to integers and then to the respective rationals from [0, 1]. In the context of ordinal regression, we are not required to fit the values y_j. Instead we require that the order of the observed values is matched by the prediction of the model. Thus, if $y_j \leqslant y_k$, we require $S_\mu(\mathbf{x}^j) \leqslant S_\mu(\mathbf{x}^k)$. As the data may contain inaccuracies or may not correspond to a Sugeno integral with respect to any capacity, we expect inconsistencies in the two orders. Our aim is to minimise the inconsistencies.

Consider the following problem:

$$\begin{aligned} &\text{minimise } F(\mu) = \sum_{j,k=1}^{J} 0 \vee (S_\mu(\mathbf{x}^j) - S_\mu(\mathbf{x}^k)) \\ &\qquad \text{for all } j, k \text{ such that } y_j \leqslant y_k \\ &\text{s.t. } \mu \text{ is monotone and } \mu(\emptyset) = 0,\ \mu(\mathcal{N}) = 1. \end{aligned} \tag{8.54}$$

Note that whenever for $y_j \leqslant y_k$ we have $S_\mu(\mathbf{x}^j) \leqslant S_\mu(\mathbf{x}^k)$, the corresponding term in the objective is 0. Also note that there are, in general, $2^n - 2$ variables $\mu(\mathcal{A})$ and $O(n2^n)$ linear inequality constraints representing monotonicity.

The objective F is not a convex function. We present some methods of solution in the next section based on its DC decomposition.

8.11.2 Optimisation Approaches

The objective function F in the preceding formulation is a piecewise linear non-convex function. Such functions require methods of non-smooth optimisation [BKM14], as they are not differentiable (especially at the critical points). Furthermore, there are many linear constraints on the feasible domain, which require specialised methods or introducing suitable penalty terms. However, in [GJB19] it was argued that the structure of these objective functions can be exploited.

To illustrate this, first we consider the problem of fitting a symmetric capacity, which we later extend to other types of capacities. In this particular case, the objective becomes:

$$\begin{aligned} \text{minimise } F(\mathbf{h}) = \sum_{j,k=1}^{J} 0 \vee \left(\max_{i=1,\dots,n} \min\{x_i^j, h_i\} - \max_{i=1,\dots,n} \min\{x_i^k, h_i\} \right) \\ \text{for all } j, k \text{ such that } y_j \leqslant y_k \\ \text{s.t. } 1 = h_1 \geqslant h_2 \geqslant \dots \geqslant h_n \geqslant 0, \end{aligned} \tag{8.55}$$

where, for brevity of notation, we have $\mu(H_i) = h_i$ and now assume that each $\mathbf{x}^j$ has been increasingly ordered in advance. The objective is a piecewise linear function, and as such it is Lipschitz and also a DC function. The latter is quite important as it allows us to apply the methods of DC optimisation [HT99] to numerically solve a challenging global optimisation problem.

DC Decomposition

We now provide a DC decomposition of the objective $F = G - H$, where G and H are both convex functions, that will facilitate the calculation of the subdifferentials and application of the DC algorithm. We shall use the following identities:

$$\max_{i=1,\dots,n} \{g_i - h_i\} = \max_{i=1,\dots,n} \{g_i + \sum_{j\neq i} h_j\} - \sum_{i=1,\dots,n} h_i,$$

$$\min\{g, h\} = 0 - \max\{-g, -h\}, \; g, h \text{ linear},$$

$$\max_{i=1,\dots,n} \min\{x_i, h_i\} = \max_{i=1,\dots,n} \sum_{j\neq i} \max\{-x_j, -h_j\} - \sum_{i=1,\dots,n} \max\{-x_i, -h_i\}.$$

Note that the right-hand side in each formula is the difference of two convex functions, so applying them recursively and using the fact that the sum of convex functions is convex, we get

$$\begin{aligned} F(\mathbf{h}) = \sum_{j,k} 0 \vee \Bigg[&\left(\max_{i=1,\dots,n} \sum_{m\neq i} \max\{-x_m^j, -h_m\} + \sum_{i=1,\dots,n} \max\{-x_i^k, -h_i\} \right) \\ &- \left(\sum_{i=1,\dots,n} \max\{-x_i^j, -h_i\} + \max_{i=1,\dots,n} \sum_{m\neq i} \max\{-x_m^k, -h_m\} \right) \Bigg]. \end{aligned} \tag{8.56}$$

Lastly, applying the formula $\max\{0, G - H\} = \max\{H, G\} - H$, we get the DC formulation

$$F(\mathbf{h}) = \sum_{j,k|y_j \leqslant y_k} \max\{H_{jk}(\mathbf{h}), G_{jk}(\mathbf{h})\} - \sum_{j,k|y_j \leqslant y_k} H_{jk}(\mathbf{h}), \tag{8.57}$$

where

$$\begin{aligned} G_{jk}(\mathbf{h}) = &\quad \max_{i=1,\ldots,n} \sum_{m \neq i} \max\{-x_m^j, -h_m\} + \sum_{i=1,\ldots,n} \max\{-x_i^k, -h_i\}, \\ H_{jk}(\mathbf{h}) = &\quad \sum_{i=1,\ldots,n} \max\{-x_i^j, -h_i\} + \max_{i=1,\ldots,n} \sum_{m \neq i} \max\{-x_m^k, -h_m\}. \end{aligned}$$

Note that $G_{jk}(\mathbf{h}) = H_{kj}(\mathbf{h})$. We can also write

$$\begin{aligned} &\max_{i=1,\ldots,n} \sum_{m \neq i} \max\{-x_m^j, -h_m\} \\ &\qquad = \sum_{i=1,\ldots,n} \max\{-x_i^j, -h_i\} - \min_{i=1,\ldots,n} \max\{-x_i^j, -h_i\}, \end{aligned}$$

keeping in mind that $\min\{-x, -y\} = -\max\{x, y\}$ and $\max\{-x, -y\} = -\min\{x, y\}$.

Finally, we can rewrite the above objective as

$$F(\mathbf{h}) = G(\mathbf{h}) - H(\mathbf{h}) \tag{8.58}$$

with

$$\begin{aligned} G(\mathbf{h}) = &\sum_{j=1}^{J-1} \sum_{k=j+1}^{J} \max\left\{ \bigvee_{i=1}^{n} \min\{x_i^j, h_i\}, \bigvee_{i=1}^{n} \min\{x_i^k, h_i\} \right\} \\ &- \sum_{j=1}^{J} (J-1) \sum_{i=1}^{n} \min\{x_i^j, h_i\} \end{aligned} \tag{8.59}$$

$$\begin{aligned} H(\mathbf{h}) = &\sum_{j=1}^{J} (j-1) \bigvee_{i=1}^{n} \min\{x_i^{\sigma(j)}, h_i\} \\ &- \sum_{j=1}^{J} (J-1) \sum_{i=1}^{n} \min\{x_i^j, h_i\}, \end{aligned} \tag{8.60}$$

where σ is a permutation such that for all j it either holds that $y_{\sigma(j)} < y_{\sigma(j+1)}$ or that both $y_{\sigma(j)} = y_{\sigma(j+1)}$ and

$$\bigvee_{i=1}^{n} \min\{x_i^{\sigma(j)}, h_i\} \geqslant \bigvee_{i=1}^{n} \min\{x_i^{\sigma(j+1)}, h_i\}.$$

The latter qualification is needed to deal with ties. However, if all observations in $\mathbf{y}$ are unique, σ is just the ordering permutation of the outputs and it might be computed once at the data preprocessing step.

DC Algorithm

DC decomposition allows one to apply specialised optimisation algorithms to locate the minima of non-convex multiextremal functions, which is a very challenging optimisation problem. Recall that convex functions have unique global minima (although not necessarily a single minimiser), and that many convex optimisation methods with proven convergence are available. The DC optimisation approach for $F = G - H$ is an iterative method, where at each iteration one replaces the convex function H with a suitable linear expression L and hence solves a convex optimisation problem for $G - L$, outlined below, by some standard technique. The DC iterations converge to a local minimum of F.

Let us recall the DC algorithm (DCA) [LLP04], which consists in iterating two steps.

Algorithm 8.3: DCA

Input: starting point $\mathbf{h}_0$ and $i = 0$
While convergence criteria not met **do**:
1. Compute $\mathbf{y} \in \partial H(\mathbf{h}_i)$.
2. Solve $\mathbf{z} = \arg\min_{\mathbf{h}} G(\mathbf{h}) - \langle \mathbf{h}, \mathbf{y} \rangle$ for a fixed $\mathbf{y}$.
   Set $\mathbf{h}_i = \mathbf{z}$ and $i = i + 1$.

Typical convergence criteria include reaching the difference between two successive iterations below a threshold, a threshold on the value of F, or a predefined number of iterations. Note that the optimisation problem at Step 2 is convex and hence is solvable by many standard methods. Calculation of an element of the subgradient of H, denoted ∂H, in Step 1 is straightforward in our case, e.g. we can simply take one-sided derivatives of the terms in H. Let us then focus on Step 2 of the DC algorithm now.

Inner Step: Linear Programming-Based Approach

The function G is piecewise linear and convex and so is the objective in the DCA. One approach is to rely on linear programming (LP), whereby we convert Step 2 into a linear programming problem. Let us introduce auxiliary variables $\mathbf{a} \in \mathbb{R}^{Jn}$, $\mathbf{b} \in \mathbb{R}^{J(J-1)/2}$ in order to convert the expression for G into a linear function

$$a_{ij} = \min\{x_i^j, h_i\},$$

$$b_{jk} = \max\{\bigvee_{i=1}^{n} a_{ij}, \bigvee_{i=1}^{n} a_{ik}\}.$$

Then the associated linear programming problem at Step 2 is

$$\begin{aligned}
\text{minimise } P(\mathbf{h}, \mathbf{a}, \mathbf{b}) &= -\sum_{i=1}^{n} y_i h_i + \sum_{j,k,j<k} b_{jk} - \sum_{i=1}^{n}\sum_{j=1}^{J}(J-1)a_{ij} \\
\text{s.t.} \quad & b_{jk} \geqslant a_{ij}, i = 1, \ldots, n, \ \forall j < k \qquad (8.61)\\
& b_{jk} \geqslant a_{ik}, i = 1, \ldots, n, \ \forall j < k \\
& a_{ij} \leqslant x_i^j, i = 1, \ldots, n, \ j = 1, \ldots, J \\
& a_{ij} \leqslant h_i, i = 1, \ldots, n, \ j = 1, \ldots, J \\
& h_i \geqslant h_{i+1}, \ i = 1, \ldots, n-1 \\
& h_1 = 1.
\end{aligned}$$

This problem is solved by the standard simplex method. Note that the same set of constraints is involved irrespective of the value $\mathbf{y}$ obtained at Step 1 of the DCA. For this reason, as DCA iterates, we have a sequence of similar LP problems at Step 2, which only differ by some of the coefficients in the objective. Hence, from the practical implementation perspective, one can set up a generic LP model and update it from iteration to iteration by changing the coefficients in the objective, rather than setting a new LP problem every time.

8.11.3 Non-symmetric Capacities

We now address the issue of fitting general capacities, where the Sugeno integral is given by Eq. (5.17), and the values $\mu(\mathcal{A})$ satisfy monotonicity and boundary conditions. Here the objective (8.58) remains essentially the same, but we change the variables $\mathbf{h}$ to $\mu \in R^{2^n}$ and also change expressions of the form $\min\{x_i^j, h_i\}$ to $\min\{x_i^j, \mu(\{\sigma(i), \sigma(i+1), \ldots, \sigma(n)\})\}$, where σ is a permutation such that $x_{\sigma(i)}^j$ is the i-th smallest element of vector $\mathbf{x}^j$.

Hence, for general capacities we have a much larger number of variables and constraints. Nevertheless the DC decomposition in (8.58) remains. Similar to the case of symmetric capacities, the DCA can be applied. The convex piecewise linear problem at Step 2 can be solved either by using a derivative free optimisation method or by using linear programming similar to formulation (8.61), with the corresponding increase in the number of variables and constraints. The latter could offer significant advantages because (a) handling a large number of linear constraints is efficient using LP methods, and (b) the system of constraints is very sparse, and LP software is well suited to handling sparse matrices.

Still, even moderate dimension n leads to an excessively large number of monotonicity constraints. As a way of reducing the complexity of capacities, the concept of k-order capacities can be used. In the framework of the Sugeno integral, k-maxitive capacities are used [BC18, MGB17].

The k-maxitive capacities, for a particular $1 \leqslant k \leqslant n$, satisfy the condition $\mu(\mathcal{A}) = \max_{\mathcal{B} \subset \mathcal{A}} \mu(\mathcal{B})$ for $|A| > k$. This implies that for at least one $\mathcal{A}$ of cardinality k we have $\mu(\mathcal{A}) = 1$. Fitting the Sugeno integral to ordinal data subject to k-maxitivity was recently performed in [BC18] using simulated annealing optimisation.

It is possible to specify the objective (8.63) under the condition of k-maxitivity, and even preserve the structure of the DC decomposition and problem formulation (8.61), but at the expense of an extra set of 0-1 variables, which model the constraints that $\mu(\mathcal{A}) = \mu(\mathcal{A} \setminus \{i\})$ for at least one $i \in \mathcal{A}$ for all $\mathcal{A}$ with cardinality greater than k [BW21], and this becomes expensive numerically.

To avoid an excessive number of monotonicity constraints, we use k-interactive capacities: the values of the capacity are fixed at subsets of cardinality larger than k in such a way as to maximize its partial entropy. The Sugeno integral with respect to such a capacity is expressed as

$$S_\mu(\mathbf{x}) = \left(\bigvee_{\mathcal{A} \subseteq \mathcal{N}, |\mathcal{A}| \leqslant k} (\min_{i \in \mathcal{A}} x_i) \wedge \mu(\mathcal{A}) \right) \bigvee x_{(n-k)}. \tag{8.62}$$

That is, the $n-k-1$ smallest components of $\mathbf{x}$ do not participate in the calculation, and no weighting is applied to the $n-k$-th smallest component.

The advantage of using this approach is a very significant reduction in the number of monotonicity constraints, and also of the number of variables. On the other hand, one can apply the DC decomposition (8.58) and solve an LP problem similar to (8.61) at Step 2 of the DC algorithm. For example, for a symmetric capacity, we only need to add the constraints $h_2 = h_3 = \ldots = h_{n-k} = 1$ in (8.61).

On the technical side, note that the same set of constraints is involved irrespective of the value $\mathbf{y}$ obtained at Step 1 of the DCA. As was the case with symmetric capacities, we have a sequence of similar LP problems at Step 2 that only differ in terms of the coefficients in the objective. Once again we can save computation time by making small updates to a generic LP model from iteration to iteration.

8.11.4 Fitting Sugeno Integral for Learning Capacities Using PAVA Isotone Regression

As earlier, fitting the values y_j exactly may be unnecessary. What is important is that the order of observed values is matched by the prediction of the model. Thus if $y_j \leqslant y_k$ we require $S_\mu(\mathbf{x}^j) \leqslant S_\mu(\mathbf{x}^k)$. Our aim is therefore to minimise the inconsistencies between the orderings. In this section we follow [BD19].

The learning problem is formulated as

$$\begin{aligned} \text{minimise } F(\mu) &= \sum_{j,k=1}^{J} D(S_\mu(\mathbf{x}^j), S_\mu(\mathbf{x}^k)) \\ &\text{for all } j, k \text{ such that } y_j \leqslant y_k \\ &\text{s.t. } \mu \text{ is monotone and } \mu(\emptyset) = 0, \mu(\mathcal{N}) = 1. \end{aligned} \tag{8.63}$$

Here the function $D : [0, 1]^2 \to [0, 1]$ is defined in such a way that it is positive only when $a > b$. For example, we can use $D(a, b) = 0 \vee (a - b)$, or $D(a, b) = 0$ if $a \leqslant b$ and 1 otherwise.

As previously, if for $y_j \leqslant y_k$ we have $S_\mu(\mathbf{x}^j) \leqslant S_\mu(\mathbf{x}^k)$, the corresponding term in the objective is 0. We assume that the data are reordered according to the outputs, that is, $y_j \leqslant y_k$ whenever $j < k$.

In the context of isotone regression [BC90, RWD88], monotonising the inputs is a commonly used strategy. Usually the data is monotonised by minimising the squared differences to the inputs. It becomes a quadratic programming problem, solved by active set methods [BC90]. A popular PAV algorithm (Pool Adjacent Violators Algorithm PAVA) is one method that provides efficient numerical solution.

PAVA consists in the following steps. A sequence z_j, $j = 1, \ldots, J$ is processed from the start. If violation of monotonicity $z_j > z_{j+1}$ is found, both values z_j and z_{j+1} are replaced with their average z'_j, and both values form a block. Since the new value z'_j is now smaller than z_j, monotonicity may become violated with respect to the datum z_{j-1}. If this is the case, the values for $j-1$, j and $j+1$ are merged into a block and their values are replaced with their average. We continue back-averaging as needed to get monotonicity.

Another approach to isotonic regression is called the minimum lower set (MLS) algorithm [BC90, RWD88]. It provides the same solution as the PAVA, but works differently. For each datum (or block), MLS selects the largest contiguous block of subsequent data with the smallest average. If this average is smaller than that of the preceding block, the blocks are merged, and the data in the block are replaced with their average. MLS is also an active set method [BC90], but its complexity is $O(J^2)$ as opposed to $O(J)$ of the PAVA and of another active set algorithm proposed in [BC90] under the name Algorithm A.

In all mentioned cases the solution to isotone regression problem can be found by solving the maximin problem

$$u_j = \max_{k \leqslant j} \min_{l \geqslant j} \hat{z}(k, l),$$

with $\hat{z}(k, l)$ being the unrestricted maximum likelihood estimator of $z_k \dots, z_l$. For a quadratic cost function $\hat{z}(k, l)$ is the mean, as in PAV and MLS algorithms; for the absolute deviations it becomes the median, and for other cost functions an M-estimator of location.

The resulting monotone sequence u_j minimises the following objective:

$$\min \sum_{j=1}^{J} (u_j - z_j)^2.$$

Now we apply PAVA to simplify the objective (8.63). We need to match the ordering of y_j, which are already sorted in increasing order, by the values $z_j = S_\mu(\mathbf{x}^j)$. By applying PAVA to the sequence z_j we obtain a monotonised sequence u_j, closest to z_j in the least squares sense, whose l_2 distance to z_j, denoted by $PAVA(\mathbf{x}, \mu)$, is now taken as the value of the objective.

Thus the method to learn the capacity consists in solving

$$\begin{aligned} &\text{minimise } F(\mu) = PAVA(\mathbf{x}, \mu) \\ &\quad \text{s.t. } \mu \text{ is monotone and } \mu(\emptyset) = 0, \mu(\mathcal{N}) = 1. \end{aligned} \tag{8.64}$$

The smallest value of the objective is 0, which is achieved if the order of the values $S_\mu(\mathbf{x}^j)$ matches that of y_j, which is the goal of ordinal regression.

8.11.5 *Fitting the Sugeno Integral by Minimising the Maximum or Median Residual Error*

As already mentioned, in the case of zero error, the output of the Sugeno integral coincides with either one of the inputs or one of the weights, $\mu(H_i)$, and so with enough observed data the capacity values for all sets can be inferred. In [BGJ20a] a bi-level optimisation approach for learning the weights $\mu(\mathcal{A})$ was proposed such that the maximum error for an observed sample is minimised, which in turn can be applied to heuristic methods aimed at minimizing other fitting criteria such as the median residual.

The approach uses a residual function, referred to as the auxiliary error and defined as follows.

Definition 8.2 For an input vector $\mathbf{x} \in [0, 1]^n$, let $\mathbf{h} \in [0, 1]^n$ denote the vector induced from a capacity μ according to the non-decreasing ordering $x_{(1)} \leqslant x_{(2)} \leqslant \dots \leqslant x_{(n)}$, i.e. $h_i = H_i = \mu(\{(i), \dots, (n)\}), i = 1, \dots, n$. For an observed output, y, let a denote the maximum index such that $x_{(a-1)} < y$ and b the minimum index such that $x_{(b)} > y$, that is,

$$a = \max\{i | x_{(i-1)} < y\}, \qquad b = \min\{i | x_{(i)} > y\},$$

with $x_{(0)} = -\infty$, $x_{(n+1)} = \infty$, $h_{n+1} = -\infty$ by convention.

The residual function $R_{\mathbf{h}}(\mathbf{x}, y) : [0, 1]^{n+1} \to [0, 1]$ associated with the Sugeno integral with respect to $\mathbf{h}$ (denoted $S_{\mathbf{h}}$) is given by

$$R_{\mathbf{h}}(\mathbf{x}, y) = \begin{cases} \max(0, y - h_a, h_b - y), & x_{(1)} \leqslant y \leqslant x_{(n)}; \\ \max(0, y - x_{(n)}, y - h_n), \ y > x_{(n)}; \\ \max(0, x_{(1)} - y, h_2 - y), \ y < x_{(1)}. \end{cases} \tag{8.65}$$

We then have the following proposition, which justifies the use of $R_{\mathbf{h}}$ as a residual function.

Proposition 8.2 *The residual $R_{\mathbf{h}}(\mathbf{x}, y) = 0$ if and only if $S_{\mathbf{h}}(\mathbf{x}) = y$.*

Example 8.1 Consider Fig. 8.3. For this vector with $n = 10$, $\mathbf{x}$ is shown as shaded bars in non-decreasing order. The corresponding values of h_i for each $x_{(i)}$ are shown as the decreasing step function. The obtained value of $S(\mathbf{x}, \mathbf{h})$ is shown as the horizontal dotted line, while an observed value of y is shown as the horizontal solid line. Since the line for y intersects the $\mathbf{x}$ distribution between $x_{(2)}$ and $x_{(3)}$, the values of a and b coincide with both equal to 3, and the value of $R_{\mathbf{h}}$ is $h_a - y = h_3 - y$.

Clearly $R_{\mathbf{h}}$ will be an overestimation of the actual error $|S_{\mathbf{h}}(\mathbf{x}) - y|$. The extent to which this is the case will depend on how close a is to the actual index used in the Sugeno integral calculation (i.e. the i such that $S_{\mathbf{h}}(\mathbf{x}) = \min(x_{(i)}, h_i)$). However, with the help of an additional parameter, we can use $R_{\mathbf{h}}$ to help us minimise the actual maximum error for any given dataset.

We first provide a linear programming formulation that minimises the maximum $R_{\mathbf{h}}$.

We let a_j, b_j denote the values of a and b with respect to the instance $(\mathbf{x}^j, y_j)$ as per Definition 8.2. Then h_{a_j} refers to the weight h_a corresponding to the instance j. Similarly, h_{b_j} is the weight corresponding to h_b for instance j. We remind that h_i corresponds with $H_i = \mu(\{(i), (i + 1), \ldots, (n)\})$. Now, let r denote the maximum $R_{\mathbf{h}}(\mathbf{x}^j, y_j)$ for all j. We have

$$\text{Minimise } r, \text{ w.r.t. } \mu, r \tag{8.66}$$

and the following linear constraints:

For all j such that $x^j_{(1)} \leqslant y_j \leqslant x^j_{(n)}$

$$h_{a_j} + r \geqslant y_j$$
$$-h_{b_j} + r \geqslant -y_j.$$

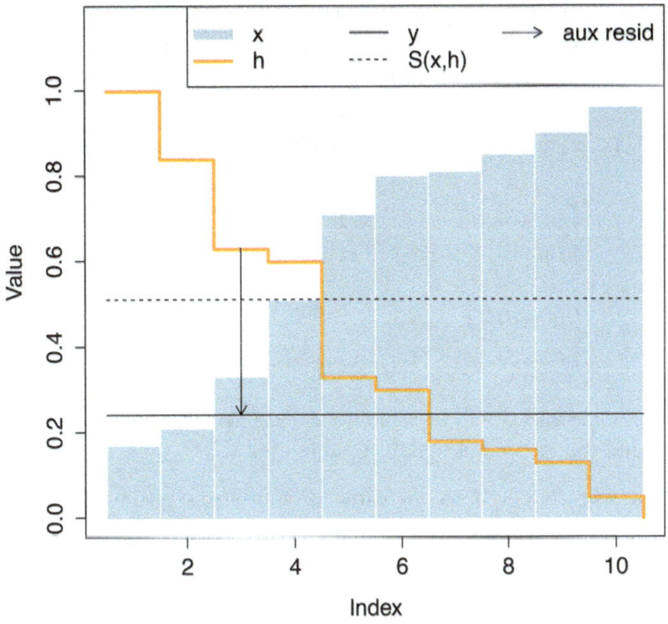

Fig. 8.3 The residual $R_{\mathbf{h}}$ (the auxiliary error) indicated by the arrow and calculated with respect to an input vector **x** and weights **h** induced by the ordering of the inputs for a given capacity. The residual is the difference between y and the relevant weight h_3, where $x_{(2)} < y < x_{(3)}$

For all j such that $y_j < x^j_{(1)}$

$$r \geqslant x^j_{(1)} - y_j$$
$$-h_{2_j} + r \geqslant -y_j.$$

For all j such that $y_j > x^j_{(n)}$

$$r \geqslant y_j - x^j_{(n)}$$
$$h_{n_j} + r \geqslant y_j.$$

Now consider a non-negative value t such that we associate the interval $[y_j - t, y_j + t]$ with the observation y_j. We let α denote the maximum index such that $x_{(\alpha-1)} < y - t$ and β the minimum index such that $x_{(\beta)} > y + t$, i.e.

$$\alpha = \max\{i | x_{(i-1)} < y - t\}, \qquad \beta = \min\{i | x_{(i)} > y + t\}$$

with $x_{(0)} = -\infty$, $x_{(n+1)} = \infty$, $h_{n+1} = -\infty$ by convention.

We can now update $R_{\mathbf{h}}$ with respect to t.

$$R_{\mathbf{h}}(\mathbf{x}, y, t) = \max(R^+_{\mathbf{h},t}(\mathbf{x}, y, t), R^-_{\mathbf{h},t}(\mathbf{x}, y, t)), \tag{8.67}$$

where $R^+_{\mathbf{h},t}(\mathbf{x}, y)$ and $R^-_{\mathbf{h},t}(\mathbf{x}, y)$ are given by

$$R^+_{\mathbf{h}}(\mathbf{x}, y, t) = \begin{cases} \max(0, h_\beta - (y+t)), & y+t \geqslant x_{(1)}; \\ \max(0, x_{(1)} - (y+t), h_2 - (y+t)), & y+t < x_{(1)}, \end{cases} \tag{8.68}$$

$$R^-_{\mathbf{h}}(\mathbf{x}, y, t) = \begin{cases} \max(0, (y-t) - h_\alpha), & y-t \leqslant x_{(n)}; \\ \max(0, (y-t) - x_{(n)}, (y-t) - h_n), & y-t > x_{(n)}. \end{cases} \tag{8.69}$$

If $t = 0$, then α and β will be equal to a and b, respectively; however, if $t > 0$, then we expand the interval at which $R_{\mathbf{h}} = 0$.

Example 8.2 Figure 8.4 shows the same $\mathbf{x}$, y and $\mathbf{h}$ as given in Fig. 8.3. Now we have $\alpha = 1$ and $\beta = 4$, and a residual value of

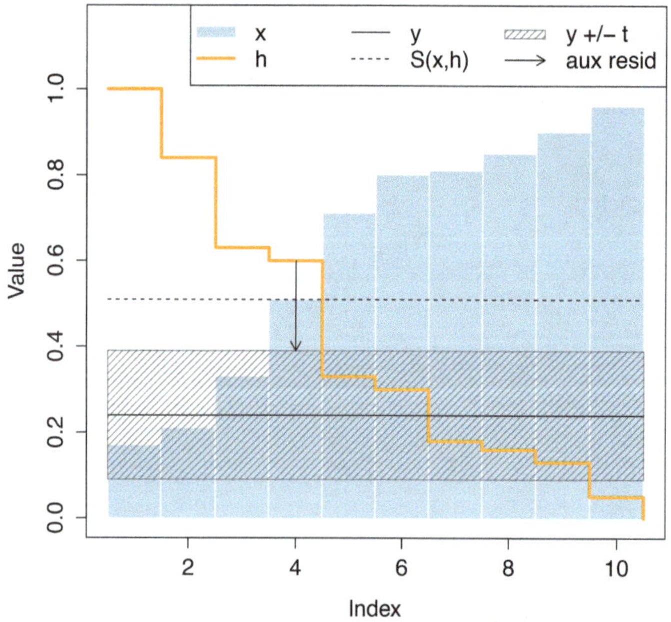

Fig. 8.4 The auxiliary error calculated with respect to the same input vector $\mathbf{x}$ and induced weights $\mathbf{h}$, where an additional parameter t is used to create an interval around y. The parameter reduces the difference and also changes the relevant weight, which is now h_4 with $x_{(1)} > y-t$ and $x_{(4)} > y+t$. Once t becomes equal to or greater than the absolute error, i.e. $t \geqslant |S_{\mathbf{h}}(\mathbf{x}) - y|$, the relevant weight will be h_5 (or h_6, h_7, etc.) and $R_{\mathbf{h}}(\mathbf{x}, y, t) = 0$

$$R_{\mathbf{h}}(\mathbf{x}, y, t) = \max(h_4 - (y+t), (y-t) - h_1) = h_4 - (y+t).$$

In summary, we have the following formulation in terms of t (along with monotonicity, boundary, or other constraints on the capacity as required):

$$\begin{aligned}
&\text{Minimise } r \quad \text{w.r.t. } \mu, r && (8.70)\\
&\text{s.t. } h_{\alpha_j} + r \geqslant y_j - t, \quad \forall j | y_j - t \leqslant x^j_{(n)}\\
&h_{n_j} + r \geqslant y_j - t, \quad \forall j | y^j - t > x^j_{(n)}\\
&r \geqslant y_j - t - x^j_{(n)}, \quad \forall j | y_j - t > x^j_{(n)}\\
&-h_{\beta_j} + r \geqslant -(y_j + t), \quad \forall j | y_j + t \geqslant x^j_{(1)}\\
&r \geqslant x^j_{(1)} - (y_j + t), \quad \forall j | y_j + t < x^j_{(1)}\\
&-h_{2_j} + r \geqslant -(y_j + t) \quad \forall j | y_j + t < x^j_{(1)}.
\end{aligned}$$

Algorithm 8.4 was also proposed for minimizing the median residual error. It relies on sampling from the observed dataset, minimising the maximum error, and then calculating the error with respect to the obtained capacity across all observed values.

Algorithm 8.4: Capacity weights minimising median residual

Input: a dataset consisting of J observations $(\mathbf{x}^j, y_j)$, s (sampling size), B (number of bisections), Q (number of iterations)
Output: μ (best fitting capacity) and r_{Med} (median residual)

1 Initialise the capacity μ as the additive and symmetric capacity
2 Calculate and initialise the best residual r_{Med}, where

$$r_{\mathsf{Med}} = \mathsf{Med}(\{|S_\mu(\mathbf{x}^j) - y_j|, j = 1, \ldots, J\})$$

3 For each of Q iterations:

3.1 Randomly sample s instances $(\mathbf{x}^j, y_j)$ from the dataset
3.2 Use bisection search (B bisections) to minimise t such that $r = 0$ in (8.70) across the s observations and obtain μ_S;
3.3 Calculate the residuals, $r^j = |S_{\mu_S}(\mathbf{x}^j) - y_j|$ and then let $r_S = \mathsf{Med}(r^1, \ldots, r^J)$;
3.3 If $r_S < r_{\mathsf{Med}}$ then replace r_{Med} with r_S and replace μ with μ_S;

4 Return μ along with r_{Med}.

The least median residual can be seen as an appropriate fitting criterion for ordinal data; however, the algorithm could be adapted so that any fitting criteria could be used.

8.12 Unsupervised Methods

In unsupervised learning there are no target output values y_j, so the capacity values need to be determined based only on the inputs $\mathbf{x}^j$.

8.12.1 A Method Based on Correlation Indices in the Data

A method by Kojadinovic [Koj07] suggests using the concept of mutual information contained in a subset $\mathcal{A} \subset \mathcal{N}$, and defines capacities by

$$\mu(\mathcal{A}) = \frac{h(\mathcal{A})}{h(\mathcal{N})},$$

where h is the entropy of subset $\mathcal{A}$, defined as the entropy of any of random vector $(x_1, x_2, \ldots, x_r)$ composed of random variables $x_1, x_2, \ldots, x_r \in \mathcal{A}$. The input data is a sample of m profiles $\mathbf{p}^{\mathcal{A}}$, and each profile vector is composed of components of inputs $\mathbf{x}^j$ that belong to $\mathcal{A}$. The resulting capacity is submodular.

A method reported in [RGL15] uses the correlation matrix R of J observations $\mathbf{x}^j$. If the variables are independent, R would be the identity matrix I_n. Denote by K^* an estimated number of non-interacting variables. The authors propose to use principal component analysis (PCA) to determine K^*. PCA calculates an ordered set of eigenvalues of R, called the principal components. Let λ_j denote the jth largest eigenvalue. Then we take

$$K^*(\mathcal{A}) = \sum_{j,\lambda_j<1} \lambda_j + |\{\lambda_j|\lambda_j \geqslant 1\}|.$$

The capacity values are then constructed as

$$\mu(\mathcal{A}) = \frac{K^*(\mathcal{A})}{K^*(\mathcal{N})}.$$

It is shown that such a definition ensures monotonicity and boundary conditions on the set function; hence, we obtain a normalised capacity. For independent variables the resulting capacity is additive and symmetric.

Some data manipulations prior to PCA are required, namely, scaling the inputs (that need not be in $[0, 1]^n$) to unit variance and subtracting the mean of each scaled component from x_i^j (thus obtaining mean-centred standardised data).

8.12.2 Similarity-Based Interaction Learning

The main idea of this type of method is based on similarity and interaction [Dua17].

To quantify the interactions among criteria, similarity or correlation coefficients are first derived from the decision matrix. These coefficients serve as a basis for estimating the strength and direction of interdependencies between criteria. The interaction indices are then defined in a manner that is consistent with the observed correlation structure, ensuring conceptual alignment between empirical data and model representation. An optimisation-based approach is subsequently developed to formalise and operationalise this alignment, enabling the systematic determination of interaction indices that best reflect the underlying relationships among criteria.

For example, in [PD24], a linear programming model is proposed with t to ensure the same ratio between interaction indices and similarity measures for all pairs of criteria,

$$
\begin{aligned}
\max_{t, I} \quad & t \\
\text{s.t.} \quad & \varphi_i - \frac{1}{2} \sum_{i' \neq i} \operatorname{sign}(-\psi_{i,i'}) \cdot I_{i,i'} \geqslant 0, \quad \forall i \in \mathcal{N} \\
& I_{i,i'} - \operatorname{sign}(-\psi_{i,i'}) \cdot t \cdot \psi_{i,i'} = 0, \quad \forall i, i' \in \mathcal{N} \\
& \sum_i \varphi_i = 1, \\
& 0 \leqslant t \leqslant 1,
\end{aligned}
$$

where φ_i is the Shapley importance of criterion i, $I_{i,i'}$ is the Shapley interaction index of criteria i and i', and $\psi_{i,i'}$ is the similarity index of decision data with respect to criteria i and i'.

8.12.3 Online Learning

In this approach proposed in [GG+25] the capacity values are not calculated for the whole power set $2^{\mathcal{N}}$ but only for one maximal chain, sufficient for calculating the Choquet integral of a particular input value. The method is oriented to online learning (the dataset is not present and the data appear one at a time) and high values of n, making it impractical to store all capacity values.

The method is based on defining capacity values by means of a weighting vector and a non-conjunctive aggregation function F as follows.

For a subset $\mathcal{A} \in 2^{\mathcal{N}}$ we define a function $\phi_A \colon [0,1]^n \to [0,1]^n$ as $\phi_A(\mathbf{x}) = (z_1, \ldots, z_n)$, where

$$z_i = \begin{cases} x_i, & \text{if } i \in \mathcal{A} \\ 0 & \text{otherwise} \end{cases} \quad \forall i \in \{1, \ldots, n\}. \tag{8.71}$$

For a fixed $\boldsymbol{\omega} = (\omega_1, \ldots, \omega_n) \in [0,1]^n$ and a non-conjunctive aggregation function F, we can construct a normalised capacity defined as

$$\mu(\mathcal{A}) = \begin{cases} 1, & \text{if } \mathcal{A} = \mathcal{N}, \\ F(\phi_A(\boldsymbol{\omega})), & \text{otherwise.} \end{cases} \tag{8.72}$$

The aggregation function F is required to be non-conjunctive to ensure that for all $\boldsymbol{\omega} \in [0,1]^n$ the induced capacity μ is not the smallest capacity vanishing for all subsets except for $\mathcal{N}$.

The authors have shown monotonicity of the resulting set function and some of its properties that result from various choices of F and $\boldsymbol{\omega}$. An important point is that $\boldsymbol{\omega}$ can be a weighting vector provided a priori, but also the data itself $\boldsymbol{\omega} = \mathbf{x}$.

References

[AKH10] D. Anderson, J. Keller, T. Havens, Learning fuzzy-valued fuzzy measures for the fuzzy-valued Sugeno fuzzy integral, in *Lecture Notes in Artificial Intelligence 6178* (2010), pp. 502–511

[AGM10] S. Angilella, S. Greco, B. Matarazzo, Non-additive robust ordinal regression: a multiple criteria decision model based on the Choquet integral. Eur. J. Oper. Res. **201**(1), 277–288 (2010)

[Ang+16] S. Angilella et al., Non additive robust ordinal regression for urban and territorial planning: an application for siting an urban waste landfill. Ann. Oper. Res. **245**(1–2), 427–456 (2016)

[BKM14] A. Bagirov, N. Karmitsa, M. Makela, *Introduction to Non-smooth Optimization: Theory, Practice and Software* (Springer, Berlin, Heidelberg, 2014)

[Bel07] G. Beliakov. `Fmtools` package, version 4. 2007–2021. http://personal-sites.deakin.edu.au/~gleb/fmtools.html

[Bel09] G. Beliakov, Construction of aggregation functions from data using linear programming. Fuzzy Sets Syst. **160**, 65–75 (2009)

[BBC16] G. Beliakov, H. Bustince, T. Calvo, *A Practical Guide to Averaging Functions* (Springer, New York, Berlin, Heidelberg, 2016)

[BD19] G. Beliakov, D. Divakov, Fitting Sugeno integral for learning fuzzy measures using PAVA isotone regression. Inform. Sci. **34**, 2863–2771 (2019)

[BGJ20a] G. Beliakov, M. Gagolewski, S. James, Robust fitting for the Sugeno integral with respect to general fuzzy measures. Inform. Sci. **514**, 449–461 (2020)

[BJL11] G. Beliakov, S. James, G. Li, Learning Choquet-integral-based metrics for semisupervised clustering. IEEE Trans. Fuzzy Syst. **19**(3), 562–574 (2011)

[BW19] G. Beliakov, J.-Z. Wu, Learning fuzzy measures from data: simplifications and optimisation strategies. Inform. Sci. **494**, 100–113 (2019)

[BW21] G. Beliakov, J.-Z. Wu, Learning k-maxitive fuzzy measures from data by mixed integer programming. Fuzzy Sets Syst. **412**, 41–52 (2021)

[Bel24] G. Beliakov, *Rfmtool: Fuzzy Measure Tools*. R package version 5.0.4. 2024. https://CRAN.R-project.org/package=Rfmtool

[BBA25] G. Beliakov, C. Beliakov Amaya, Efficient methods for verifying monotonicity of 2-additive fuzzy measures. Comp. Appl. Math. **44**, 25 (2025)

[BJW25] G. Beliakov, S. James, J.-Z. Wu, Capacities satisfying the buoyancy property on average. Theory Decis. (2025). https://doi.org/10.1007/s11238-025-10039-x

[BC90] M. Best, N. Chakravarti, Active set algorithms for isotonic regression; a unifying framework. Math. Program. **47**, 425–439 (1990)

[BV04] S. Boyd, L. Vandenberghe. *Convex Optimization* (Cambridge University Press, Cambridge, 2004)

[BC18] Q. Brabant, M. Couceiro, k-Maxitive Sugeno integrals as aggregation models for ordinal preferences. Fuzzy Sets Syst. **343**, 65–75 (2018). ISSN: 0165-0114

[CJ89a] A. Chateauneuf, J.-Y. Jaffray, Some characterizations of lower probabilities and other monotone capacities through the use of Möbius inversion. Math. Soc. Sci. **17**(3), 263–283 (1989)

[CJ89b] A. Chateauneuf, J. Jaffray, Some characterizations of lower probabilities and other monotone capacities through the use of Möbius inversion. Math. Soc. Sci. **17**, 263–283 (1989)

[CGI16] S. Corrente, S. Greco, A. Ishizaka, Combining analytical hierarchy process and Choquet integral within non-additive robust ordinal regression, Omega **61**, 2–18 (2016)

[Cor+13] S. Corrente, S. Greco, M. Kadziński, R. Słowiński, Robust ordinal regression in preference learning and ranking. Mach. Learn. **93**(2–3), 381–422 (2013)

[Dua17] L.T. Duarte, A novel multicriteria decision aiding method based on unsupervised aggregation via the Choquet integral. IEEE Trans. Eng. Manag. **65**(2), 293–302 (2017)

[Fle00] R. Fletcher, *Practical Methods of Optimization*, 2nd edn. (Wiley, New York, 2000)

[GJB19] M. Gagolewski, S. James, G. Beliakov, Supervised learning to aggregate data with the Sugeno integral. IEEE Trans. Fuzzy Syst. **27**, 810–815 (2019)

[GJ18] M. Gagolewski, S. James, Fitting symmetric fuzzy measures for discrete Sugeno integration, in *Advances in Fuzzy Logic and Technology 2017*, ed. by J. Kacprzyk et al., vol. 642. Advances in Intelligent Systems and Computing (Springer, Berlin, 2018), pp. 104–116

[GW03] E.M. Gertz, S.J. Wright, Object-oriented software for quadratic programming. ACM Trans. Math. Software **29**, 58–81 (2003)

[GG+25] X. Gonzalez-Garcia, L. Horanska, G. Beliakov, H. Bustince, Efficient online generation of fuzzy measures via aggregation functions. Inform. Fusion **125**, 103452 (2026)

[GKM08] M. Grabisch, I. Kojadinovic, P. Meyer, A review of methods for capacity identification in Choquet integral based multi-attribute utility theory: applications of the Kappalab R package. Eur. J. Oper. Res. **186**(2), 766–785 (2008)

[GKM15] M. Grabisch, I. Kojadinovic, P. Meyer, `kappalab` package, version 0.4 (2015). https://CRAN.R-project.org/package=kappalab

[GMS00] M. Grabisch, T. Murofushi, M. Sugeno (eds.), *Fuzzy Measures and Integrals. Theory and Applications* (Physica–Verlag, Heidelberg, 2000)

[Gra95] M. Grabisch, A new algorithm for identifying fuzzy measures and its application to pattern recognition, in *Fuzzy Systems, 1995. International Joint Conference of the Fourth IEEE International Conference on Fuzzy Systems and The Second

International Fuzzy Engineering Symposium, Proceedings of 1995 IEEE Int (IEEE, Piscataway, 1995), pp. 145–150

[GN94] M. Grabisch, J.-M. Nicolas, Classification by fuzzy integral: performance and tests. Fuzzy Sets Syst. **65**(2), 255–271 (1994). Fuzzy Methods for Computer Vision and Pattern Recognition. ISSN: 0165-0114

[GB17] M. Grant, S. Boyd. *CVX: Matlab Software for Disciplined Convex Programming* (2017). https://cvxr.com/cvx/

[GMS14] S. Greco, V. Mousseau, R. Słowiński, Robust ordinal regression for value functions handling interacting criteria. Eur. J. Oper. Res. **239**(3), 711–730 (2014)

[HT99] R. Horst, N. Thoai, DC programming: overview. J. Optim. Theory Appl. **103**, 1–43 (1999)

[Koj07] I. Kojadinovic, Minimum variance capacity identification. Eur. J. Oper. Res. **177**(2), 498–514 (2007)

[KG05] I. Kojadinovic, M. Grabisch. *Non Additive Measure and Integral Manipulation Functions, R package version 0.2* (2005). http://www.polytech.univnantes.fr/kappalab

[KM07] I. Kojadinovic, J.-L. Marichal, Entropy of bi-capacities. Eur. J. Oper. Res. **178**, 168–184 (2007)

[KMR05] I. Kojadinovic, J.-L. Marichal, M. Roubens, An axiomatic approach to the definition of the entropy of a discrete Choquet capacity. Inform. Sci. **172**, 131–153 (2005)

[LLP04] H.A. Le Thi, H.M. Le, D.T. Pham, New and efficient DCA based algorithms for minimum sum-of-squares clustering. Pattern Recognit. **47**, 388–401 (2004)

[Mar00] J.-L. Marichal, Behavioral analysis of aggregation in multicriteria decision aid, in *Preferences and Decisions Under Incomplete Knowledge*, ed. by B.D.B. J. Fodor, P. Perny (Physica-Verlag, Heidelberg, 2000), pp. 153–178

[Mar02] J.-L. Marichal, Entropy of discrete Choquet capacities. Eur. J. Oper. Res. **137**, 612–624 (2002)

[Mar04] J.-L. Marichal, Tolerant or intolerant character of interacting criteria in aggregation by the Choquet integral. Eur. J. Oper. Res. **155**, 771–791 (2004)

[Mar07] J.-L. Marichal, k-intolerant capacities and Choquet integrals. Eur. J. Oper. Res. **177**, 1453–1468 (2007)

[MR00] J.-L. Marichal, M. Roubens, Determination of weights of interacting criteria from a reference set. Eur. J. Oper. Res. **124**(3), 641–650 (2000)

[MGL11] B. Mayag, M. Grabisch, C. Labreuche, A representation of preferences by the Choquet integral with respect to a 2-additive capacity. Theory Decis. **71**(3), 297–324 (2011)

[MR05] P. Meyer, M. Roubens, Choice, ranking and sorting in fuzzy multiple criteria decision aid, in *Multiple Criteria Decision Analysis: State of the Art Surveys*, ed. by J. Figueira, S. Greco, M. Ehrogott (Springer, New York, 2005), pp. 471–503

[MGB17] J. Murillo, S. Guillaume, P. Bulacio, k-maxitive fuzzy measures: a scalable approach to model interactions. Fuzzy Sets Syst. **324**, 33–48 (2017)

[NW99] J. Nocedal, S. Wright. *Numerical Optimization* (Springer, New York, 1999)

[PD24] G.D. Pelegrina, L.T. Duarte, A novel unsupervised capacity identification approach to deal with redundant criteria in multicriteria decision making problems. IEEE Trans. Fuzzy Syst. **32**(12), 7196–7201 (2024)

[PRS09] H. Prade, A. Rico, M. Serrurier, Elicitation of Sugeno integrals: a version space learning perspective, in *Lecture Notes in Computer Science 5722* (2009), pp. 392–401

[R D25] R Development Core Team, R: a language and environment for statistical computing, in *R Foundation for Statistical Computing* (2025). http://www.R-project.org

[RWD88] T. Robertson, F. Wright, R. Dykstra. *Order Restricted Statistical Inference* (Wiley, Chichester, New York, 1988)

[Rou96] M. Roubens, Interaction between criteria and definition of weights in MCDA problems, in *44th Meeting of the European Working Group Multicriteria Aid for Decisions* (Brussels, 1996)

[RGL15] H.V. Rowley, A. Geschke, M. Lenzen, A practical approach for estimating weights of interacting criteria from profile sets. Fuzzy Sets Syst. **272**, 70–88 (2015)

[Teh+12] A. Tehrani, W. Cheng, K. Dembczynski, E. Hüllermeier, Learning monotone nonlinear models using the Choquet integral. Mach. Learn. **89**(1–2), 183–211 (2012)

[Tor22] V. Torra, (Max,)-transforms and genetic algorithms for fuzzy measure identification. Fuzzy Sets Syst. **451**, 253–265 (2022)

[Tor23a] V. Torra, Optimal transport and the Wasserstein distance for fuzzy measures: an example, in *International Conference on Intelligent and Fuzzy Systems* (Springer, Berlin, 2023), pp. 39–44

[Tor23b] V. Torra, The transport problem for non-additive measures. Eur. J. Oper. Res. **311**(2), 679–689 (2023).

[VBL14] H. Vu, G. Beliakov, G. Li, A Choquet integral toolbox and its application in customer preference analysis. in *Data mining Applications with R*, vol. 9. (Elsevier, Amsterdam, 2014), pp. 247–272. ISBN: 9780124115200

[WB18] J.-Z. Wu, G. Beliakov, Nonadditivity index and capacity identification method in the context of multicriteria decision making. Inform. Sci. **467**, 398–406 (2018)

[WB20] J.-Z. Wu, G. Beliakov, Marginal contribution representation of capacity based multicriteria decision making. Int. J. Intell. Syst. **35**(3), 373–400 (2020)

[WPS18] J.-Z. Wu, E. Pap, A. Szakal, Two kinds of explicit preference information oriented capacity identification methods in the context of multicriteria decision analysis. Int. Trans. Oper. Res. **25**, 807–830 (2018)

[Wu+15] J.-Z. Wu, S. Yang, Q. Zhang, S. Ding, 2-additive capacity identification methods from multicriteria correlation preference information. IEEE Trans. Fuzzy Syst. **23**(6), 2094–2106 (2015)

[WZ10] J.-Z. Wu, Q. Zhang, 2-order additive fuzzy measure identification method based on diamond pairwise comparison and maximum entropy principle. Fuzzy Optim. Decis. Making **9**(4), 435–453 (2010)

[Wu+14] J.-Z. Wu, Q. Zhang, Q. Du, Z. Dong, Compromise principle based methods of identifying capacities in the framework of multicriteria decision analysis. Fuzzy Sets Syst. **246**, 91–106 (2014)

[YK96] B. Yuan, G. Klir, Constructing fuzzy measures: a new method and its application to cluster analysis, in *Proceedings of the NAFIPS'96* (1996), pp. 567–571

Chapter 9
Random Sampling of Capacities

Abstract Random sampling of capacities is an important component of various approaches to learning and optimisation, especially in problems involving complex constraints and objectives. Being able to generate unbiased random capacities is also crucial when it comes to validation and evaluating comparative performance of newly proposed capacity identification methods on randomly generated datasets. In this chapter we summarise the main challenges in random sampling, as well as a number of useful approaches to random capacity generation. Some of these methods apply to general capacities, while others are specialised towards sparse capacities and specific types.

9.1 General Methods, Linear Extensions, Order Polytopes

Random generation of capacities plays a pivotal role in solving optimisation problems by introducing stochastic elements into the problem-solving process. It facilitates exploration of the solution space, aiding the algorithm in escaping local optima and diversifying its search for better solutions and empowers optimisation algorithms to address uncertainty, complex constraints, and nonlinearity effectively [BW21]. Methods based on evolutionary heuristics depend heavily on generation of random solutions in a feasible domain.

Another application of randomly generated capacities is sensitivity testing of solutions based on learning capacities. Often there is no unique solution to a capacity learning problem, and a whole set of feasible capacities is obtained and needs to be sampled. On the other hand, having defined a capacity in some way, one may try to verify its suitability by performing rankings of decision alternatives (e.g. obtained by calculating fuzzy integrals) and checking how sensible these rankings are to small variations to capacity values.

The focus of this chapter will be on pseudorandom generation of capacities of some class with uniform distribution. The reason is that other distributions can be obtained (practically or in principle) from the uniform distribution by using the acceptance-rejection approach.

G. Beliakov et al., *Choquet Capacities and Fuzzy Integrals*, Theory and Applications of Computability, https://doi.org/10.1007/978-3-031-97070-2_9

9.1.1 Challenges for Generating Capacities

It turns out that random generation of capacities with uniform distribution is a very challenging problem. The difficulties are related to the high dimensionality of the space containing capacities ($dim = 2^n$) and a complicated structure of the set of capacities. It is known [CDM13] that the set of all capacities is a polytope with an extremely large number of vertices related to the Dedekind numbers $M(n)$ (the OEIS[1] sequence A000372: 2, 3, 6,20, 168, 7581, 7828354, 2414682040998, ...), and hence, it is infeasible to use convex combinations of its vertices, for example, for random sampling.

There are some general methods of uniformly sampling polytopes [Smi84, Rub84], including partitioning and acceptance-rejection. However, these methods are impractical in such a high dimension. On the other hand, the polytope of capacities P has a special structure, and it is called an order polytope [CM06, CDM13, Gra16]. This structure allows one to perform decomposition of P into simplices (of equal volume) [CSD19], then randomly pick up one simplex, and finally generate a random point within that simplex (the last step is almost trivial [Dev86, Rub84]). The catch here is the extremely large cardinality of the simplicial partition, given by the OEIS sequence A046873, which reads (starting from $n = 1$) 1, 2, 48, 1680384, 14807804035657359360,.... Hence, these simplices cannot be explicitly enumerated for $n > 4$.

Recent research has introduced efficient techniques for randomly generating capacities and their specific types, including random node generation [HP17, CDM13], topological sort generation [BW21], linear extension-based generation [CDM13, CSD19, WB23], and vertex-convex combination generation [CM06, Wu+24].

It should be noted that random generation of capacities can be performed in any representation related to μ linearly, such as Möbius, interaction, nonadditivity, and nonmodularity, mentioned in Chap. 4. It is due to the fact that a uniform distribution on some set remains uniform under a linear transformation; hence, some approaches use, for instance, Möbius representation. However, order polytopes in one representation are no longer order polytopes in another, and hence, specialised methods are not equally applicable in different representations.

9.1.2 Main Approaches

We present here some of the main approaches to random generation of capacities and their specific families, which work in a high dimension by making use of

[1] The Online Encyclopedia of Integer Sequences, http://oeis.org

specific features of the respective subsets of capacities. These approaches are as follows:

1. Linear extension-based method
2. Topological sorting-based method
3. Random walk

We will mention one basic method of selecting randomly a point on a simplex (a polytope in $\mathbb{R}^d$ determined by d non-coplanar points). The unit (or standard) simplex S_0 is determined by the points
$(1, 0, \ldots, 0)$,$(0, 1, 0, \ldots, 0)$,$(0, 0, 1, 0, \ldots, 0) \ldots (0, \ldots, 0, 1)$.

It can always be transformed to another simplex of the same dimension by a linear change of variables. Another useful simplex is $S_1 = \{\mathbf{x} :\in [0, 1]^d | x_i \geqslant x_{i+1}\}$.

Now generate $d - 1$ uniformly distributed values from $[0, 1]$, and sort them in increasing order [Dev86]. Add to this collection the values 0 and 1. We have $d + 1$ values y_i listed in increasing order. Take the pairwise differences $x_i = y_{i+1} - y_i$. The vector $\mathbf{x} = (x_1, \ldots, x_d)$ belongs to S_0. The vector $\mathbf{x} = (y_{d-1}, \ldots, y_1)$ belongs to S_1 (of dimension $d - 1$).

Repeating the mentioned steps M times, we obtain M uniformly distributed random points on S_0 (or S_1). The computational cost is $O(d \log d)$ (the sorting step). See [Dev86] for some alternative methods.

For symmetric capacities, this approach is sufficient, and the OWA weighting vectors are randomly generated in that way.

When it comes to randomly generating capacities of a specific class, besides monotonicity we can enforce a few other conditions. Sub- and supermodularity are two very important conditions which arise in applications (see types of capacities in Chap. 2). They require verifying many more constraints than general capacities. This chapter also deals with random generation of supermodular capacities (submodular capacities are obtained as their duals). We also treat buoyant and antibuoyant capacities as special subsets of sub- and supermodular capacities.

We end this section with a few simple methods to generate capacities randomly from very restricted classes, namely, belief measures and balanced capacities, which satisfy $\mu(\mathcal{A}) \leqslant \mu(\mathcal{B})$ when $|\mathcal{A}| < |\mathcal{B}|$.

Belief measures (Definition 2.15) are characterised by the conditions $M(\mathcal{A}) \geqslant 0$ and $\sum M(\mathcal{A}) = 1$ in Möbius representation. We see that the set of belief measures is a simplex, and hence, we easily generate the required Möbius values as coordinates of a point $\mathbf{x} \in S_0$. If required, we can convert to standard representation.

Balanced capacities are also generated based on the unit simplex, by generating $2^n - 2$ values in $[0, 1]$ and sorting. For each subset cardinality $i = |\mathcal{A}|$, starting from singletons, we assign the $\binom{n}{i}$ smallest values to $\mu(\mathcal{A})$ in some random order (e.g. using a random permutation of indices representing the ith cardinality). We exclude the already assigned smallest values and continue until $i = n - 1$. After adding $\mu(\emptyset) = 0$, $\mu(\mathcal{N}) = 1$, the resulting random capacity is balanced by construction.

9.2 Linear Extension-Based Generation Method

The set of all capacities, which is an order polytope P, is determined by the $d = 2^n - 2$ non-negative values $\mu(\mathcal{A})$ and the $n2^{n-1}$ non-redundant monotonicity constraints. Note that the values $\mu(\mathcal{A})$ constitute a poset, which corresponds to the Boolean lattice B_n. It is known that this polytope has a simplicial partition, and each simplex corresponds to a linear extension of $\mathcal{P}(\mathcal{N})$.

Definition 9.1 (Linear Extension) Given a partially ordered set (poset) $\mathcal{P}$, a linear extension is a linear order (i.e. a chain) compatible with the order relations in $\mathcal{P}$.

For example, the ordering

$$\{1\} \precsim \{2\} \precsim \{1, 2\} \precsim \{3\} \precsim \{2, 3\} \precsim \{1, 3\} \precsim \{1, 2, 3\}$$

is one compatible linear extension when $n = 3$.

To randomly generate a capacity using that ordering, generate a random point $\mathbf{x}$ on S_1 (dimension $d = 2^n - 2$) and assign $\mu(\{1\}) = x_1$, $\mu(\{2\}) = x_2$, $\mu(\{1, 2\}) = x_3$, ..., $\mu(\mathcal{N}) = 1$.

One generic method we mention here is a *Markov chain* random walk on the set of indices. For computational purposes a capacity is encoded into a vector μ of length 2^n following the binary ordering (see Chap. 4). Let us fill an array $\mathbf{A} \in Z^{2^n}$ with $\mathbf{A}[i] = i, i = 0, \dots, 2^n - 1$. This array corresponds to the linear extension

$$\emptyset \precsim \{1\} \precsim \{2\} \precsim \{1, 2\} \precsim \{3\} \precsim \{1, 3\} \dots \precsim N.$$

Recall that the value $\mathbf{A}[i]$ in binary encodes a subset $\mathcal{A}$ (each bit indicates presence or absence of the respective element in $\mathcal{A}$). We will denote by $\mathbf{A}[i] \precsim \mathbf{A}[j]$ the fact that $\mathcal{A} \subseteq \mathcal{B}$.

Perform M times the following Markov steps:

- Select randomly $i \in \{1, \dots, 2^n - 3\}$.
- Exchange $\mathbf{A}[i]$ and $\mathbf{A}[i + 1]$ if $\mathbf{A}[i] \not\precsim \mathbf{A}[i + 1]$.

After each Markov step the array $\mathbf{A}$ still encodes a linear extension of B_n (the partial order has not been broken). Therefore, after M steps we obtain another linear extension of B_n.

Now, a random capacity consistent with that linear extension is generated by taking a random value $\mathbf{x} \in S_1$ and assigning $\mu(\mathbf{A}[i]) = x_i$ (the values $\mu(0) = 0$, $\mu(2^n - 1) = 1$ are always fixed).

Each linear extension of B_n corresponds with a particular simplex in the simplicial partition of the polytope of capacities P, and all such simplices have equal volume. Therefore, randomly (and uniformly) selecting a linear extension and then generating a (uniform) random point in the respective simplex will result in a uniform distribution of such random points in P and hence uniform random generation of capacities.

Thus, we have obtained one generic method for randomly generating capacities with uniform distribution, as long as we take enough Markov steps M.

The main problem with such a generic method is the huge number of linear extensions of B_n (given by the mentioned OEIS sequence A046873), which makes the random walk mixing time (number of steps M required to stabilise at the stationary uniform distribution) prohibitively large. Thus, for small $n \leqslant 4$ it is a simple and viable alternative, but for larger n we need to develop more efficient methods than a simple random walk. Having said this, performing a few thousands of Markov steps at the end of other algorithms is a little inexpensive touch to improve the randomness and uniformity of those algorithms. So we usually add simple Markov steps at the end without mentioning this explicitly.

Another simple technique to improve the uniformity of the randomly generated capacities is symmetrisation. Note that we expect the distribution to be independent of the way the inputs are labelled. Whatever the permutation of the indices of the inputs $\{1, 2, \ldots, n\}$, the distribution must be the same. However, the way we encode the linear extensions through the numbers $0, 1, 2, 3 \ldots$ in the array $\mathbf{A}$, and the implicit ordering of those numbers, $0 \leqslant 1 \leqslant 2 \leqslant \ldots$, introduces asymmetry in the resulting distribution.

To illustrate this, suppose we use the simple Markov chain method above, starting the initial linear extension with $\mathbf{A}[i] = i$. Even though the inputs with labels $1, 2,$ and 3 need to be treated symmetrically, the distance between $\mathbf{A}[1]$ and $\mathbf{A}[2]$ in the array $\mathbf{A}$ is smaller than the distance between $\mathbf{A}[1]$ and $\mathbf{A}[3]$. It is actually less likely that the values $\mathbf{A}[1]$ and $\mathbf{A}[3]$ will be eventually exchanged compared with $\mathbf{A}[1]$ and $\mathbf{A}[2]$ (our Markov steps exchange neighbouring values only). Thus, our asymmetric encoding scheme introduces an implicit bias.

To eliminate such a bias we use a symmetrisation technique, which picks an arbitrary permutation σ of the array $(1, 2, \ldots, n)$ and then relabels the inputs $i \to \sigma(i)$ in every subset $\mathcal{A} \subset \mathcal{N}$ (not just singletons). This final step in the linear extension-based methods eliminates the bias implied by the natural ordering of integers and makes the output of any random generation algorithm nonsensitive to the way the inputs are labelled.

9.2.1 MinimalsPlus Method

The huge number of linear extensions of the Boolean lattice B_n also makes it prohibitively expensive to construct the partition explicitly. Some (approximately) random algorithms for picking linear extensions have been proposed [GLS23, CSD19].

The MinimalsPlus approach [CSD19] employs random walks to select simplices in a non-enumerative manner, making it suitable for both general capacities and p-symmetric capacities. For 2-additive capacities, a method based on the simplicial partition in Möbius representation has been developed [MGS20]. An alternative technique utilises Markov chains to approximate linear extensions, resulting in a two-layer approximation that significantly reduces time consumption compared to traditional Markov chain methods [GLS23].

For a poset $\mathcal{P}$, the set of minimal elements is

$$M = \{a \in \mathcal{P} | \forall b \in \mathcal{P}, b \nprec a\}.$$

Randomly pick an element x from M and place it at the start of the linear extension of $\mathcal{P}$. Remove x from $\mathcal{P}$ and update M. Continue this process iteratively placing randomly selected minimal elements x one after another until all elements from $\mathcal{P}$ have been used. This is the essence of the *Minimals* algorithm.

Unfortunately the Minimals algorithm above does not produce linear extensions distributed uniformly. It is enhanced by choosing elements of M with probabilities which depend on the relative numbers of elements in $\mathcal{P}$ which succeed x. A further enhancement, called MinimalsPlus, uses a more sophisticated procedure to select $x \in M$ by weighting them according to the numbers of positioned antichains [CSD19].

Another approach is a Markov chain-based approximate method, a two-layer approximation method that generates a subset of linear extensions meanwhile eliminating those with very low probability. The method has similar performance compared to the Markov chain but is much less time-consuming [GLS23]. The work in [SGL22] adopted the distributions of measure values of subsets and probabilities of linear extensions for capacities to improve the random node generation method under the monotonicity condition and additional conditions.

9.2.2 Topological Sort

Another way of picking a linear extension of a poset is by topological sorting [Kah62] which considers poset $\mathcal{P}$ as a directed acyclic graph (DAG) where an edge between vertices u and v means $u \leqslant v$ in that ordering. The oldest topological sort method is due to Kahn [Kah62], which is implemented in most Unix-based computer systems, as a useful routine to perform object linking. But that method is not particularly random. A different algorithm is based on depth-first search from the nodes of the DAG in arbitrary order [Bel+22].

We construct the DAG represented by the list of pairs of vertices related by inclusion in the power set $\mathcal{P}(\mathcal{N})$, and excluding those pairs easily obtained by transitivity. We ensure that the algorithm is always presented with the nodes in arbitrary order by randomising the input pairs. This way we hope to obtain a variety of linear extensions as the outputs of the depth-first topological sort.

While indeed the linear extensions generated in that way seem random, by no means they are representative of the whole set of linear extensions. The topological sort algorithm used here has a definite preference for a particular (although large) subset of linear extensions, as reported in [BW21] through experiments.

Algorithm 9.1: Depth-first topological sort

Input: L: Empty list that will contain the sorted elements, $graph$: list of edges representing the graph, S : Set of all nodes with no incoming edge
Output: L: a topologically sorted order

1 **While** S is not empty **do:**

  1.1 remove a node n from S
  1.2 add n to L
  1.3 **For** each node m with an edge e from n to m **do:**
    remove edge e from $graph$
    If m has no other incoming edges then insert m into S

2 **If** $graph$ is not empty then return $error$ (graph has at least one cycle)
  Else return L

However, when augmented with both the additional Markov chain steps and symmetrisation, the topological sort method yielded fairly uniformly distributed capacities [BW21]. Compared with the pure Markov step method, topological sorting provided a wide variety of linear extensions as starting points for a Markov chain, reducing the mixing time M significantly.

9.2.3 K-Interactive Capacities

Let us now look at k-interactive capacities. Recall that the capacity values for $|\mathcal{A}| > k$ are fixed in some way, in particular as $\mu(\mathcal{A}) = K + (1 - K)\frac{|\mathcal{A}|-k-1}{n-k-1}$ for some number $K \in (0, 1]$ selected in advance. One special case is when $K = 1$, which corresponds with $(k + 1)$ tolerant capacities. If we replace the values $\mu(\mathcal{A}), |\mathcal{A}| = k + 1, \ldots, n$ with their respective equivalence classes, we can construct the partial order illustrated in Fig. 9.1.

This partial order is suitable to the to the linear extension-based methods [Bel+22], although we do not require randomness for the values $\mu(\mathcal{A}), |\mathcal{A}| = k + 1, \ldots, n$, which are predetermined. Hence, we take only the subset of $\mu(\mathcal{A}), |\mathcal{A}| = 0, 1, \ldots, k+1$ as the input to MinimalsPlus/topological sort (excluding the chain in the upper part of Fig. 9.1), generate the capacity values, and then scale them linearly so that $\mu(\mathcal{A}) = K$ when $|\mathcal{A}| = k + 1$. We note that this order corresponds to an appropriately scaled down version of an order polytope.

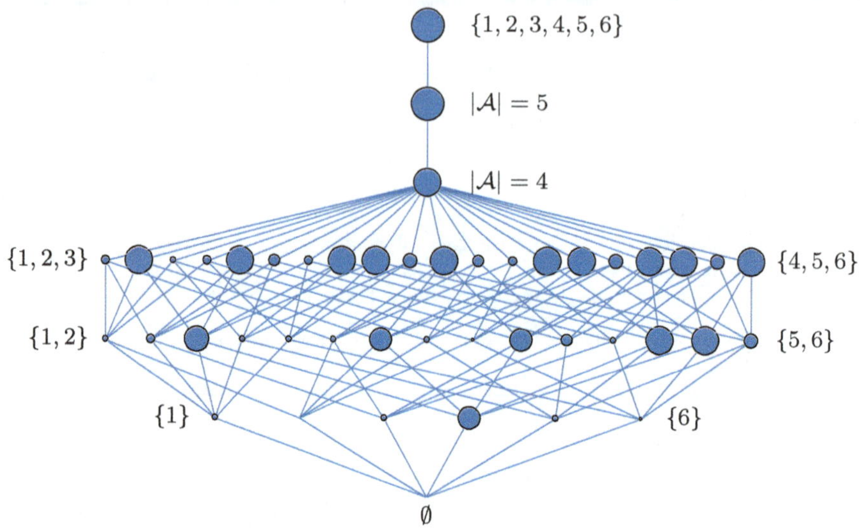

Fig. 9.1 A 3-interactive capacity for $n = 6$ (also presented in Fig. 7.1) and the implied partial order. The sizes of the circles reflect the values of the respective capacities with all $\mu(\mathcal{A})$ equal when the cardinality is fixed at $|\mathcal{A}| = 4, 5, 6$, represented as single nodes since they are treated as equivalent

9.2.4 Symmetric and p-Symmetric Capacities

We already mentioned that symmetric capacities, which can be represented by just one vector of their values of length n, are generated by taking random points on the unit simplex S_1.

The p-symmetric capacities [MGG02] are generated by using a linear extension approach (MinimalsPlus/topological sort) by supplying as input the appropriate partial order [CSD19]. An example of that partial order is shown in Fig. 9.2. Here the capacity values in each of the marked groups are the same, and it is sufficient to randomly select values for each group, which obey the partial order based on the number of elements from each class in the respective subsets. One can code each group by pairs (a, b), where a is the number of elements of the first type (first subset of indifference) and b is the number of elements of the second type in each set. For example, {1, 2, 3} is coded as (2, 1).

9.3 Random Walk and Sorting-Based Methods

A different approach to linear extensions is to pick up a starting capacity and perform random walk in the polytope of capacities P itself, rather than on the array

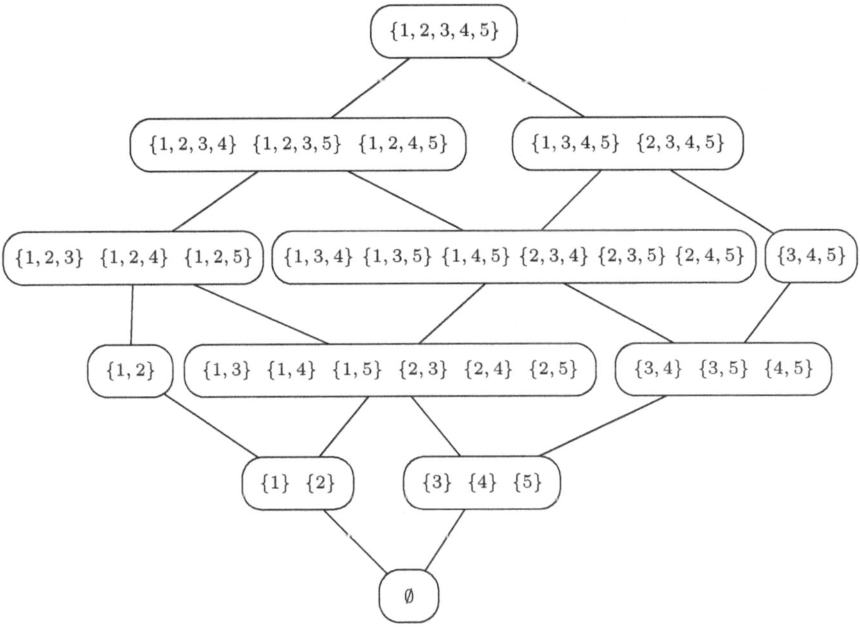

Fig. 9.2 Partial order for the 2-symmetric capacity given in Example 7.6

of indices. As capacities are viewed as geometrical points in P, we will refer to them as *points* here. The coordinates of each point provide the values of the capacity at the respective subsets $\mathcal{A}$.

The random walk is performed by iterating two steps:

- Generate randomly a point $\mathbf{y}$ in a neighbourhood of the current point $\mathbf{x}$.
- If $\mathbf{y} \in P$, then with some probability select $\mathbf{y}$ as the next point, i.e. assign $\mathbf{x} \leftarrow \mathbf{y}$.

The mentioned neighbourhood should be relatively small compared to the size of P to avoid many rejections due to selections outside P and yet relatively big so as not to stall in the same area. Selecting a random point $\mathbf{y}$ is a simple task (although we later mention some caveats related to satisfying additional constraints); the challenge is to perform efficient checks that $\mathbf{y} \in P$ or otherwise.

A direct way for doing so is to check all monotonicity constraints, which requires $O(n2^n)$ operations (each constraint has only two values to compare). We present here a better approach which reduces the checking time.

The method we present here is from [BW24, BJW24] and it is particularly suitable to the cases where candidate set functions are generated in relatively small neighbourhoods of a given capacity. This is a particular requirement of the random walk method (see above) as well as a feature in stochastic optimisation algorithms (e.g. in evolutionary computing). This method is based on various sorting algorithms and exhibits, on average, $O(2^n)$ running time, which is a substantial improvement.

9.3.1 Monotonicity Check

Consider an array of pairs $(\mu(\mathcal{A}), \mathcal{A})$:

$$\mathbf{v} = ((\mu(\emptyset), 0), (\mu(\{1\}), 1), \ldots, (\mu(N), 2^n - 1)),$$

where the second element of each pair represents the composition of the set $\mathcal{A}$ (through the bits of the respective integer, e.g. $11 = 1011_2$ corresponds with the set $\{1, 2, 4\}$). Using the notation $\mathbf{v}[i].\mu$ to denote the first element of the i-th pair and $\mathbf{v}[i].\mathcal{A}$ to denote the second, let us sort this array so that the two conditions

$$\mathbf{v}[i].\mu \leqslant \mathbf{v}[i+1].\mu \text{ and } \mathbf{v}[i].\mathcal{A} \not\supseteq \mathbf{v}[i+1].\mathcal{A}$$

are satisfied.

If such sorting can be performed, μ is a monotone set function; otherwise, it is not. In the former case, the sorted array $\mathbf{v}$ will encode a linear extension of the poset $\mathcal{P}(\mathcal{N})$.

To perform a check we adapt the classical *merge sorting* algorithm, in which the comparison operation sets an indicator variable whenever the two sorting conditions are contradictory. The merge sort starts with adjacent pairs of elements of the array, then proceeds to 4-tuples by merging pairs, and so on. An important feature is that the k-tuples are maintained when sorted, so that any contradiction between the sorting conditions are detected during k-tuple merging (unlike in other algorithms like quicksort). Note that the algorithm assumes that $\mathbf{v}[i].\mathcal{A} \not\supseteq \mathbf{v}[i+1].\mathcal{A}$ at the start.

The worst-case running time (and also the best and average times) of the merge sort algorithm is $O(d \log d) = O(n2^n)$, the same as for the standard monotonicity check. However, the constant in $O()$ is larger because we also perform memory write operations. In the experiments in [BW24, BJW24] the merge sort was approximately 1.5 times slower across all values of n tested.

Let us now move to the *insertion sort* algorithm, which sorts elements one at a time by making comparisons with one or more of the sorted elements. This algorithm has its best running time $O(d)$, but is quadratic on average and in worst-case scenarios. It works best when the array is partially or almost sorted, and this is precisely the niche we work in when random set functions are generated in the neighbourhood of a capacity.

Suppose we have a capacity μ and its representation as a sorted array of pairs $\mathbf{v}$ as described previously. Let us perturb the values of $\mu(\mathcal{A})$ and obtain a new set function ν whose monotonicity we need to check. Its representation is a partially sorted array $\mathbf{v}$ if the perturbations are not big. The condition $\mathbf{v}[i].\mathcal{A} \not\supseteq \mathbf{v}[i+1].\mathcal{A}$ needs to be verified.

If we use either the standard method or merge sort-based method, we still have to perform the same $O(n2^n)$ operations. But with insertion sort, which also checks for contradictions of the two sorting conditions, we have the expected runtime $O(2^n)$

(because it corresponds with the best-case sorting scenario). Furthermore, when the array is already sorted, no write operations are necessary.

Let us consider the situation when the monotonicity condition fails (i.e. the perturbed array cannot be sorted). Instead of exposing its worst runtime as in traditional sorting of the array, here our algorithm exits prematurely, indicating non-monotonicity. Therefore, we expect its runtime to be even better in this case, as illustrated in numerical experiments in [BW24, BJW24].

As an additional bonus, for a monotone set function, insertion sort produces a new fully sorted array that can be used as another capacity around which to perform perturbation at the next iteration of the underlying optimisation algorithm, or in a Markov chain random walk.

Let us now outline the steps of a stochastic optimisation algorithm (shown as Algorithm 9.2) and the places where the monotonicity check plays its role.

Algorithm 9.2: Random sampling and monotonicity check

Input: n (dimensionality), $d = 2^n$, μ (given capacity), represented by array of pairs **v**
Output: ν (modified capacity)

1 Use merge sort for sorting array **v** if not already sorted.
2 Sampling loop. **For k samples do:**

   2.1 Randomly perturb the values of μ
   2.2 Check monotonicity by insertion sort
   2.3 **If** μ is monotone, check its quality for the underlying application,
   Else break;

3 End loop. Set ν = best μ.
4 **Return** ν in the array **v** (already sorted).

It starts with a given monotone set function μ. In case the algorithm does not find any perturbed function, which is monotone, it returns the original μ. Otherwise it returns the best perturbed monotone set function. We have omitted some of the housekeeping details for the sake of brevity. Note that Step 1 is not performed whenever the perturbed ν is monotone, so its cost is amortised by the algorithm. In the sampling loop we use the fast insertion sort check.

By checking the quality of the capacity, we mean the application-defined criteria which make one or another capacity preferable. In an optimisation method, it is the value of the objective function, for example, fitting the desired data in a capacity learning application such as [Uri+23].

9.3.2 Convexity Checks

In this section, we apply the same methodology of sorting to efficiently verify the convexity of set functions, also referred to as supermodularity or 2-monotonicity [Gra16].

We recall the marginal contribution representation (or discrete derivatives) of set functions from Chap. 4.

Definition 9.2 (Marginal Contribution) For a capacity μ, the marginal contribution of i to the set $\mathcal{A} \subseteq \mathcal{N}$ is

$$\Delta_i(\mathcal{A}) = \mu(\mathcal{A} \cup \{i\}) - \mu(\mathcal{A} \setminus \{i\}).$$

Note that it need not hold that $i \in \mathcal{A}$. The marginal contribution representation requires more variables; however, the monotonicity property is easily verified as non-negativity of each $\Delta_i(\mathcal{A})$.

The necessary and sufficient condition for supermodularity is expressed in terms of Möbius values as [CJ89] (we reproduce Eq. (2.7)):

$$\sum_{\mathcal{B} \subseteq \mathcal{A} | i, j \in \mathcal{B}} M(\mathcal{B}) \geqslant 0, \ \forall i, j \in \mathcal{A} \subseteq \mathcal{N}, |\mathcal{A}| \geqslant 2. \tag{9.1}$$

There are altogether $\frac{1}{8}2^n n(n-1)$ non-redundant inequalities [BJL11]. Let us calculate the number of terms in these inequalities. For each pair $\{i, j\}$ in $\mathcal{N}$, there are $\sum_{k=0}^{n-2} \binom{n-2}{k}$ supersets, which gives us all the sets $\mathcal{A}$ on which this inequality must hold. Then for each of these, there will be $\sum_{l=0}^{k} \binom{k}{l} = 2^l$ terms used. Hence, the total number of terms required for checking convexity is equal to

$$\binom{n}{2} \sum_{l=0}^{n-2} 2^l \binom{n-2}{l} = \binom{n}{2} 3^{n-2}.$$

Thus, we need $O(n^2 3^n)$ operations in Möbius representation. To check supermodularity directly from its definition requires $O((2^n)^2) = O(4^n)$ operations. The supermodularity condition (4.39) in terms of discrete derivatives involves the same number of constraints as in Eq. (9.1); however, each of these will only have two terms, resulting in $\binom{n}{2} 2^{n-1} = \frac{1}{4} 2^n n(n-1)$ in total, i.e. $O(n^2 2^n)$. But it also requires calculating all marginal contributions from a set function μ in $O(n2^n)$ operations. Next, we can either verify all conditions (4.39) in $O(n^2 2^n)$ operations or do that by merge sort, also in $O(n^2 2^n)$ operations.

Now, assuming that we already have a presorted array of marginal contributions (with their labels), and that a perturbed set function ν needs to be checked for supermodularity (the monotonicity check is simply non-negativity of $\Delta_i(\emptyset)$, then it will be implied by supermodularity), we calculate the perturbed marginal contributions and sort them by insertion sort instead, similarly to the previous section. Here, however, we expect the complexity to be $O(n2^n)$, as the altered marginal contributions are expected to be partially sorted, hence involving the best-case sorting scenario.

Therefore, similarly to the monotonicity check by sorting, after one initial step with $O(n^2 2^n)$ operations, every step in the sampling loop (Algorithm 9.3) involves only $O(n2^n)$ operations (compared with the baseline algorithm $O(n^2 3^n)$).

Algorithm 9.3: Random sampling and supermodularity check

Input: n (dimensionality), $m = 2^n$, μ (given capacity) in the array **v**, marginal contributions in array of pairs **w**

Output: ν (modified capacity), $\Delta\nu$ (its marginal contributions) in **w**

1 Use merge sort for sorting array **w** if not already sorted.

2 Sampling loop. **For k samples do:**

2.1 Randomly perturb the values of μ

2.2 Calculate marginal contributions **w**

2.3 Check monotonicity as non-negativity of **w**

2.4 Check supermodularity by insertion sort of **w**

2.5 **If** μ is supermodular, check its quality for the application,
Else break;

3 End loop. Set $\nu =$ best μ.

4 **Return** ν in the array **v** and marginal contributions in **w** (already sorted).

At Step 2.4, when we perform insertion sort, assuming that now we have an array of triples **w**, so that every element $\mathbf{w}[k] = (\Delta_i \mu(\mathcal{A}), \mathcal{A}, i)$, we ensure that

$$\mathbf{w}[k].\Delta \leqslant \mathbf{w}[k+1].\Delta \text{ and not } (\mathbf{w}[k+1].\mathcal{A} \subset \mathbf{w}[k].\mathcal{A} \text{ and } \mathbf{w}[k].i = \mathbf{w}[k+1].i).$$

If the sorting algorithm completes with no contradiction between these conditions, the capacity is supermodular; in case of a contradiction at any stage, sorting cannot be performed and the capacity is not supermodular.

Note 9.1 The Möbius values of all pairs of a supermodular capacity need to be non-negative by (9.1).

9.3.3 *Verifying Antibuoyancy*

As we mentioned earlier, antibuoyancy is a stronger property than supermodularity, and it involves more constraints in the form of (4.40). It finds its application in ensuring compatibility of the Choquet integral with the Pigou-Dalton principle of progressive transfers [BJ21].

Yet it turns out that to verify this condition we can apply the same sorting approach as checking supermodularity in the same number of operations. The difference will be in using condition (4.40) instead of (4.39). Assuming that the marginal contributions are stored in the array of triples $\mathbf{w}$, the sorting procedure needs to ensure that

$$\mathbf{w}[k].\Delta \leqslant \mathbf{w}[k+1].\Delta \text{ and not}$$

$$(\mathbf{w}[k+1].A \subset \mathbf{w}[k].A \text{ and } \mathbf{w}[k+1].i \in \mathbf{w}[k].A \text{ and } \mathbf{w}[k].i \neq \mathbf{w}[k+1].i).$$

As we see, there is one extra operation at the comparison step which does not change the overall complexity of the algorithm.

9.3.4 Sparse Capacities in Möbius Representation

For moderate and large n the complexity of capacities, exponential in n, makes them hard to operate on and interpret their coefficients. For this reason, simplifications of capacities, such as k-additivity and k-interactivity, are applied to limit the number of nonzero coefficients in Möbius representation. One can take a step further and only allow a selected set of Möbius values to be nonzero. This leads to sparse capacities.

Whenever a capacity is modified in Möbius representation, by altering the values assumed to be nonzero, its monotonicity and/or convexity needs to be verified. While this can be done directly, e.g. using (9.1), to reduce complexity we present here an alternative approach based on the sorting methods developed in the previous sections. To do so, we need to convert from Möbius representation to the standard μ, which adds to the complexity of the algorithm.

Assume now that the set of nonzero Möbius values is M and its cardinality is $|M|$. Also assume that whenever the Möbius values are modified, only $|\hat{M}|$ values are changed, which we store in a sparse array $\mathbf{\Delta m}$. Since the values of μ (array $\mathbf{v}$) are obtained through a matrix-vector operation from $\mathbf{m}$, $\mathbf{v} = \mathbf{Zm}$, where $\mathbf{Z}$ is a triangular matrix [Bel21], we also have $\mathbf{\Delta v} = \mathbf{Z\Delta m}$. Let us then apply sparse matrix-vector multiplication, which takes $O(|\hat{M}|2^n)$ operations (2^{n-1} is the number of nonzero elements in $\mathbf{Z}$). After this the values of μ are updated and sorting algorithms are applied. The overall complexity of this step is $O(|\hat{M}|2^n)$ for the monotonicity check and $O((|\hat{M}|+n)2^n)$ for checking convexity by insertion sort. Both values are better than verifying monotonicity and convexity in Möbius representation directly.

The linear transformation between $\mathbf{v}$ and $\mathbf{w}$ is achieved by matrix-vector multiplication $\mathbf{v} = \mathbf{Aw}$, where $\mathbf{A}$ is a suitable 0-1 matrix. Here we underline that having the uniform distribution of random variates in one representation automatically implies the uniformity in another equivalent representation related by a linear change of variables [Dev86, Rub84]. For instance, a uniform distribution in marginal contribution representation implies uniformity in the standard representation μ, as well as in Möbius representation. The interaction indices [GMR00] will also be distributed uniformly.

Tracking Negative Möbius Values

To further reduce the complexity of direct verification of convexity by (9.1), we focus specifically on the negative Möbius values in the convexity constraints. Recall that non-negativity of Möbius values implies convexity, and the values of singletons and pairs need to be non-negative. Therefore, only the Möbius values of k-tuples ($k > 2$) can be negative.

We can therefore focus only on the negative Möbius values and develop a heuristic to reduce the number of subsets $\mathcal{A}$ in (9.1) to check.

Suppose we have the set of L Möbius values, of which the values of pairs and singletons are non-negative by construction, and there is a subset Neg of subsets $\mathcal{A}$ with negative Möbius values, $|Neg| = \tilde{L}$ and the respective subset with non-negative values Pos, $Neg \cup Pos = \mathcal{P}(\mathcal{N})$. Consider now condition (9.1). Obviously, we need to check for all $\mathcal{A} \in Neg$. We also need to verify the conditions for all $\mathcal{A} = \cup_i \mathcal{C}_i, \mathcal{C}_i \in Neg$. However, if $\mathcal{A}$ is not a union of one or more subsets from Neg, condition (9.1) is automatically satisfied whenever it holds for subsets in Neg and their unions and hence need not be verified.

To see this, let $\mathcal{A} \in Pos$ and let $\mathcal{D} = (\cup_k \mathcal{C}_k), \mathcal{C}_k \in Neg, \mathcal{C}_k \subset \mathcal{A}$. If $\mathcal{A} = \mathcal{D}$ we verify (9.1) directly. Otherwise the sum (9.1) is written as

$$\sum_{\mathcal{B}\subseteq\mathcal{D}|i,j\in\mathcal{B}} M(\mathcal{B}) + \sum_{\mathcal{B}\not\subseteq\mathcal{D},\mathcal{B}\subseteq\mathcal{A}|i,j\in\mathcal{B}} M(\mathcal{B}).$$

The second sum contains only $\mathcal{B} \in Pos$, but the first sum is not negative because we verified (9.1) for $\mathcal{D}$ (playing the role of $\mathcal{A}$ in (9.1)). Thus, (9.1) holds for $\mathcal{A} \neq \mathcal{D}$.

Therefore, out of 2^n possible subsets $\mathcal{A}$ in (9.1), we only need to consider those that are unions of elements of Neg. As long as $\tilde{L}$ is much smaller than L, it makes sense to preprocess the Möbius values and select those key subsets.

Thus, we add to the algorithm a preprocessing step to select $\tilde{L}$ subsets and form their unions (altogether L^* subsets $\mathcal{A}$ in (9.1)). The selection process is as follows. Instead of forming all possible unions of $\tilde{L}$ subsets (of which there are only L^*, but the construction process takes $2^{\tilde{L}}$ operations), we go through all 2^n subsets and verify if any of them can be represented as unions of subsets $\mathcal{C}_i \in Neg$. For this we take conditional set differences $\mathcal{A} \setminus \mathcal{C}_1 \setminus \mathcal{C}_2 \setminus \ldots \setminus \mathcal{C}_{\tilde{L}}$, whenever $\mathcal{C}_i \subseteq \mathcal{A}$. If the resulting difference is the empty set, then (and only then) $\mathcal{A}$ is the union of a collection of subsets $\mathcal{C}_i$ from Neg. This operation is done in $\tilde{L}$ time for each $\mathcal{A}$, so overall to select the required set unions we need $O(\tilde{L}2^n)$ time. The overall time complexity of verification of convexity by (9.1) is therefore $O(\tilde{L}2^n)$ (preprocessing) plus $O(n^2 L L^*)$ (verifying (9.1) L^* times). Obviously, $\tilde{L} \leqslant L^* \leqslant 2^n$.

Compared to the baseline $O(n^2L2^n)$ we have potential improvement, as long as only a few Möbius values are negative. Even if $L^* = 2^n$ the preprocessing step does not increase the (asymptotic) complexity of the algorithm, as in this case we have worst runtime $O(n^2L2^n)$ (the preprocessing term $\tilde{L}2^n \leqslant L2^n \leqslant n^2L2^n$).

We should mention that the above reasoning and methods apply to any sparse capacity (many 0 Möbius values, not necessarily based on set cardinality), having just L nonzero Möbius values. It is also valuable for k-additive capacities as a special case.

9.3.5 K-Additive and k-Additive Convex Capacities

All k-additive capacities form polytopes, although these are no longer order polytopes. The sets of k-additive capacities exhibit more intricate structures, as discussed in [CM06, Gra97, MGS20]. Although these structures have lower dimensionality, some vertices are not 0-1 measures.

It is natural to work in Möbius representation and fix many values at 0. We can employ random walks for random generation of k-additive capacities. The monotonicity at every step can be verified directly using (1.10) or by converting to the standard representation and using sorting-based verifications. Similarly, supermodularity can be checked. Note that despite the reduction of the dimensionality of the space containing k-additive capacities, the numbers of monotonicity and supermodularity constraints remain the same (the special case of 2-additive capacities is an exception). Thus, the cost of verifying monotonicity in either way is rather high.

9.3.6 Generation of Starting Points

Generation of each type of capacity by random walks requires as a starting point: an a priori monotone or convex capacity. For *general* capacities in Möbius representation, we take a linear convex combination of two random measures (k-additive or not), a belief measure and its dual plausibility measure, which are easily generated. The result is a general capacity to start the random walk.

For *supermodular* k-additive capacities we use two types of starting points. The first one is a randomly generated k-additive belief measure. The second one is a 2-additive belief measure. The reason is that in the first case the random walk spends most of the time in the simplex of belief measures, so it is hard to obtain a supermodular measure with negative Möbius values. The second option is to place the starting point on the boundary of the set of belief measures rather than in its interior, so that negative Möbius values are somewhat favoured.

In terms of random steps in Möbius representation, to keep the condition $\sum_{\mathcal{A}} M(\mathcal{A}) = 1$, we generate perturbations on an L-simplex $\Delta(\mathcal{A}) \geqslant 0, \sum_{\mathcal{A}} \Delta(\mathcal{A}) = 1$ and then subtract $\frac{1}{L}$ from each coordinate to satisfy $\sum_{\mathcal{A}} \Delta(\mathcal{A}) = 0$. Therefore, the perturbed values add to one: $\sum_{\mathcal{A}} M(\mathcal{A}) + \varepsilon \Delta(\mathcal{A}) = 1$. Here ε is a sufficiently small number, which indicates the size of a step in the random walk, and is selected for each dimension and k to avoid too large steps leading to frequent rejections of the moves. In practice the value ε should be chosen adaptively (too many rejections leading to decreasing the value of ε and few rejections leading to increasing its value).

For k-interactive submodular capacities the initial point is also a belief measure; however, for supermodular ones we need a starting dual submodular capacity.

9.4 2-Additive Capacities

The 2-additive capacities are a special subset of k-additive capacities, but as opposed to k-additive capacities in general, they have a very special structure, which allows one to perform monotonicity and convexity checks for random walks very efficiently. Furthermore, random generation of 2-additive capacities can be performed directly in just $O(n^2)$ operations.

They can be specified in Möbius representation by using $d = n + \frac{n(n-1)}{2}$ values (singletons and pairs), although the number of inequalities that determine half-space representation remains $n2^{n-1}$. For convenience, in this section we will denote our variables by $m_i = M(\{i\})$ and $m_{ij} = M(\{i, j\})$.

9.4.1 Monotonicity Conditions

A key result is that the polytope of 2-additive capacities has vertices which are all 0-1 capacities, and their total number is n^2 [CM06, MGS20]. There are exactly n^2 such 0-1 capacities of three types:

1. 0-1 probability measures $m_i = 1$ for some $i \in \mathcal{N}$ and the rest are 0.
2. Unanimity games $m_{ij} = 1$ for some pair i, j and 0 otherwise.
3. Dual unanimity games with $m_{ij} = -1$, $m_i = m_j = 1$ for one pair i, j, and the rest 0.

Also we denote the three mentioned types of 0-1 capacities as p_i (probability measure), b_{ij} (belief measure focused on $\{i, j\}$), and pl_{ij} (plausibility measure focused on $\{i, j\}$).

Thus, the polytope of 2-additive capacities has a more compact vertex representation as opposed to hyperplane representation with $n2^{n-1}$ inequalities. This feature was used in algorithms for identifying 2-additive capacities from data [HT13].

Another feature of that polytope P is that its composition is the convex hull of 2-additive belief measures $B = \{b_{ij}, i, j \in \mathcal{N}, i > j\}$, its dual plausibility measures $Pl = \{pl_{ij}, i, j \subset \mathcal{N}, i > j\}$, and probabilities $Pr = \{p_i, i \in \mathcal{N}\}$.

Since the set of belief measures is a simplex (all Möbius values are non-negative and add to 1), its dual is also a simplex. Thus, P is the convex hull of three relatively simple subsets.

For our purposes we quote the relevant facts from [MGS20]:

1. Any capacity μ can be written as a unique convex linear combination of d vertices, such that whenever b_{ij} has a nonzero coefficient, the respective coefficient of pl_{ij} is null and vice versa.
2. For any collection C of vertices of P, the convex hull of C is a face of P if and only if $p_i, p_j \in C \iff b_{ij}, pl_{ij} \in C$.
3. Two vertices μ_1, μ_2 of P are adjacent unless $\mu_1 = p_i, \mu_2 = p_j$ or $\mu_1 = b_{ij}$, $\mu_2 = pl_{ij}$.
4. The collection of all polytopes with d vertices specified in point 1 above forms a simplicial partition of P.

The first fact above gives a particularly simple way to check whether a set function is a capacity. Note that the converse of that implication is also true (by item 4 above): if a set function cannot be written as the mentioned convex combination, then it is not in P and hence not monotone.

Theorem 9.1 ([BBA25]) *A 2-additive set function is a capacity if and only if for all $i, j \in N$:*

$$0 \leqslant a_i = m_i + \sum_{j \neq i | m_{ij} < 0} m_{ij} \leqslant 1, \tag{9.2}$$

$$0 \leqslant m_i \leqslant 1, \quad -1 \leqslant m_{ij} \leqslant 1, \tag{9.3}$$

and

$$\sum_i m_i + \sum_{j \neq i} m_{ij} = 1. \tag{9.4}$$

Proof The conditions (9.3) are the standard bounds on the Möbius values. We have by fact 1 above

$$\mu = \sum_{i=1,\dots,n} a_i p_i + \sum_{i \neq j | m_{ij} > 0} m_{ij} b_{ij} + \sum_{i \neq j | m_{ij} < 0} |m_{ij}| pl_{ij}, \tag{9.5}$$

where a_i are some positive coefficients previously specified. Consider the coefficients of the terms pl_{ij} for a fixed i. These entries contribute to the value of the singleton m_i by $m_i = a_i + \sum_{i \neq j | m_{ij} < 0} |m_{ij}|$. Since $0 \leqslant m_i \leqslant 1$ and $a_i \geqslant 0$ we obtain (9.2). So the representation of a capacity as a convex linear combination of belief, plausibility, and probability measures implies (9.2). The converse of that implication follows from facts 1 and 4 above, i.e. if (9.2) does not hold, i.e. some $a_i < 0$, the set function cannot be represented as the mentioned convex combination and hence is not a capacity.

Note that conditions (9.2)–(9.3) have $O(n^2)$ terms and can be verified in $O(n^2)$ steps by the following Algorithm 9.4, which is significantly more efficient than the baseline. There are only n inequalities in (9.2) (the second inequality $\leqslant 1$ follows automatically) and $n^2/2$ trivial inequalities in (9.3).

Algorithm 9.4: Verification of monotonicity of a 2-additive set function

Input: $m \in \mathbb{R}^d$ (given 2-additive set function in Möbius representation), represented by arrays of singletons m_i and pairs m_{ij}

Output: True or False

1 Verify all $m_i \leqslant 1$. If it fails return *False*

2 Main loop. **For** $i, j \in N, j > i$**:**

2.1 Verify (9.3) for pairs, if fails return *False*.

2.2 If $m_{ij} < 0$:

2.2.1 Set $m_i = m_i + m_{ij}$

2.2.2 Set $m_j = m_j + m_{ij}$

3 Verify all $m_i \geqslant 0$. If not, return *False*

4 Return *True*.

Thus, for random walks we have a very efficient algorithm for checking monotonicity of 2-additive capacities. As far as convexity is concerned, a 2-additive capacity is supermodular if and only if it is a belief measure; hence, we only need to check the non-negativity of pairs and singletons.

9.4.2 Random Generation of 2-Additive Capacities

In this section we refer to the simplicial partition of P and an algorithm for randomly generating 2-additive capacities with uniform distribution proposed in [MGS20]. The algorithm relies on the fact that all simplices in that partition have the same volume, and therefore, one can randomly pick up a simplex and then generate a random point in that simplex using a sorting procedure. It is claimed in [MGS20] that the complexity of that process is $O(n^4)$ because it involves matrix multiplication (matrix of size $n^2 \times n^2$) to convert coordinates of the standard unit simplex to the one selected from the partition.

It was then shown in [BBA25] that this multiplication is not needed and the complexity is in fact $O(n^2 \log n)$. To see this note that random points on the standard unit simplex $S_0 = \{v \in \mathbb{R}^d | v_i \geqslant 0, \sum v_i = 1\}$ with uniform distribution are generated by sorting (Sect. 9.1.2). Clearly the complexity of that step is $O(d \log d) = O(n^2 \log n)$.

Next, following [MGS20] we randomly determine a simplex of the partition of P (randomly deciding whether we take m_{ij} as b_{ij} or pl_{ij}). Then we set either $m_{ij} = v_k$ or $m_{ij} = -v_k$, where k corresponds with the position of the pair ij in the array of singletons and pairs of length d. We then also set $m_i = v_i + \sum_{i \neq j | m_{ij} < 0} |m_{ij}|$. The latter step is done whenever m_{ij} is assigned, similar to Steps 2.2 of Algorithm 9.4, so it requires almost no extra effort. Since the assignment of values v_k step takes $O(n^2)$ operations, the overall complexity of the random generation algorithm is $O(n^2 \log n)$, and the matrix multiplication is done in place.

9.5 Verifying Convexity of k-Interactive Capacities

In this section we employ another type of capacities which limit interactions to sets of at most k elements, but differently to k-additivity. The k-interactive capacities were introduced in [BW19]. We recall their definitions from Chap. 7.

Definition 9.3 (k-interactive Capacity) [BW19] A capacity is called k-interactive if for some chosen $K \in [0, 1]$ and $1 \leqslant k \leqslant n$

$$\mu(\mathcal{A}) = K + \frac{|\mathcal{A}| - k - 1}{n - k - 1}(1 - K), \text{ for all } \mathcal{A}, |\mathcal{A}| > k.$$

It was shown in [BW19] that these k-interactive capacities are duals to those whose Möbius values are nonzero for sets of cardinality $|\mathcal{A}| \geqslant n - k$ and the singletons. Equivalently, k-interactive capacities have co-Möbius values zero $\check{m}(\mathcal{A}) = 0$ when $|\mathcal{A}| = 2, 3, \ldots, n - k - 1$.

We will use the notation $M^* = M_{\mu^*}$, denoting the Möbius values of the dual capacity μ^*. Note that $M^*(\mathcal{A}) = 0$, $|\mathcal{A}| = 2, 3, \ldots, n - k - 1$ and it makes sense to perform verifications for k-interactive capacities in Möbius representation by using their duals M^*.

The advantage of k-interactivity is that both monotonicity and convexity checks involve only L inequalities (recall L is the number of nonzero Möbius values) in $O(L^2)$ time regardless of n. The monotonicity check is translated into the following conditions in terms of the Möbius values of the dual capacity M^*:

$$M^*(\{i\}) + \sum_{\mathcal{B} \subseteq \mathcal{A}, i \in \mathcal{B}, |\mathcal{B}| \geqslant n-k} M^*(\mathcal{B}) \geqslant 0, \ \forall \mathcal{A} \subseteq \mathcal{N}, |\mathcal{A}| \geqslant n - k, i = 1, \ldots n. \tag{9.6}$$

Generation of submodular k-interactive capacities involves (9.1) (applied to M^*) and non-negativity of singletons, with the restrictions to all $\mathcal{A}, \mathcal{B} : |\mathcal{A}| \geqslant |\mathcal{B}| \geqslant n - k$, that is, in $O(L^2)$ time. This way we obtain a supermodular capacity μ^* whose dual is the one we look for, a k-interactive submodular capacity.

In contrast to k-additive capacities, to generate supermodular k-interactive capacities merely by taking the duals of k-interactive submodular ones does not work, because unlike k-additivity, k-interactivity is not preserved under duality. Instead we check for submodularity of the dual capacities μ^*, of a k-interactive one, and then we obtain a supermodular k-interactive capacity through duality. For this we verify the submodularity conditions for μ^*:

$$\sum_{\mathcal{B} \subseteq \mathcal{A}, i, j \in \mathcal{B}, |\mathcal{B}| \geqslant n-k} M^*(\mathcal{B}) \leqslant 0, \ \forall \mathcal{A} \subseteq \mathcal{N}, |\mathcal{A}| \geqslant n-k, i, j = 1, \ldots n \tag{9.7}$$

which correspond with non-positivity of the second derivative.

Now, unlike the case of supermodular capacities, where monotonicity follows from the non-negativity of singletons, here we also need to check monotonicity separately using (9.6), which can be done at the same time with a few extra operations.

9.6 Buoyant and Antibuoyant Capacities

Buoyancy and antibuoyancy are relatively restrictive properties, such that even if we have access to uniformly distributed capacities, an acceptance-rejection approach would not be practical in most contexts. Methods for random generation proposed in [BGJ22] include adapting the linear extension approach and the use of randomly generated vertices of the buoyancy/antibuoyancy polytope.

Suppose we have a linear extension, which we will denote by $\mathbf{L}$ such that each $L_i, i = 1, \ldots, d$ is associated with one of the sets $\mathcal{A} \subseteq \mathcal{N}$ and $L_i \prec L_{i+1}$ is consistent with the partial ordering. Then for a randomly generated vector $\mathbf{r}$ sorted such that $r_i \leqslant r_{i+1}$, we can assign values such that r_i correspond with the marginal contributions Δ_j for each set. The resulting capacity will be antibuoyant since the marginal contributions will be increasing with set cardinality (see Algorithm 9.5).

This will result in fairly extreme capacities, since the differences will be increasing according to the vector $\mathbf{r}$. Algorithm 9.6 is one option for alleviating this effect, by essentially fitting a randomly generated capacity from Algorithm 9.5 to an antibuoyant and symmetric capacity by taking a linear combination with the additive and symmetric capacity.

Another option for generating random antibuoyant or buoyant capacities is using random linear combinations of vertices of the antibuoyant capacity polytope. The polytope itself is quite complicated, with it being impractical to enumerate all the vertices for even modest sized n; however, some vertices can be generated algorithmically and these can potentially be used alongside random capacities obtained from either Algorithm 9.5 or Algorithm 9.6.

We let $\mathcal{U} = \{\mu^1, \ldots, \mu^u\}$ denote a subset of capacities, with $|\mathcal{U}| = u$.

Algorithm 9.5: Random generation of an antibuoyant capacity

Input: **L** (Linear extension), and **r** (sorted vector of differences with $r_i \leqslant r_{i+1}$)
Output: μ
1 Initialise $\mu(\mathcal{A}) \leftarrow 0$, edgesInto$(\mathcal{A}) \leftarrow 0$, for all $\mathcal{A}$
2 **For** $i = 1, 2, \ldots, d$

2.1 $\mathcal{A} \leftarrow L_i$

2.2 $\text{minVal}(\mathcal{A}) \leftarrow \max\left\{\max_{\mathcal{B} \subseteq \mathcal{N}} \mu(\mathcal{B}), \max_{\mathcal{B} \subseteq \mathcal{A}} (\mu(\mathcal{B}) + \text{edgesInto}(\mathcal{B}))\right\}$

2.3 $\mu(\mathcal{A}) \leftarrow \text{minVal}(\mathcal{A}) + r_i$

2.4 **If** $|\mathcal{A}| = 1$ edgesInto$(\mathcal{A}) \leftarrow \mu(\mathcal{A})$
Else $\text{edgesInto}(\mathcal{A}) \leftarrow \mu(\mathcal{A}) - \min_{\substack{\mathcal{B} \subset \mathcal{A} \\ |\mathcal{B}| = |\mathcal{A}| - 1}} \mu(\mathcal{B})$

3 $\mu(\mathcal{A}) \leftarrow \mu(\mathcal{A})/\mu(\mathcal{N})$, for all $\mathcal{A}$

Algorithm 9.6: Random generation of an antibuoyant capacity with cardinality index adjustment

Input: μ^r (a randomly generated capacity using Algorithm 9.5), **w** (an antibuoyant weighting vector satisfying $w_1 \leqslant w_2 \leqslant \cdots \leqslant w_n$), and μ^a (the additive and symmetric capacity)
Output: μ
1 Initialise μ^s
2 **For** $i = 1, 2, \ldots, n$, and all $|\mathcal{A}| = i$,
$\mu(\mathcal{A}) \leftarrow \sum_{j=1}^{i} w_j$
3 $c_1, c_2 \leftarrow \arg\min_{c_1, c_2} \|\mu^s - (c_1\mu^a + c_2\mu^r)\|$
4 $\mu \leftarrow (c_1\mu^a + c_2\mu^r)$

Then, any convex combination will also be antibuoyant, given by

$$\mu = \sum_{i=1}^{u} c_i \mu^i$$

with $\sum_{i=1}^{u} c_i = 1$ and $c_i \geqslant 0$.

A method for generating antibuoyant vertices that correspond with p-symmetric capacities is shown in Algorithm 9.7 with a demonstration of the approach in Example 9.1.

Example 9.1 Let $\mathcal{A} = \{1, 2, 3\}$ for $n = 5$. Then we have a $(2+1) \times (3+1)$ matrix P. Results from the following steps are shown in Fig. 9.3:

Step (1) We know that all subsets $\mathcal{B} \subseteq \mathcal{A}$ will satisfy $\mu(\mathcal{B}) = 0$.

Step 1)

	0	1	2	3
0	0	0	0	0
1				
2				

Step 2)

	0	1	2	3
0	0	0	0	0
1				1/2
2				1

Step 3)

	0	1	2	3
0	0	0	0	0
1			1/4	1/2
2			5/8	1

Step 4)

	0	1	2	3
0	0	0	0	0
1	1/16	1/8	1/4	1/2
2	7/32	3/8	5/8	1

Fig. 9.3 Calculation of the entries in the matrix P from Example 9.1

Algorithm 9.7: Generation of a p-symmetric antibuoyant vertex

---

Input: n, $\mathcal{A}$ (the largest set such that $\mu(\mathcal{A})$ is set to 0 along with its subsets)
Output: μ
1 $a \leftarrow |\mathcal{A}|$
2 $b \leftarrow n - a$
3 Initialise $(b+1) \times (a+1)$ matrix P, with $P_{i,j} = 0, \forall i, j$
4 **For** $i = 2, 3, \ldots, b+1$
  $P_{i,a+1} \leftarrow (i-1)/b$ (Values assigned to supersets of $\mathcal{A}$)
5 **For** $j = a, a-1, \ldots, 1, i = 2, 3, \ldots, b+1$
  $P_{i,j} \leftarrow (P_{i-1,j} + P_{i,j+1})/2$ (Mean of entries to the right and above)
6 **For** $\mathcal{B} \subseteq \mathcal{N}$

  6.1 $i \leftarrow |B \setminus A| + 1$
  6.2 $j \leftarrow |B \cap A| + 1$
  6.3 $\mu(B) \leftarrow P_{i,j}$

Step (2) We then fill down the last column with an arithmetic sequence ending in 1.
Step (3) Starting on the second row and the second last column (the cell corresponding with two elements from $\mathcal{A}$ and one element not from $\mathcal{A}$), we assign the average of the cell to the right and above.
Step (4) This continues until we complete the entire matrix.

After the matrix is obtained, the corresponding fuzzy measure values are determined based on how many of the elements come from $\mathcal{A}$ and how many come from $\mathcal{N} \setminus \mathcal{A}$.

Once we have a desirable set of capacities, they can be used for random generation or even for fitting [BGJ22].

9.7 Random Generation Subject to Equality Constraints

This section is based on [BSW25]. We are interested in random generation of capacities constrained by a number of linear equalities. For example, one can fix some of the capacity values $\mu(\mathcal{A}) = c$. Or one can fix the values of the Choquet integral at some $\mathbf{x}$, or fix some of the interaction indices. In all cases we fix any linear combination of capacity values:

$$\sum_{\mathcal{A}} a_{\mathcal{A}}\mu(\mathcal{A}) = c,$$

and obtain a hyperplane in $\mathbb{R}^d$. More than one linear combination defines the intersection of the respective hyperplanes, which is itself an affine space H in $\mathbb{R}^d$.

Our object of interest is the intersection of the polytope of capacities P with H which we presume non-empty. This affine space will be represented by the set of (consistent, non-redundant) linear constraints in matrix form:

$$\mathbf{Av} = \mathbf{c}.$$

The polytope P will be represented by the monotonicity constraints

$$\mathbf{Mv} \geqslant \mathbf{0}.$$

The vector $\mathbf{v}$ contains the capacity values and represents it as a point in $\mathbb{R}^d$.

9.7.1 *Fixed Capacity Values*

In this section we present a basic case of some $\mu(\mathcal{A}_k) = c_k$. Consider the linear extension-based methods, where we have identified a random linear extension (a simplex in $\mathbb{R}^d$) and need to generate a random point in the intersection of that simplex with our hyperplane H.

Hence, we have the problem of generating uniform random vectors on a simplex with some values fixed. We now have two arrays: the labels of subsets $\mathcal{A}$ ordered according to the chosen linear extension and the values $\mathbf{v}[i] = \mu(\mathcal{A})$. Assign the fixed values c_k to the elements of $\mathbf{v}$ that correspond to $\mathcal{A}_k, k = 1, \ldots, K$. Now we have a partition of $1, \ldots, d$ into $K + 1$ contiguous subsets of indices. Denote the number of unassigned values in each subset by $d_k, k = 1, \ldots, K + 1$. Next, for each subset of indices generate d_k uniform random numbers from $[c_{k-1}, c_k]$ and sort them. Assign the resulting sorted values to the respective $\mathbf{v}[i]$; see Fig. 9.4.

The reason this approach generates a uniform distribution on the intersection of a simplex with the affine space H is that this intersection is the Cartesian product of smaller simplices that correspond to the partition of the variables into $K+1$ groups.

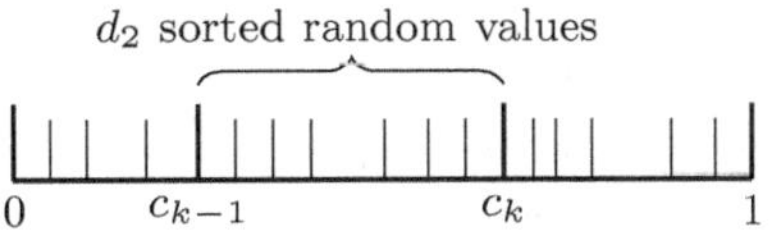

Fig. 9.4 Uniformly distributed random values are generated in each subinterval $[c_{k-1}, c_k]$

Hence, as long as the values of the variables in each group are uniformly distributed, the overall distribution on the Cartesian product will be uniform.

9.7.2 Generation of Capacities Subject to Constraints $\mathbf{Av} = \mathbf{c}$

For the general case the authors of [BSW25] present several approaches.

Explicitly Resolving $\mathbf{Av} = \mathbf{c}$ for Some Components

We consider the general case of linear equality constraints which define an affine space H, whose intersection with P, the polytope of capacities, is our region of interest PH. We assume w.l.o.g. $\mathbf{A} \in \mathbb{R}^{m \times d}$ is of full row-rank m.

Using Linear Extension (LE)-Based Methods

In this approach we express explicitly m dependent components through $d - m$ independent components of $\mathbf{v}$ by doing Gauss elimination and writing

$$\mathbf{w} = \mathbf{Bu}.$$

The independent components $\mathbf{u} \in \mathbb{R}^{d-m}$ have to satisfy monotonicity constraints and can be generated randomly and uniformly by the mentioned topological sort. We can apply efficient methods based on linear extensions of partial orders.

Then we generate the dependent components $\mathbf{w}$ and insert them, based on their values and monotonicity, into $\mathbf{v}$. For each such value only n comparisons are needed; hence, the overall cost of checking monotonicity is $O(mn)$. Overall this method uses the acceptance-rejection approach, selecting $\mathbf{v}$ in the affine plane determined by $\mathbf{Av} = \mathbf{c}$ and then verifying monotonicity of μ for the dependent components.

The preprocessing step to express $\mathbf{w} = \mathbf{Bu}$ takes $O(m^2 d)$ time. It is in fact closely related to determining a suitable basis in the nullspace of $\mathbf{A}$, $NS(\mathbf{A})$ [GH87, Ber+85, CP87]. Since every vector $\mathbf{v}$ in the affine space H can be written as

$$\mathbf{v} = \mathbf{v}_0 + \mathbf{By},$$

where $\mathbf{y} \in NS(\mathbf{A})$ and $\mathbf{v}_0$ is any solution to $\mathbf{Av} = \mathbf{c}$, we are interested in expressing $\mathbf{y}$ so that $d-m$ of its components are also components of $\mathbf{v}$ (because we already have an efficient method for their random generation subject to monotonicity constraints). If the columns of $\mathbf{B}$ are given by a basis of $NS(\mathbf{A})$, we have

$$\mathbf{Av} = \mathbf{Av}_0 + \mathbf{ABy} = \mathbf{Av}_0 = \mathbf{c}.$$

A standard way of constructing a basis for $NS(\mathbf{A})$ is through orthogonal factorisation, namely,

$$\mathbf{A} = \mathbf{QR} = \mathbf{Q}[\mathbf{R}_1|\mathbf{R}_2],$$

where $\mathbf{Q}$ is $m \times m$ orthogonal, $\mathbf{R}_1$ is an $m \times m$ upper rectangular, and $\mathbf{R}_2$ is $m \times (d-m)$ [CGP06]. Then we can take

$$\mathbf{B} = \begin{bmatrix} \mathbf{R}_1^{-1}\mathbf{R}_2 \\ \mathbf{I}_{d-m} \end{bmatrix} \text{ and } \mathbf{v}_0 = \begin{bmatrix} \mathbf{w} \\ \mathbf{0}_{d-m} \end{bmatrix}$$

where $\mathbf{w} = \mathbf{R}_1^{-1}\mathbf{Q}^T\mathbf{c}$. In this case we can take $\mathbf{y} = \mathbf{u}$ and $\mathbf{By}$ will have the last $n-m$ elements as $\mathbf{u}$, and consequently so will $\mathbf{v}$. The cost is $(m^2 d)$, and only $O(md)$ elements are nonzero.

The QR factorisation method with pivoting is done by any standard linear algebra library. If $\mathbf{A}$ is sparse, SPQR routine from SuiteSparse is a better choice to account for sparseness.

Using Random Walks in PH

Rather than generating the independent components of $\mathbf{u}$ separately, we can explicitly define the polytope PH through the set of modified monotonicity constraints. Let the monotonicity constraints be expressed as $\mathbf{Mv} \geqslant \mathbf{0}$, where $\mathbf{M} \in \mathbb{R}^{n\frac{d}{2} \times d}$ (sparse) matrix. We also have $\mathbf{w} = \mathbf{Bu}$ and $\mathbf{v} = \mathbf{v}_0 + \mathbf{Bu}$.

Now we express the inequality constraints as follows:

$$\mathbf{M}(\mathbf{v}_0 + \mathbf{Bu}) = (\mathbf{MB})\mathbf{u} + \mathbf{c} \geqslant \mathbf{0}.$$

Thus, we have arrived at a description of the polytope PH in the lower dimensional subspace H. The matrix $(\mathbf{MB}) \in \mathbb{R}^{n\frac{d}{2} \times (d-m)}$ has the same number of monotonicity constraints. However, this is no longer an order polytope and is less sparse than $\mathbf{M}$. In saying this, we expect many inequalities to become redundant as m becomes larger, and thus, a preprocessing step to eliminate redundant constraints is warranted.

Generation of random vectors in PH is then performed by random walks, as outlined in the next section. Dikin walk can be used to avoid checking inequalities every time, but the cost is passed onto determining the parameters of an ellipsoid at each step [BW21].

As an alternative, instead of considering PH for a random walk, we may consider only one element S of the simplicial partition that corresponds with a fixed LE. That is, we select an LE and then do a random walk in the intersection of S and H that we call SH. A simplex has a smaller number of monotonicity constraints, namely, $d+1$, which are also expressed in sparse matrix format as $\mathbf{M}_S\mathbf{v} \geqslant \mathbf{0}$. Then we will have a polytope SH defined by the half-spaces

$$(\mathbf{M}_S\mathbf{B})\mathbf{u} + \mathbf{c} \geqslant \mathbf{0},$$

which is a smaller system than for PH and hence a faster walk. A disadvantage is that SH may be empty (this can be checked at the set-up stage), and in general different SH have different volumes and therefore lead to more non-uniformity. However, the expected number of points per LE (i.e. per SH) is one; hence, smaller volume does not automatically imply higher density.

9.7.3 Random Walks in H

In this approach we start with a random point already in PH (e.g. obtained by an orthogonal projection) and perform a random walk in the affine subspace H (for which we generate vectors in the nullspace of $\mathbf{A}$). For this we construct an orthogonal basis in $NS(\mathbf{A})$ following the Gram-Schmidt orthogonalisation process.

Using Half-Space Representation

For a random walk we need to generate random points on the unit sphere of dimension $d_{NS} = dim(NS(\mathbf{A})) = d - m$, and then take a linear combination of the orthogonal basis vectors weighted by the coordinates of the points on the unit sphere.

Random unit vectors on a sphere are generated by taking d_{NS} iid values with normal distribution and normalising them to 1. An important question is the cost of calculating a suitable basis in $NS(\mathbf{A})$. We distinguish the case of relatively small d_{NS} (large m) and large d_{NS} (small m).

A textbook method for finding a basis for the $NS(\mathbf{A})$ is to form an augmented matrix (identity matrix I_d of dimension d)

$$[\mathbf{A}^T|\mathbf{I}_d]$$

and reduce it to row-echelon form. Taking the rows with pivot elements 0 (i.e. the last $d-m$ rows of the modified $\mathbf{I}_d$), we obtain a basis for $NS(\mathbf{A})$, which we then orthonormalise.

In the previous section we already mentioned a few methods for constructing bases for $NS(\mathbf{A})$ based on orthogonal and sparse LU factorisation [GH87, Ber+85, CP87]. These bases are sparse but not orthogonal.

The Gram-Schmidt process takes $O(d_{NS}^2 d)$ time, whereas row reduction takes $O(m^2 d)$ time, together giving the cost of set-up $O((m^2 + d_{NS}^2)d)$. Then generation of random vectors in $NS(\mathbf{A})$ takes $O(dd_{NS})$ time.

We note that random walks would be more efficient if the basis for $NS(\mathbf{A})$ is sparse (i.e. having just over d_{NS} nonzero terms). Then generation takes $O(d_{NS})$ time. To do this one can perform sparsity-preserving LU or LL^T factorisation [Kow24, GT08],

$$\mathbf{A}^T = \mathbf{L}\mathbf{L}^T.$$

When m is small, either row reduction of $[\mathbf{A}^T | \mathbf{I}_d]$ or QR factorisation takes only a few steps and leaves the basis for the nullspace sparse. That sparsity will be gone after the Gram-Schmidt steps, but for large d it is almost orthogonal by itself.

Next, we need to ensure that at each step of the random walk the resulting points belong to P. This can be done by verifying monotonicity constraints in $O(n2^n)$ time. However, we can also use a more efficient method based on sorting mentioned earlier, which takes on average $O(2^n)$ time (i.e. linear in d). Thus, the overall time in this method is $O(Sd\, d_{NS} + Q)$, where S is the number of random steps and Q is the complexity of the orthogonal projection and generating a random vector in H.

Using Vertex Representation

This approach uses an alternative representation of a polytope based on its vertices. Since PH is a polytope, assume we identified all or most of its vertices. Now, starting with a random point in PH, perform a random walk as follows:

Take a random subset of vertices V_i of PH and form their linear combination (not necessarily convex). Then make a step of randomly chosen length in that direction (i.e. in some affine subspace of H). The new point will stay in H, but needs to be verified for monotonicity constraints. The last step can be executed in $O(d)$ by using insertion sort as mentioned above [BJW24].

Compared with the random walk in the nullspace of $\mathbf{A}$, here we only use a few random vertices V_i to determine the direction, which takes $O(d)$ time. For example, if we take only three vertices, the random step will be taken in the two-dimensional plane spanned by the three points, and we then generate a random direction in that plane (a point in the unit circle $(\cos t, \sin t)$ with just one random value $t \in [0, 2\pi]$ generated) and make a step of random length in that direction. It will help if the three vertices have many components at 0 (sparsity).

When the vectors in a plane are not orthonormal, [BSW25] presents a few methods to overcome the algorithm's preference in certain directions. Its set-up takes $O(d\, d_{NS}^2)$ time (computation of pairwise scalar products), and generation steps take $O(d)$ each (even less if the directional vectors are sparse).

Of course, one needs to know the vertices of PH for that method to work, which are generally hard to determine. But in some special cases these vertices are available, for example, for k-additive capacities we know that some vertices are 0-1 capacities, which are easily found.

Our specific focus in this section is on the challenge of high dimensionality of the polytope of capacities and hence the computational cost of various alternatives. This cost is governed not only by the dimensionality $d = 2^n - 2$ and the number of constraints m but also by the sparsity of the matrix of constraints. Indeed it is highly beneficial to use sparsity to reduce the computational cost, but it is also challenging, as constructing the nullspace of the matrix of constraints and making orthogonal projections onto it require special sparsity-preserving methods.

We notice that the cost of the presented methods differs, depending on whether we have a large number of simple constraints, such as k-additive capacities, where the dimension of the nullspace is small compared to d, or a small number of more complex constraints involving the values of the Choquet integral or Shapley values.

9.8 Some Heuristics

We conclude this chapter with some heuristic methods. Table 9.1 presents a list and some features of the those mentioned.

Table 9.1 Some capacity generation methods

Methods	Basic principle	References	Features
Random node generation	Randomly select a subset (node) from N first and identify a random value within its allowable range	[HP17, CDM13]	Simple to implement but does not produce uniform distribution
Topological sort generation	Considers poset $\mathcal{P}$ as a directed acyclic graph (DAG) where an edge between vertices u and v means $u \leqslant v$ in that ordering	[BW21, Bel+22]	Efficient, produces linear extensions distributed uniformly
Linear extension-based generation	View the random generation of capacities as a process of assigning ordered random values to the linear extensions of all subsets in $\mathcal{P}(N)$	[CDM13, CSD19, GLS23, MGS20]	Produces linear extensions distributed uniformly, needs many Markov steps
Vertex-convex combination generation	The convex combination of boundary or vertex measures can provide a better coverage of the feasible domain	[CM06, BGJ22]	Better coverage but for uniformity needs additional steps

9.8.1 Random Node Generation Method

A straightforward but naive approach to generating a random capacity on set $\mathcal{N}$ is to randomly select a subset $\mathcal{A}$ (node) from $\mathcal{N}$ first and then to identify a random value within its allowable range [CM06, HP17]. One can then assign that value to $\mu(\mathcal{A})$ and update the allowable ranges for the remaining subsets using monotonicity conditions. This need not be done immediately but only when a new node is selected.

The most notable advantage of this method lies in its low complexity and fast execution speed. However, from a theoretical standpoint, the capacities it generates are not uniformly distributed. This non-uniformity stems from two factors: the dependence of a subset's allowable range on its cardinality and rank and the distribution of generated values deviating significantly from a uniform distribution [SGL22]. Moreover, efforts have been made to enhance random node generation by leveraging distribution information of capacity values and probabilities of different linear extensions, particularly under monotonicity and additional conditions [SGL22].

9.8.2 Vertex-Convex Combination Generation Method

The set of capacities is a convex set (which also applies to various interesting subsets of capacities). Let us call a (convex) subset of capacities of interest a feasible set, and elements of this set feasible capacities. Convex combinations of feasible capacities will also be feasible. The convex combination of capacities $\mu_1, \ldots, \mu_t$, $t > 1$, is defined as [CM06]:

$$\lambda_1\mu_1 + \cdots + \lambda_t\mu_t, \quad \sum_{i=1}^{t} \lambda_i = 1, \quad \lambda_i \geqslant 0. \tag{9.8}$$

This formula allows any number of feasible capacities to serve as the parent feasible capacities for generating new capacities between them. However, one of the ultimate goals of a random generation method is to cover the entire feasible range as much as possible. In this context, the use of extreme capacities [CM06], which usually correspond to the boundary or vertex points of the polytope of capacities, can provide a better coverage of the feasible domain [Wu+24].

If one restricts themselves only to the boundary and monotonicity conditions, the extreme values of $\mu(\mathcal{A})$ can reach 0 and 1 in two opposite directions, and one basic fact is that any capacity is a convex combination of 0-1 valued extreme capacities [Gra16]. The number of extreme capacities that take only 0 or 1 as their values is given by $M(n)-2$, where $M(n)$ is a Dedekind number that grows very quickly with n. As n becomes large, it becomes computationally impractical to generate all such extreme points. However, one can randomly pick subsets of extreme capacities to generate their random convex combinations, and then use those as starting points for a random walk.

Another method proposed in [BW24] is to use a random combination of 0-1 capacities not to be taken as the required random capacity but merely to identify the particular linear extension it corresponds to. In other words, we take a point in the polytope of capacities P and determine a particular simplex in the simplicial partition it belongs to. After that, a new random point in that simplex is generated, which is not necessarily a convex combination of the originally taken vertices. That point is the randomly generated capacity, which has a better distribution than a convex combination of vertices.

9.8.3 Generation of Special Capacities

A method for generating supermodular capacities based on linear extensions has been proposed in [Bel22]. It used MinimalsPlus, randomised topological sort, and Markov chain random walk, combined with a smaller set of linear constraints and amenable to acceptance-rejection. This approach translates into an efficient algorithm for generating supermodular capacities, and by duality, also submodular ones. Still it does not scale up beyond $n = 6$.

Relying on the fact that convex combinations of (anti)buoyant capacities remain within the same class, the authors of [BGJ22] constructed various subsets of vertices for (anti)buoyant capacities based on their desired cardinalities and linear extensions. These subsets were further utilised to create possible representations of (anti)buoyant capacities by employing their convex combinations (see Sect. 9.6 above).

In [Bel+22], the MinimalsPlus and topological sort methods were employed to generate k-interactive capacities, and a random walk (Markov chain) was used on a modified Boolean lattice corresponding to k-interactive capacities to ensure more uniform distribution of outputs.

References

[Bel21] G. Beliakov, On the derivatives of set functions in matrix representation. Inform. Sci. **546**, 469–480 (2021)

[Bel22] G. Beliakov, On random generation of supermodular capacities. IEEE Trans. Fuzzy Syst. **30**(1), 293–296 (2022)

[BGJ22] G. Beliakov, M. Gagolewski, S. James, Reduction of variables and constraints in fitting antibuoyant fuzzy measures to data using linear programming. Fuzzy Sets Syst. **451**, 266–284 (2022)

[BJ21] G. Beliakov, S. James, Choquet integral based measures of economic welfare and species diversity. Int. J. Intell. Syst. **37**, 2849–2867 (2021)

[BJW24] G. Beliakov, S. James, J.-Z. Wu, Efficient monotonicity and convexity checks for randomly sampled fuzzy measures. IEEE Trans. Fuzzy Syst. **32**(12), 6767–6774 (2024)

[BJL11] G. Beliakov, S. James, G. Li, Learning Choquet-integral-based metrics for semisupervised clustering. IEEE Trans. Fuzzy Syst. **19**(3), 562–574 (2011)

[BSW25] G. Beliakov, P. Sun, J.-Z. Wu, Random projections of constrained fuzzy measures. Fuzzy Sets Syst. Under review (2025)

[BW19] G. Beliakov, J.-Z. Wu, Learning fuzzy measures from data: simplifications and optimisation strategies. Inform. Sci. **494**, 100–113 (2019)

[BW21] G. Beliakov, J.-Z. Wu, Random generation of capacities and its application in comprehensive decision aiding. Inform. Sci. **577**, 424–435 (2021)

[BBA25] G. Beliakov, C. Beliakov Amaya, Efficient methods for verifying monotonicity of 2-additive fuzzy measures. Comp. Appl. Math. **44**, 25 (2025)

[Bel+22] G. Beliakov, F.J. Cabrerizo, E. Herrera-Viedma, J.-Z. Wu, Random generation of k-interactive capacities. Fuzzy Sets Syst. **430**, 48–55 (2022)

[BW24] G. Beliakov, J.-Z. Wu, An efficient algorithm for sampling fuzzy measures. IEEE Trans. Fuzzy Syst. **32**(7), 4205–4208 (2024)

[Ber+85] M. Berry et al., An algorithm to compute a sparse basis of the null space. Numerische Mathematik **47**(4), 483–504 (1985)

[CGP06] S. Chandrasekaran, M. Gu, T. Pals, A fast ULV decomposition solver for hierarchically semiseparable representations. SIAM J. Matrix Anal. Appl. **28**(3), 603–622 (2006)

[CJ89] A. Chateauneuf, J.-Y. Jaffray, Some characterizations of lower probabilities and other monotone capacities through the use of Möbius inversion. Math. Soc. Sci. **17**(3), 263–283 (1989)

[CP87] T.F. Coleman, A. Pothen, The null space problem II. Algorithms. SIAM J. Algebraic Discrete Methods **8**(4), 544–563 (1987)

[CDM13] E.F. Combarro, I. Díaz, P. Miranda, On random generation of fuzzy measures. Fuzzy Sets Syst. **228**(4), 64–77 (2013)

[CSD19] E.F. Combarro, J. Hurtado de Saracho, I. Díaz, Minimals Plus: an improved algorithm for the random generation of linear extensions of partially ordered sets. Inform. Sci. **501**, 50–67 (2019)

[CM06] E.F. Combarro, P. Miranda, Identification of fuzzy measures from sample data with genetic algorithms. Comput. Oper. Res. **33**(10), 3046–3066 (2006)

[Dev86] L. Devroye. *Non-uniform Random Variate Generation* (Springer, New York, 1986)

[GH87] J.R. Gilbert, M.T. Heath, Computing a sparse basis for the null space. SIAM J. Algebraic Discrete Methods **8**(3), 446–459 (1987)

[GT08] C. Gotsman, S. Toledo, On the computation of null spaces of sparse rectangular matrices. SIAM J. Matrix Anal. Appl. **30**(2), 445–463 (2008)

[Gra16] M. Grabisch. *Set Functions, Games and Capacities in Decision Making* (Springer, Berlin, New York, 2016)

[Gra97] M. Grabisch, K-order additive discrete fuzzy measures and their representation. Fuzzy Sets Syst. **92**(2), 167–189 (1997)

[GLS23] M. Grabisch, C. Labreuche, P. Sun, An approximation algorithm for random generation of capacities. Order **41**(2), 281–306 (2023)

[GMR00] M. Grabisch, J.-L. Marichal, M. Roubens, Equivalent representations of set functions. Math. Oper. Res. **25**(2), 157–178 (2000)

[HP17] T.C. Havens, A.J. Pinar, Generating random fuzzy (capacity) measures for data fusion simulations, in *2017 IEEE Symposium Series on Computational Intelligence (SSCI)* (IEEE, Piscataway, 2017), pp. 1–8

[HT13] E. Hüllermeier, A.F. Tehrani, Efficient learning of classifiers based on the 2-additive Choquet integral, in *Computational Intelligence in Intelligent Data Analysis* (Springer, Berlin, 2013), pp. 17–29

[Kah62] A.B. Kahn, Topological sorting of large networks. Commun. ACM **5**, 558–562 (1962)

[Kow24] P. Kowal. *Null Space of a Sparse Matrix, MATLAB Central* (2024)

[MGS20] P. Miranda, P. García-Segador, Combinatorial structure of the polytope of 2-additive measures. IEEE Trans. Fuzzy Syst. **28**, 2864–2874 (2020)

[MGG02] P. Miranda, M. Grabisch, P. Gil, p-symmetric fuzzy measures. Int. J. Uncertain. Fuzziness Knowl.-Based Syst. **10**(supp01), 105–123 (2002)

[Rub84] P.A. Rubin, Generating random points in a polytope. Commun. Stat.-Simul. Comput. **13**(3), 375–396 (1984)

[Smi84] R.L. Smith, Efficient Monte Carlo procedures for generating points uniformly distributed over bounded regions. Oper. Res. **32**(6), 1296–1308 (1984)

[SGL22] P. Sun, M. Grabisch, C. Labreuche, An improvement of random node generator for the uniform generation of capacities, in *International Conference on Scalable Uncertainty Management* (Springer, Berlin, 2022), pp. 202–216

[Uri+23] M. Uriz, D. Paternain, H. Bustince, M. Galar, A supervised fuzzy measure learning algorithm for combining classifiers. Inform. Sci. **622**, 490–511 (2023)

[WB23] J.-Z. Wu, G. Beliakov, Generating fuzzy measures from additive measures (2023). arXiv preprint arXiv:2309.15399

[Wu+24] J.-Z. Wu, G. Beliakov, J. Simon, M. Gagolewski, Random generation of linearly constrained fuzzy measures and domain coverage performance evaluation. Inform. Sci. **659**, 120080 (2024)

Chapter 10
Optimisation Models Based on Fuzzy Integrals

Abstract Some contexts give rise to optimisation problems in which the objective involves minimising or maximising the Choquet integral with respect to a set of decision variables. This allows for interaction between variables to be accounted for, resulting in either convex or non-convex optimisation problems depending on the capacity used. In this chapter, we present some of the different scenarios that can lead to Choquet integral objectives along with methods for formulating and solving the resulting optimisation problems.

10.1 Optimisation with Fuzzy Integrals

This chapter examines optimisation (operations research, OR) problems in which the objective function accounts for variable dependencies. We focus on the Choquet integral replacing linear objectives in linear programming and mixed integer programming. Depending on the type of capacity, we may obtain a convex or non-convex optimisation problem, which are treated in the respective sections.

To account for input dependencies, we first formulate a general optimisation problem with the Choquet integral-based objective. We are interested in:

$$\begin{aligned} \text{maximise} \quad & C_\mu(\mathbf{x}) = \sum_{\mathcal{A} \subseteq \mathcal{N}} M(\mathcal{A}) \min_{\mathcal{A}}(\mathbf{x}) \\ \text{s.t.} \quad & \mathbf{C}\,\mathbf{x} \leqslant \mathbf{d}. \end{aligned} \tag{10.1}$$

The system of linear inequality constraints with matrix $\mathbf{C} \in \mathbb{R}^{m \times n}$ defines the feasible domain. The function $\min_A$ denotes the minimum over the subset of components $x_i, i \in \mathcal{A}$. Note that we use Möbius representation of a capacity here for two reasons. Firstly, it helps linearise the objective explicitly and convert (10.1) to a more traditional form used in OR. Secondly, it helps in simplifying the problem in the case of employing k-order capacities, in which many Möbius values are zero.

Such problems for supermodular capacities (the objectives become concave functions) were considered previously [Gra16, GL10, BJW19, MWM10, Bot+18,

G. Beliakov et al., *Choquet Capacities and Fuzzy Integrals*, Theory and Applications of Computability, https://doi.org/10.1007/978-3-031-97070-2_10

LMP13, BJ21a], [BJ21b, Tim12, MP20, Bel21, Isl+19, GPS10]. In this setting (or, equivalently, minimising a convex objective when the capacity is submodular), the optimisation problem has a single optimum, which can be found directly or by converting the problem to a linear program (LP). The supermodularity of the objective models synergies between the decision variables. The contributions of pairs, triples, and other tuples of variables are no smaller than the sum of individual contributions. It fits the context of maximising the total value of a portfolio (options, projects, items) when there are positive synergies leading to even higher values if some decision variables are jointly increased.

The minimisation problems with convex Choquet integral objectives (submodular capacities) model discounts, when the price paid for groups of items is smaller than the sum of individual prices.

In general though, the objective function may be neither concave nor convex, and problem (10.1) becomes a challenging nonlinear non-smooth optimisation problem. We deal with this problem in Sect. 10.2.

The objective in (10.1) is a piecewise linear function; it is non-smooth. In addition the system of constraints can be quite large. While generic non-smooth optimisation algorithms can be used directly, in this chapter we show how such problems can be reformulated to use more efficient linear programming techniques.

10.1.1 Linear Programming Formulation

Consider a *supermodular* capacity μ, which implies concavity of the Choquet integral. Following [Bel21] we translate (10.1) into

$$\begin{aligned} \text{maximise} \quad & \min_{\sigma} \sum_{i=1}^{n} w_i^{\sigma} x_i \\ \text{s.t.} \quad & \mathbf{C}\,\mathbf{x} \leqslant \mathbf{d}, \end{aligned} \tag{10.2}$$

where w^{σ} are the vectors of coefficients on each of the linear segments of the objective (they correspond to the canonical simplices S_{σ}).

Let us now translate (10.2) into a linear programming problem:

$$\begin{aligned} \text{maximise} \quad & T \\ \text{s.t.} \quad & \mathbf{C}\,\mathbf{x} \leqslant \mathbf{d}, \\ & \sum_{j=1}^{n} w_j^{\sigma} x_j \geqslant T \text{ for all } \sigma. \end{aligned} \tag{10.3}$$

While the linear structure of problem (10.3) is clear, its size (the $n!$ number of constraints) makes its numerical solution impractical even for moderate n.

Next, we will employ k-order fuzzy measures which limit high-order interactions and reduce problem complexity.

10.1.2 k-Additive Capacities

The k-additive capacities limit interactions to subsets of cardinality at most k by setting the respective Möbius values to 0; see Definition 7.5. In Möbius representation using auxiliary variables $T_{\mathcal{A}} = \min_{\mathcal{A}}(\mathbf{x})$, we have

$$\begin{aligned} \text{maximise} \quad & \sum_{\mathcal{A}:|\mathcal{A}|\leqslant k} M(\mathcal{A})T_{\mathcal{A}} \\ \text{s.t.} \quad & \mathbf{C}\,\mathbf{x} \leqslant \mathbf{d}, \\ & x_j \geqslant T_{\mathcal{A}} \text{ for all } j \in \mathcal{A} \text{ and all } \mathcal{A} \subseteq \mathcal{N}, |\mathcal{A}| \leqslant k. \end{aligned} \tag{10.4}$$

The special case of 2-additive capacities, which take into account only pairwise interactions, is particularly simple. Supermodular capacities imply all Möbius values are non-negative and sum to 1, and we have $n+\binom{n}{2}$ variables in this case. There are $O(n^2)$ variables and constraints, but with a very sparse structured matrix.

Thus, for 2-additive capacities problem (10.4) translates into

$$\begin{aligned} \text{maximise} \quad & \sum_i M(\{i\})x_i + \sum_{i,j:i>j} M(\{i,j\})T_{ij} \\ \text{s.t.} \quad & \mathbf{Cx} \leqslant \mathbf{d}, \\ & x_i \geqslant T_{ij} \text{ for all } i, j \in \mathcal{N}, i > j \\ & x_j \geqslant T_{ij} \text{ for all } i, j \in \mathcal{N}, i > j. \end{aligned} \tag{10.5}$$

In the portfolio optimisation context, special 2-additive capacities were used in [MWM10, MP20].

10.1.3 Symmetric and Antibuoyant Capacities with a Single Constraint

We consider problem (10.1) in which there is only one constraint. For supermodular capacities there is not much simplification to (10.3), as all the permutations σ need to be considered, making the LP large. However, in the case of symmetric or antibuoyant capacities, we have some significant results.

In the case of a single constraint, we aim to solve

$$\begin{aligned} \text{maximise} \quad & C_\mu(\mathbf{x}) \\ \text{s.t.} \quad & \sum_{j=1}^{n} a_i x_i \leqslant 1. \end{aligned} \tag{10.6}$$

For symmetric capacities, the Choquet integral objective is equivalent to an OWA, resulting in

$$\begin{aligned} \text{maximise} \quad & \sum_{i=1}^{n} w_i x_{(i)} \\ \text{s.t.} \quad & \sum_{i=1}^{n} a_i x_i \leqslant 1 \end{aligned} \tag{10.7}$$

with $a_i, w_i > 0$.

It was shown in [CF03, CF18, CF19] that the solution can be obtained directly, given by

$$\max_{k \in \{1,\dots,n\}} \frac{w_1 + \dots + w_k}{a_{\sigma_1} + \dots + a_{\sigma_k}},$$

where $0 < a_{\sigma_1} \leqslant \dots \leqslant a_{\sigma_n}$ and the corresponding k can be used to determine $\mathbf{x}$, so that:

$$\begin{aligned} x_{\sigma_1} &= \dots = x_{\sigma_k} = \frac{1}{a_{\sigma_1} + \dots + a_{\sigma_k}} \\ x_{\sigma_{k+1}} &= \dots = x_{\sigma_n} = 0. \end{aligned}$$

Notably, the constraint induces the relative ordering of variables in the solution vector, which can also be found using standard linear programming solvers with additional constraints enforcing $x_{\sigma_1} \geqslant \dots \geqslant x_{\sigma_n}$.

10.1.4 Asymmetric Antibuoyant Capacities with a Single Constraint

When the Choquet integral is defined with respect to a general capacity, the weights will differ depending on the simplex used; however, it was established in [BJ21b] that the optimum will have a similar form.

Proposition 10.1 *When optimising* (10.6) *for all* $a_i > 0$*, the (not necessarily unique) maximum will lie at one of the points* $\mathbf{x}^*$ *where*

$$x_i^* = \frac{1}{\sum\limits_{j \in \mathcal{A}} a_j}, \textit{ for each } i \in \mathcal{A}$$

$$x_i^* = 0, \textit{ otherwise,}$$

for some $\mathcal{A} \subseteq \mathcal{N}$.

The following example taken from [BJ21b] demonstrates that the solution can be found through enumeration.

Example 10.1 Consider maximising a Choquet integral with respect to the capacity,

$$\mu = \begin{pmatrix} & 1 & \\ 0.5 & 0.6 & 0.6 \\ 0.1 & 0.3 & 0.3 \end{pmatrix},$$

subject to the constraint, $10x_1 + 11x_2 + 29x_3 = 1$.

We enumerate the following potential solutions:

$\mathcal{A}$	$\mathbf{x}$	$C_\mu(\mathbf{x})$	$\mathcal{A}$	$\mathbf{x}$	$C_\mu(\mathbf{x})$
$\{1\}$	$(\frac{1}{10}, 0, 0)$	1/100	$\{1, 3\}$	$(\frac{1}{39}, 0, \frac{1}{39})$	1/65
$\{2\}$	$(0, \frac{1}{11}, 0)$	3/110	$\{2, 3\}$	$(0, \frac{1}{40}, \frac{1}{40})$	3/200
$\{3\}$	$(0, 0, \frac{1}{29})$	3/290	$\{1, 2, 3\}$	$(\frac{1}{50}, \frac{1}{50}, \frac{1}{50})$	1/50
$\{1, 2\}$	$(\frac{1}{21}, \frac{1}{21}, 0)$	1/42			

and hence given that $3/110$ is the largest output, we determine that this solution at $\mathbf{x} = (0, \frac{1}{11}, 0)$ is our optimum.

This simplifies even further in the case of antibuoyant capacities, where we have the following proposition from [BJ21b].

Proposition 10.2 *Let* μ *be a supermodular capacity and* $0 < a_1 \leqslant a_2 \leqslant \ldots \leqslant a_n$. *Then the solution to* (10.6) *is always on the boundary of the canonical simplex* $S_1 = \{\mathbf{x} \in [0, M]^n | \, x_1 \geqslant x_2 \geqslant \ldots \geqslant x_n\}$ *if and only if* μ *is antibuoyant.*

Mirroring the result for OWA objectives, in the case of antibuoyant capacities, problem (10.6) reduces to a standard linear optimisation formulation with weights determined by the ordering on the constraint and solution vector found with additional constraints that ensure that $\mathbf{x}$ is located on S_1, i.e. the problem becomes

$$\text{maximise} \sum_{i=1}^{n} w_i^1 x_i \tag{10.8}$$

$$\text{s.t.} \sum_{i=1}^{n} a_i x_i \leqslant 1,$$

$$\mathbf{x} \in S_1.$$

In the event that the constraint coefficients are equal, we also have the following.

Corollary 10.1 *If all a_i are equal, the optimal solution to* (10.6) *is on the diagonal of the cube* $[0, M]^n$, $x^* = (t, t, t, \dots, t)$.

We now reformulate (10.8) as a simplified problem. Consider the cumulative sums $w_i^{(c)} = \sum_{k=1}^{i} w_k$ and $a_i^{(c)} = \sum_{k=1}^{i} a_k$. Also define the new non-negative variables $\mathbf{y} \in \mathbb{R}^n$, $y_i = x_i - x_{i+1}$, with $x_{n+1} = 0$ by convention. This way the original variables can be found through reverse cumulative sums $x_1 = \sum_{i=1}^{n} y_i$, $x_2 = \sum_{i=2}^{n} y_i, \dots, x_n = y_n$. Now we have an equivalent problem:

$$\text{maximise} \sum_{i=1}^{n} w_i^{(c)} y_i \tag{10.9}$$

$$\text{s.t.} \sum_{i=1}^{n} a_i^{(c)} y_i \leqslant 1.$$

We observe the following. Problem (10.9) is an LP with a single budget constraint. Beside its own interest it can be thought of as a relaxation of the knapsack problem (which is a mixed integer programming problem). As such it can provide a putative optimal solution to the knapsack problem and also serve as an upper bound on the solution. The transformation of variables from $\mathbf{x}$ to $\mathbf{y}$ simplified other constraints into that of non-negativity. Such constraints are natural in the simplex method of LP and hence need not be explicitly enforced.

Problem (10.9) also provides a simple solution to a complicated resource allocation problem with dependencies expressed as (10.6), as long as the capacity is antibuoyant. It then suffices to identify the order in which the coefficients a_i of the constraint form a non-decreasing sequence, relabel the variables accordingly, and solve (10.9).

10.2 General Non-convex Problems

In this section we consider general capacities, neither sub- nor supermodular. The Choquet integral in the objective is neither convex nor concave, making the problem multimodal. Therefore, we hope at most for convergence of algorithms to local optima. Multistart can be used to locate deep local, if not global optima. A multistart scheme requires efficient local optimisation algorithms which we discuss in the sequel.

Some generic numerical approaches to solve this problem directly can be found in [Pow15]. The main challenges for generic direct approaches are as follows: (1) a large number of stationary points due to non-convexity, (2) the size of the problem in terms of the exponential number of terms in the objective, and (3) complexity arising from monotonicity and other additional constraints, which all make the generic approaches non-scalable.

We now reformulate problem (10.1) as a DC (difference of convex) problem in which the inner problems are LPs, so that we can rely on the proven techniques of linear programming. Let us instantiate (10.1) in the cases of differences of *concave* Choquet integrals.

10.2.1 DC Problem Formulation

To optimise the Choquet integral-based objective which is neither concave nor convex, we apply the DC algorithm [AT05, HT99], which is a generic iterative method that uses a decomposition of the objective into convex and concave parts (i.e. as a difference of concave parts for maximisation problems). Any capacity can be written as a difference of two supermodular capacities (non-normalised) μ and ν [GS94, Gra16] (see Sect. 10.2.2), and therefore, the Choquet integral can be expressed as a difference of concave Choquet integrals,

$$C(\mathbf{x}) = C_\mu(\mathbf{x}) - C_\nu(\mathbf{x}). \tag{10.10}$$

The DC algorithm is employed to maximise the function C in Eq. (10.10), subject to the constraints $\mathbf{Cx} \leqslant \mathbf{d}$, and is formulated as follows:

Algorithm 10.1: DCA

Input: starting point $\mathbf{x}_0$ and $k = 0$
While convergence criteria are not met **do**:

1. Compute $\mathbf{y} \in \partial C_\nu(\mathbf{x}_k)$.
2. Solve $\mathbf{z} = \arg\max_{\mathbf{x}} C_\mu(\mathbf{x}) + \langle \mathbf{x}, \mathbf{y} \rangle$ (for a fixed $\mathbf{y}$), s.t. $\mathbf{Cx} \leqslant \mathbf{d}$.
3. Set $\mathbf{x}_{k+1} = \mathbf{z}$ and $k = k + 1$.

The convergence criteria are standard and involve reaching a given threshold on pairwise differences of the optima of two consecutive iterations.

The expressions ∂C_ν denotes the Clarke's subdifferential of the function $(-C_\nu)$, and $\mathbf{y}$ is any of its subgradients, which is then fixed at Step 2. The optimisation problem at Step 2 is of the same type as (10.3), or (10.4) if we employ k-additive capacities, i.e. it is an LP.

By taking one-sided derivatives of the functions $\min_{\mathcal{A}}(\mathbf{x})$, we obtain an explicit formula for a subgradient of a piecewise linear convex function $(-C_\nu)$:

$$y_i = M_\nu(\{i\}) + \sum_{\mathcal{A}} M_\nu(\mathcal{A})\delta(x_i, \mathcal{A})$$

with $\delta(x_i, \mathcal{A}) = 1/|\mathcal{B}_{\mathcal{A}}|$ if $x_i = \min_{\mathcal{A}}(\mathbf{x})$, $\mathcal{B}_{\mathcal{A}} = \{j : x_j = \min_{\mathcal{A}}(\mathbf{x})\}$ and 0 otherwise. The expression $M_\nu(\mathcal{A})\delta(x_i, \mathcal{A})$ indicates that the coefficient $M_\nu(\mathcal{A})$ is "shared" among the $|\mathcal{B}_{\mathcal{A}}|$ variables x_j whenever we have a tie.

Proposition 10.3 *The values of $C(\mathbf{x}_k)$ in the DC algorithm form a non-decreasing sequence and after a finite number of steps $C(\mathbf{x}_{k+1}) = C(\mathbf{x}_k)$.*

Proof The proof is given in the context of general piecewise linear functions in [PDTHLT17], Section 4. It uses the fact that for any choice of $\mathbf{y} \in \partial C_\nu(\mathbf{x}_k)$, the value $C(\mathbf{x}_k)$ can be reduced finitely often and eventually $\mathbf{x}_{k+1} = \mathbf{x}_k$.

From the practical point of view, the LPs at Step 2 in different iterations of the DCA are all the same, except for some of the coefficients in the objective. Implementations of this instance of DCA can avoid the overheads associated with setting up the linear problem in (10.3) (or (10.4)), and only modify its objective, which will lead to significant computational savings.

10.2.2 Approximation by k-Additive Convex and Concave Capacities

While the general scheme of DCA and solution methods for each of its steps are straightforward, the challenge is to find a suitable decomposition of the Choquet integral in (10.10) into concave and convex parts.

This section presents various possible ways to decompose a capacity for DCA. We present both exact and approximate methods. The exact methods in the next two subsections do not modify the overall objective $C(\mathbf{x})$, but share the same computational complexity as the original capacity, i.e. exponential in the number of variables. They are suitable for rather small n. The approximate methods in the remaining subsections reduce the complexity by limiting one or both resulting convex capacities to sparse or k-order capacities, so that the overall number of coefficients is polynomial. This is done at the cost of ignoring some higher-order interactions among the variables, therefore modifying the original objective $C(\mathbf{x})$. Note that the roles of μ and ν in DCA are different, so we do not treat them symmetrically. By minimising the fitting criteria, we obtain simplified decompositions which approximate the original capacity, and hence the objective, best.

We recall the conditions for supermodularity in Möbius representation [CJ89]:

$$\sum_{\mathcal{B}\subseteq\mathcal{A}|i,j\in\mathcal{B}} M(\mathcal{B}) \geqslant 0,\ \forall \mathcal{A}\subseteq\mathcal{N},\ \forall\{i,j\}\subseteq\mathcal{A}, i\neq j. \tag{10.11}$$

Explicit Decomposition

There are many distinct decompositions of the DC functions. First, we note that a solution to the convex decomposition problem always exists. For this, following [GS94], consider the positive and negative parts of a capacity in Möbius representation $M = m^+ - m^-$, where $m^+(\mathcal{A}) = \max(0, M(\mathcal{A})), m^-(\mathcal{A}) = \max(0, -M(\mathcal{A}))$. The set functions m^+, m^- are totally monotone (i.e. belief measures) but may fail the normalisation condition, which is easily accommodated:

$$\begin{aligned} M &= D_1\hat{m}^+ - D_2\hat{m}^- \\ \hat{m}^+ &= \tfrac{1}{D_1}m^+, \quad \hat{m}^- = \tfrac{1}{D_2}m^-, \\ D_1 &= \textstyle\sum_{\mathcal{A}} m^+(\mathcal{A}), \quad D_2 = \sum_{\mathcal{A}} m^-(\mathcal{A}), \end{aligned} \tag{10.12}$$

as long as $D_1, D_2 > 0$, for otherwise either $m^- = 0$ or $m^+ = 0$ and the decomposition is trivial. Note that $D_1 + D_2 = 1$. Thus, $\hat{m}^+, \hat{m}^-$ are normalised belief measures.

Formula (10.12) provides an explicit decomposition of any capacity into the weighted difference of belief measures.

Decomposition by Solving an LP

We now formulate other ways of finding a similar decomposition. Consider the requirement of having the capacity ν as close as possible to an additive measure. The reason for this is that the DCA method uses linearisation of the Choquet integral C_ν, so the more accurate this linearisation compared to ν is, the better DCA will perform. Thus, our next decomposition model is to solve

$$\begin{aligned} \text{minimise} \quad & \sum_{\mathcal{A},|\mathcal{A}|>1} |M_\nu(\mathcal{A})| \\ \text{s.t.} \quad & M_\mu(\mathcal{A}) - M_\nu(\mathcal{A}) = M(\mathcal{A}), \\ & M_\mu, M_\nu \text{ both satisfy (10.11).} \end{aligned} \tag{10.13}$$

Note that now the values of M_μ, M_ν are not required to be non-negative, i.e. we are not restricting them to be belief measures. Nor do they require normalisation, because it is immaterial for our optimisation problem. Problem (10.13) is converted into an LP and solved by standard methods.

Decomposition by k-Additive Capacities

The goal of our next model is to simplify μ by restricting it to k-additive supermodular capacities. The rationale is to take advantage of the simplifications (10.4), (10.5). Notice that for general capacities, we are required to solve (10.3), which becomes impractical for $n > 10$ due to $n!$ permutations σ. In contrast, k-additive capacities introduce only a polynomial number of constraints. To find such a decomposition, we solve

$$\begin{aligned} \text{minimise } & \sum_{\mathcal{A}} |M_\mu(\mathcal{A}) - M_\nu(\mathcal{A}) - M(\mathcal{A})| \qquad (10.14)\\ \text{s.t. } & M_\mu(\mathcal{A}) = 0 \text{ if } |\mathcal{A}| > k,\\ & M_\mu, M_\nu \text{ both satisfy (10.11).} \end{aligned}$$

Notice that we do not expect the condition $M = M_\mu - M_\nu$ to be satisfied exactly, as the problem may be infeasible, although whenever M itself is a k-additive capacity, an exact k-additive decomposition is possible. Instead we minimise the l_1-norm of the difference between M and its k-additive convex decomposition. Here we enforce k-additivity of μ only, because it has a larger effect on the complexity of DCA in Step 2, whereas ν affects Step 1 only, which is a summation. However, one can also enforce k-additivity of ν if desired.

For 2-additive capacities, the solution is very efficient, but for larger $k > 2$ the number of constraints (10.11) grows exponentially with n, which limits the applicability of this approach.

Sparse Decomposition

Another way to reduce complexity in (10.1) is by using sparse capacities, where we require many Möbius values to be 0. This way the problem translates into (10.4) where instead of condition $|\mathcal{A}| \leqslant k$ we simply take $\mathcal{A} : M(\mathcal{A}) \neq 0$. We obtain such capacities by following parameter reduction strategies used in shrinkage methods like Lasso [Tib96].

$$\begin{aligned} \text{minimise } & \sum_{\mathcal{A}, |\mathcal{A}|>1} |M_\mu(\mathcal{A})| \qquad (10.15)\\ \text{s.t. } & M_\mu(\mathcal{A}) - M_\nu(\mathcal{A}) = M(\mathcal{A}),\\ & M_\mu, M_\nu \text{ both satisfy (10.11).} \end{aligned}$$

The importance here is to simplify μ for Step 2 of the DCA, but we can also combine the objectives from (10.13) and (10.15) (with different weights) to make both μ and ν sparse. Of course, we can also move the equality constraint into the objective like in (10.14) if we accept an approximation of the original capacity.

Decomposition by k-Interactive Capacities

For larger n, problems (10.13)–(10.15) involve a large number of supermodularity constraints (10.11), of the order of $n2^n$, except for the 2-additive capacities. The next method is based on k-interactive capacities [BCMH22, BW19], which also limit interactions to smaller groups but in a different way (see Sect. 7.7).

Recall that the duals of k-interactive capacities have Möbius values $M^*(\mathcal{A}) = 0$ for all subsets with cardinalities $1 < |\mathcal{A}| \leqslant n - k - 1$, which reduces the number of parameters to fit similarly to k-additive capacities. However, because the nonzero values are for the larger subsets (and singletons), the number of supermodularity constraints in (10.11) is the same as the number of such subsets (times their cardinality), that is, polynomial in terms of n, which is a significant advantage.

Note that we fit the dual of μ in this way using duals of k-interactive convex capacities (which are themselves concave). Therefore, we obtain a decomposition of μ into a difference of concave (submodular) capacities. For the objective in the optimisation problem (10.1), this is not an issue if we just exchange the roles of μ and ν. Let M^* be the dual capacity of M (in Möbius representation).

Then the problem we solve is

$$\begin{aligned} \text{minimise } & \sum_{\mathcal{A}} |M_\mu(\mathcal{A}) - M_\nu(\mathcal{A}) - M^*(\mathcal{A})| \\ \text{s.t. } & M_\mu(\mathcal{A}) = M_\nu(\mathcal{A}) = 0, \\ & 1 < |\mathcal{A}| \leqslant n - k - 1 \\ & M_\mu, M_\nu \text{ both satisfy (10.11).} \end{aligned} \tag{10.16}$$

Since the Choquet integral satisfies $\mathcal{C}(\mathbf{x}) = 1 - \mathcal{C}^*(1 - \mathbf{x})$, to maximise $\mathcal{C}(\mathbf{x})$ we maximise $\mathcal{C}_\nu(1 - \mathbf{x}) - \mathcal{C}_\mu(1 - \mathbf{x})$, where both parts are concave functions.

As we mentioned, the advantage of (10.16) over the models (10.13)–(10.15) is a polynomial number of supermodularity constraints and a polynomial number of nonzero Möbius values, which translate into a polynomial number of constraints in (10.4). This property makes it particularly efficient when dealing with relatively large-scale problems, e.g., for $8 \leqslant n \leqslant 25$, where it can effectively capture interactions among pairs and triples of variables. However, for even larger values of n, the preferable approach is to adopt 2-additive or 2-interactive convex decompositions in (10.14)–(10.16) to strike a balance between numerical efficiency and precision.

10.3 Knapsack

In this section we give an overview of the special case of knapsack optimisation problems where the Choquet integral is used as the objective.

10.3.1 Knapsack Problem

We start the discussion with the standard knapsack problem.

A decision-maker selects a number of projects m, not fixed a priori, from n available options. Each project i has an associated cost c_i and a value for the customer v_i. The portfolio selection problem is to choose a set of projects such that the total budget is met and the total value is maximised:

$$\begin{aligned} \text{maximise} \quad & \sum_{i=1}^{n} v_i x_i \\ \text{s.t.} \quad & \sum_{i=1}^{n} c_i x_i \leqslant C \\ & x_i \text{ is integer.} \end{aligned} \tag{10.17}$$

It is assumed that the values and costs are non-negative numbers and the constant C is the allocated budget. The single linear constraint is a specific feature of the knapsack problem. More than one item of a kind can be included, and a special case of at most one item is called 0–1 knapsack. The following references discuss alternative formulation of knapsack problems including the multichoice knapsack, multiple knapsacks, the max-min knapsack problem, and others [Lin98, KPP04, Sch+21, Gal+23].

The knapsack problem (10.17) has been extensively studied; it is an instance of an NP-complete problem, which means that unless $P = NP$, there is no prospect of polynomially scalable (with n) solution methods. The solution strategies for large-scale knapsack problems involve relaxations, heuristics, and meta-heuristics, including general methods for mixed integer problems.

The additivity in the linear objective indicates that the projects in (10.17) are treated as if they are independent. This is a crude approximation to reality: there is little value putting a camping stove into your knapsack without a gas canister, but together these two items provide significant value. It illustrates nonadditivity of the objective: the combined value of the items is more than the sum of their individual values. Similarly in an investment portfolio, the value of shares from particular industries (e.g., in agriculture, hi-tech or travel domains) are co-dependent, as illustrated by droughts, .com crash, and the impacts of COVID-19 restrictions.

10.3.2 Quadratic Knapsack Problem

One way to approach nonadditivity of the objective in the knapsack problem is change it to a quadratic function, thus leading to the quadratic knapsack problem, introduced in [GHS80]; see also [KPP04].

$$\begin{aligned} \text{maximise} \quad & \sum_{i=1}^{n} v_i x_i + \sum_{i,j=1,j>i}^{n} a_{ij} x_i x_j \\ \text{s.t.} \quad & \sum_{i=1}^{n} c_i x_i \leqslant C \\ & x_i \in \{0, 1\}. \end{aligned} \tag{10.18}$$

Here the coefficients a_{ij} are the non-negative values of the pairs of items (i, j) in the knapsack/portfolio. The non-negativity implies synergy, added value, and the product $x_i x_j$ is nonzero only when both items are present.

One can write problem (10.18) as

$$\begin{aligned} \text{maximise} \quad & \sum_{i,j=1}^{n} p_{ij} x_i x_j \\ \text{s.t.} \quad & \sum_{i=1}^{n} c_i x_i \leqslant C \\ & x_i \in \{0, 1\}, \end{aligned} \tag{10.19}$$

where p_{ii} corresponds to the linear values v_i and $p_{ij} + p_{ji} = a_{ij}$, and the matrix $\mathbf{P}$ is also symmetric. As stated, the quadratic knapsack (QKP) is strongly NP-hard [Pis07].

10.3.3 Formulation of the Knapsack Problem

We now proceed with the general knapsack problem formulation with dependencies, which we later instantiate for some special types of capacities. By replacing the linear objective with an extension of pseudo-Boolean functions, we have

$$\text{maximise} \quad F_\mu(\mathbf{x}) = \sum_{\mathcal{A} \subseteq \mathcal{N}} M(\mathcal{A}) h_A(\mathbf{x}) \tag{10.20}$$

$$\text{s.t.} \quad \sum_{i=1}^{n} c_i x_i \leqslant C$$
$$x_i \in D_i.$$

Here h_A denotes some function of $\mathbf{x}$ which also depends on the subset $\mathcal{A}$ and D_i is the domain of each variable. Notably we can use a subset of non-negative integers or, in the case of 0–1 knapsack, $D_i = \{0, 1\}$. It turns out that when we take Owen's multilinear extension of pseudo-Boolean functions [Owe82] with respect to a 2-additive supermodular capacity (and hence $h_{\mathcal{A}}$ is the product over $\mathcal{A}$) this problem becomes precisely the quadratic knapsack, where the coefficients of the quadratic objective correspond with the Shapley values and Shapley pairwise interaction indices (which are equal to the Möbius values in this case).

If we take instead the Lovász extension, we then take $h_{\mathcal{A}}(\mathbf{x}) = \min_{\mathcal{A}} x_i$, and it leads to the Choquet integral.

The Choquet integral results in a piecewise linear problem we later convert to a linear MIP. It can model not only pairwise but higher order interactions within the same framework, which would not be feasible with Owen's extension. The Choquet integral is a homogeneous function, which allows a straightforward extension from 0–1 to integer variables.

It is interesting to observe that when restricting the variables to 0–1, problem (10.20) can be stated as the combinatorial problem,

$$\text{maximise} \quad \mu(\mathcal{A}) \tag{10.21}$$
$$\text{s.t.} \quad \sum_{i \in \mathcal{A}} c_i x_i \leqslant C.$$

For additive capacities this is just the standard knapsack. Otherwise, with no specific structure in μ, it can be solved by enumerating all subsets $\mathcal{A}$. Evidently this has complexity $O(2^n)$ which limits its practicality. We aim at employing the methods of mixed integer programming, including branch and bound, relaxations, and various heuristics, as well as making use of particular features of capacities in order to avoid the exponential cost of numerical solution.

Let us translate problem (10.20) into a linear MIP. We do this in two ways. By using the Möbius representation and functions $h_A(\mathbf{x}) = \min_{i \in A} x_i$ as well as auxiliary variables $T_A = h_A(\mathbf{x})$, we obtain

$$\text{maximise} \quad C_\mu(\mathbf{x}) = \sum_{\mathcal{A} \subseteq \mathcal{N}} M(\mathcal{A}) T_{\mathcal{A}} \tag{10.22}$$
$$\text{s.t.} \quad \sum_{i=1}^{n} c_i x_i \leqslant C$$
$$T_A \leqslant x_i, \ \forall i \in \mathcal{A} \text{ and } \forall \mathcal{A} \subseteq \mathcal{N}$$

$$x_i \in D_i.$$

We note here that even though the capacity μ is supermodular (and hence the objective in (10.22) is concave), some of the Möbius values can be negative.

Alternatively, we note that the Choquet integral with respect to a supermodular capacity is a piecewise linear *concave* function (see, e.g. [BJL11, MP20]), and that any such function can be written as the minimum of linear functions. In our case we have

$$C_\mu(\mathbf{x}) = \min_\sigma L_\sigma(\mathbf{x}) = \min_\sigma \sum_{i=1}^{n} w_i^\sigma x_i, \tag{10.23}$$

where σ are permutations of the set $\mathcal{N}$ and w_i^σ denote the respective weights obtained from the pairwise difference of capacity values μ on each canonical simplex. Then we translate problem (10.20) into

$$\begin{aligned} \text{maximise} \quad & T \\ \text{s.t.} \quad & \sum_{i=1}^{n} c_i x_i \leqslant C \\ & \sum_{j=1}^{n} w_i^\sigma x_i \geqslant T \text{ for all } \sigma \\ & x_i \in D_i. \end{aligned} \tag{10.24}$$

Problems (10.22) and (10.24) are linear MIP problems, and therefore, they are solvable by many methods of integer programming. Thus, we have obtained a MIP formulation of the knapsack problem with all possible dependencies by using the Choquet integral. This formulation was also recently explored in [MP20]. The catch is the size of the problem. As usual, there are n variables x_i (μ appear in the expressions for w_i^σ as pairwise differences), and $n!$ constraints involving T in (10.24) (and $2^n - n - 1$ non-trivial constraints in (10.22)). It is practical only for small n. Having said this, there are some interior point methods for convex optimisation which would allow one to deal with the convex MIP (10.20) directly, but regardless, this formulation is not scalable due to an exponential number of possible dependencies.

In [MP20] an efficient method for solving (10.24) was proposed based on constraint generation. It is based on the following observation: Suppose that $\mathbf{x}^*$ is a solution to a reduced version of (10.24), where not all but only some permutations σ define the constraints. Then if in addition $\mathbf{x}^*$ satisfies the constraint $\sum_{i=1}^{n} w_i^\pi x_i \geqslant T$, where π is the permutation that corresponds to the sorting order of the components of $\mathbf{x}^*$, then $\mathbf{x}^*$ solves (10.24). The reason is that the ordering of the components of $\mathbf{x}^*$ determines precisely for which permutation σ in (10.23) the minimum is reached, and hence one needs only to check whether $\mathbf{x}^*$ satisfies that particular constraint in (10.24).

The constraint generation algorithms in [MP20] work as follows. The reduced MIP problem (10.24) is set up for just one randomly chosen constraint (permutation σ^1). Its solution $\mathbf{x}^*$ needs to be checked against the other constraints, but in fact only one of them is essential, the constraint σ^2 which corresponds to the ordering of the components of $\mathbf{x}^*$. If $\mathbf{x}^*$ satisfies that constraint, problem (10.24) is solved (the rest of the constraints are automatically satisfied), if not, constraint σ^2 is added to the reduced problem (10.24) which is solved again. The new solution $\mathbf{x}^*$ is checked against its own constraint σ^3 which is determined by the ordering in $\mathbf{x}^*$ and the process repeats.

This constraint generation approach was studied numerically in [MP20] (in the slightly different setting of multiagent 0–1 knapsack). Even though in the worst case all $n!$ constraints need to be generated, this case did not occur in the experiments conducted, which show very modest numbers of constraints that are generated (less than 200 even for $n = 500$ on average for randomly generated capacities). However, despite the apparent scalability of the numerical solution, the question of eliciting all those $2^n - 2$ capacity values remains.

We now look into several capacity simplification strategies to translate (10.20) into a simplified but scalable problem. There are several options at our disposal.

10.3.4 k-Additive Capacities

The k-additive capacities limit interactions to subsets of cardinality at most k by setting the respective Möbius values to 0; see Definition 7.5. We have seen that the quadratic knapsack can be formulated using 2-additive capacities. In Möbius representation using auxiliary variables $T_{\mathcal{A}} = \min_{\mathcal{A}}(\mathbf{x})$, we have

$$
\begin{aligned}
\text{maximise} \quad & \sum_{\mathcal{A} | \, |\mathcal{A}| \leqslant k} M(\mathcal{A}) T_{\mathcal{A}} && (10.25)\\
\text{s.t.} \quad & \sum_{i=1}^{n} c_i x_i \leqslant C \\
& x_i \geqslant T_{\mathcal{A}} \text{ for all } i \in \mathcal{A} \text{ and all } \mathcal{A} \subseteq \mathcal{N}, |\mathcal{A}| \leqslant k \\
& x_i \in D_i .
\end{aligned}
$$

There are $\sum_{i=1}^{k} \binom{n}{k}$ nonzero parameters $M(\mathcal{A})$, significantly smaller than 2^n.

However, the special case of 2-additive capacities, which take into account only pairwise interactions, is particularly simple. Supermodular capacities imply all Möbius values are non-negative and sum to 1, and we have $n + \binom{n}{2}$ variables in this case, the same number of terms as in the quadratic knapsack problem. But now the problem is a classical linear MIP, with a straightforward LP relaxation (by removing the restriction $x_i \in D_i$), and is amenable to standard MIP solution methods and software. The key here is that the relaxed LP problem solved at the intermediate steps of many MIP algorithms is significantly less challenging than a QP problem. There are $O(n^2)$ variables and constraints, but with a very sparse structured matrix.

Thus, for 2-additive capacities problem (10.25) translates into

$$\begin{aligned} \text{maximise} \quad & \sum_i M(\{i\})x_i + \sum_{i,j|i>j} M(\{i,j\})T_{ij} \\ \text{s.t.} \quad & \sum_{i=1}^{n} c_i x_i \leqslant C \\ & x_i \geqslant T_{ij} \text{ for all } i, j \in N, i > j \\ & x_j \geqslant T_{ij} \text{ for all } i, j \in N, i > j \\ & x_i \in D_i. \end{aligned} \tag{10.26}$$

The variables T_{ij} are also limited to $D_i \cap D_j$, but this constraint needs not be specified as it will be automatically satisfied at the optimum. We also mention that when some of the variables are independent (i.e. the respective $M(\{i,j\}) = 0$), it is straightforward to incorporate this knowledge into (10.26) by simply excluding the corresponding variables T_{ij} and constraints, thus reducing the size of the problem explicitly. That is, sparsity of the problem can be accommodated to reduce the cost of the solution. Also note the similarity in form to QKP (10.18).

10.3.5 k-Interactive Capacities

An alternative to k-additive case is the k-interactive capacities (Definition 9.3). The respective Choquet integral is given by (8.20), which we now translate into a MIP as follows. We recall problem (10.24) and the Choquet integral expression (8.20).

Now in contrast to (10.24) the number of distinct maximal chains is not $n!$ but $n \cdot (n-1) \ldots \cdot (n-k+1)$, one case being illustrated in Figs. 7.1/9.1. Thus, the number of linear constraints here is of the same order as that in the case of k-additive capacities (for $k << n$).

We also note that as opposed to (10.25), we do not use Möbius representation here but the standard representation μ. It means that the coefficients w_i^σ need to satisfy a different set of linear constraints that ensure supermodularity. These constraints are important when eliciting capacity values from user preferences, but in the knapsack problems w_i^σ are fixed.

In the case of 0–1 knapsack and k-interactive capacities, for problem (10.24) there are some further insights. Formulas (5.1) and (8.20) yield two cases. In the first case, suppose that the knapsack constraint allows m items. With $m > k$

take $\mathbf{x}$ with m ones and the rest zeros. Then the only nonzero term in (5.1) is $\mu(\{(n-m+1),\ldots,(n)\})$, but the latter has a fixed value which depends on K and k only, as per Definition 9.3. The implication is that in this case the optimum is reached when as many items are placed in knapsack as possible (even if they are of smaller value): as long as $m > k$ the objective function depends only on m. The values and interactions between these items m do not count, because the values μ of groups of more than k items do not depend on their composition (for these types of capacities). Thus, problem (10.20) converts to the simplified traditional knapsack problem whose objective is to maximise $\sum x_i$ (no weights!).

For the second case, $m \leqslant k$ brings us back to problem (10.24) (but with a reduced number of permutations σ) which can be approached by using the constraint generation method in [MP20].

Another special class of capacities is those that are antibouyant [BJ21a, BJ21b]. These capacities further restrict the notion of supermodularity and allow very efficient numerical solution to the Choquet integral optimisation problem, but they are yet to be studied in the integer programming setting.

10.3.6 Using the Shapley Values

We now present another alternative which is handy for very large problems $m > 10^5$. Here the number of possible interactions, and hence of the variables, becomes too large for MIP methods. We can convert the problem into the standard knapsack (10.17) by using Shapley values φ_i as the coefficients v_i. The Shapley values computed by (3.3) take into account not only the values of individual items but of their subsets and hence in some way account for interactions. On the other hand, this is the standard knapsack problem and hence can be solved using existing software.

Shapley values, along with the Banzhaf values, which are well-known solutions to a game, can be seen as the best approximations of a capacity in the least squares sense by an additive set function [HH92, MM11]; see also [Gra16] Section 2.16 and references therein. Banzhaf values arise when all capacity values are approximated with the same weights, whereas Shapley values correspond to a particular set of weights depending on subset cardinality. That is, given a capacity μ for which problem (10.20) is too large for linear MIP, we can convert it to the standard knapsack by picking the best representative *additive* set function (given by the Shapley or Banzhaf values), thus accounting at least partially for variable dependencies.

Thus, given a capacity μ, the first step is to determine its approximation by an additive capacity by computing the Shapley values. Then the standard knapsack problem based on the computed Shapley values is solved. The resulting putative solution can be taken as the output or can be plugged into the original problem (10.20) to be verified for feasibility/optimality, for example, following the approach in [MP20].

Note that there are also some methods of constructing 2-additive capacities approximating a given capacity, e.g. [HH92, GR20], by solving quadratic or linear programming problems, thus providing more sophisticated treatment of dependencies at the cost of extra parameters leading to (10.26), which seems to be a promising area.

10.3.7 *Illustrative Example*

Suppose a grant-funding agency considers a number of proposals n with costs $c_i, i = 1, \ldots, n$, and benefits $\mu(i)$ expressed on some common scale. The decision involves funding or not funding some proposals ($x_i \in \{0, 1\}$), and also a number of projects of particular type ($x_j \in \{0, 1, \ldots, m\}$). For example, local councils may consider building or not a particular road and also funding some number of playgrounds or roundabouts.

While the majority of projects are considered independent, the assessment group has identified and valued the combined benefit of some grouped proposals. For instance, it is considered that projects P_1 (funding a swimming pool) and P_2 (funding a gym) have a higher value together than their sum. Similarly, projects P_3 (extending a tram line), P_4 (building a car park near a terminal station), and P_5 (building a shopping precinct near the car park) together had significant value but almost no value separately (it is expected that a car park near the extended tram line will prompt people to use more public transport, which requires funding both projects and so on).

Similar valuations were performed for various other grouped projects and recorded as the combined values $\mu(\mathcal{A})$. The suggested decision to allocate funding can then be found by solving the k-additive knapsack problem (10.25), which also takes advantage of the sparsity of the Möbius representation, as terms (and constraints) involving independent groups $\mathcal{A}$ are omitted. We refer the reader to [BGJ22] for visual representation of sparse hierarchical structures.

Such resource allocation scenarios are common in other domains where the projects can be (but need not be) grouped to provide together a higher value, see, e.g. [MP20]. The number of projects can range between a few and many hundreds, the decision variables can be 0–1 or integers, and not only pairwise but higher-order dependencies may be present. The Choquet integral with respect to k-order and sparse capacities offers good modelling capabilities, and the corresponding knapsack problems can be solved by using standard linear MIP techniques.

10.4 Optimisation of the Sugeno Integral

We recall Definition 5.3, formula (5.16) reproduced below

$$S_\mu(\mathbf{x}) = \bigvee_{\mathcal{A}\subset\mathcal{N}} \left((\min_{i\in\mathcal{A}} x_i) \wedge \mu(\mathcal{A}) \right), \tag{10.27}$$

where $\vee$ is the maximum and $\wedge$ is the minimum.

The Sugeno integral is neither convex nor concave. Solving problem

$$\begin{aligned} \text{maximise} \quad & \mathcal{S}_\mu(\mathbf{x}) = \bigvee_{\mathcal{A}\subseteq\mathcal{N}} \min\{\mu(\mathcal{A}), \min_{\mathcal{A}}(\mathbf{x}))\\ \text{s.t.} \quad & \mathbf{Cx} \leqslant \mathbf{d} \end{aligned} \tag{10.28}$$

can be achieved by using generic non-smooth optimisation methods, combined with random start.

As an alternative to [DO+16], here we reformulate optimisation problem (10.28) as a mixed integer linear programming problem, using auxiliary variables $z_{\mathcal{A}} \in \{0, 1\}$. We solve

$$\begin{aligned} \text{maximise} \quad & T \\ \text{s.t.} \quad & T \geqslant p_{\mathcal{A}}, \ \forall \mathcal{A} \\ & p_{\mathcal{A}} \leqslant \mu(\mathcal{A}), \ \forall \mathcal{A} \\ & p_{\mathcal{A}} \leqslant x_i, \ \forall \mathcal{A}, i \in \mathcal{A} \\ & T \leqslant p_{\mathcal{A}} + (1 - z_{\mathcal{A}})M, \ \forall \mathcal{A} \\ & \sum_{\mathcal{A}} z_{\mathcal{A}} \geqslant 1 \\ & \mathbf{Cx} \leqslant \mathbf{d}. \end{aligned} \tag{10.29}$$

Here M is some large constant, and $z_{\mathcal{A}} = 1$ indicates that $T = p_{\mathcal{A}}$ for that particular $\mathcal{A}$. The second last constraint ensures this equality holds for at least one $\mathcal{A}$.

MIP problems are NP-hard, so the addition of $2^n - 2$ binary variables results in significant computational burden for $n > 10$. But for smaller n the use of well-designed MIP software can be an advantage.

Minimisation problems are reformulated in a similar way using an alternative Sugeno integral representation (5.18).

Optimisation of the Sugeno integral with respect to k-maxitive or k-minitive capacities brings an advantage in reducing the number of independent capacity values $\mu(\mathcal{A})$; hence, the constraints need only be checked for a smaller subset of $\mathcal{A}$. Here we use the fact that for a k-maxitive capacity

$$\bigwedge_{i\in\mathcal{A}} x_i \leqslant \bigwedge_{i\in\mathcal{B}\subset\mathcal{A}} x_i,$$

$$\min\{\bigwedge_{i\in\mathcal{A}} x_i, \mu(\mathcal{A})\} \leqslant \min\{\bigwedge_{i\in\mathcal{B}\subset\mathcal{A}} x_i, \mu(\mathcal{B})\}$$

for some $\mathcal{B} \subset \mathcal{A}$. Hence the problem is reformulated as (10.29) with the exception that all constraints run over $\mathcal{A}$, $|\mathcal{A}| \leqslant k$.

In [DO+16] the authors propose a polynomial algorithm based on bisection and the maxflow algorithm (at every bisection step). In essence, $S_\mu(\mathbf{x}) \geqslant t$ iff

$$\exists \mathcal{A}, |\mathcal{A}| \leqslant k, s.t. (\wedge_{i\in\mathcal{A}} x_i \wedge \mu(\mathcal{A})) \geqslant t.$$

Then for a fixed t one can determine whether such an $\mathcal{A}$ exists in $O(\binom{n}{k})$ operations (even by exhaustive enumeration of $\mathcal{A}$), and $\mathbf{x}$ is found using the maxflow algorithm in $O(n^3)$ time, solving

$$\begin{aligned} &\text{maximise } T \\ &\text{s.t. } T \leqslant x_i, \forall i \in \mathcal{A} \\ &T \leqslant \mu(\mathcal{A}) \\ &\mathbf{Cx} \leqslant \mathbf{d}. \end{aligned} \tag{10.30}$$

The bisection algorithm to determine t is logarithmic (relative to the desired accuracy). So overall the complexity is polynomial in n.

Alternatively, still using bisection to select the largest achievable value of S_μ, one can solve a feasibility problem for each $\mathcal{A}$, $|\mathcal{A}| \leqslant k$:

$$\begin{aligned} &\text{maximise } 1 \\ &\text{s.t. } T \leqslant x_i, \forall i \in \mathcal{A} \\ &T \leqslant \mu(\mathcal{A}) \\ &\mathbf{Cx} \leqslant \mathbf{d} \\ &T \geqslant t. \end{aligned} \tag{10.31}$$

For a fixed t, $\mathcal{A}$ this problem is solved by setting $T = t$, $x_i = t$ for $i \in \mathcal{A}$ and the rest $x_i = 0$, and verifying the constraints $\mathbf{Cx} \leqslant \mathbf{d}$ and $\mu(\mathcal{A}) \geqslant t$ by direct computation. Thus, the complexity of each step is linear in n. The algorithm is organised by checking $\mathcal{A}$ in the order of decreased cardinality or in the order of decreasing $\mu(\mathcal{A})$.

References

[AT05] L. An, P. Tao, The DC (difference of convex functions) programming and DCA revisited with DC models of real world nonconvex optimization problems. Ann. Oper. Res. **133**, 23–46 (2005)

[Bel21] G. Beliakov, Knapsack problems with dependencies through non-additive measures and Choquet integral. Eur. J. Oper. Res. **301**, 277–286 (2021)

[BCMH22] G. Beliakov, T. Cao, V. Mak-Hau, Aggregation of interacting criteria in land combat vehicle selection by using fuzzy measures. IEEE Trans. Fuzzy Syst. **30**(9), 3979–3989 (2022)

[BGJ22] G. Beliakov, M. Gagolewski, S. James, Hierarchical data fusion processes involving the Möbius representation of capacities. Fuzzy Sets Syst. **433**, 1–21 (2022)

[BJ21a] G. Beliakov, S. James, Choquet integral based measures of economic welfare and species diversity. Int. J. Intell. Syst. **37**, 2849–2867 (2021)

[BJ21b] G. Beliakov, S. James, Choquet integral optimisation with constraints and the buoyancy property for fuzzy measures. Inf. Sci. **578**, 22–36 (2021)

[BJL11] G. Beliakov, S. James, G. Li, Learning Choquet-integral-based metrics for semisupervised clustering. IEEE Trans. Fuzzy Syst. **19**(3), 562–574 (2011)

[BJW19] G. Beliakov, S. James, J.-Z. Wu, *Discrete Fuzzy Measures: Computational Aspects* (Springer, Berlin, Heidelberg, 2019)

[BW19] G. Beliakov, J.-Z. Wu, Learning fuzzy measures from data: simplifications and optimisation strategies. Inf. Sci. **494**, 100–113 (2019)

[Bot+18] M. Bottero, V. Ferretti, J. Figueira, S. Greco, B. Roy, On the Choquet multiple criteria preference aggregation model: Theoretical and practical insights from a real-world application. Eur. J. Oper. Res. **271**, 120–140 (2018)

[CJ89] A. Chateauneuf, J.-Y. Jaffray, Some characterizations of lower probabilities and other monotone capacities through the use of Möbius inversion. Math. Soc. Sci. **17**(3), 263–283 (1989)

[CF03] L. Coroianu, R. Fullér, A note on constrained OWA aggregation. Fuzzy Sets Syst. **139**, 543–546 (2003)

[CF18] L. Coroianu, R. Fullêr, On the constrained OWA aggregation problem with single constraint. Fuzzy Sets Syst. **332**, 37–43 (2018)

[CF19] L. Coroianu, R. Fullêr, Minimum of constrained OWA aggregation problem with a single constraint, in *Fuzzy Logic and Applications, Lecture Notes in Computer Science, vol. 11291, WILF 2018* (Springer, Cham, 2019), pp. 183–192

[DO+16] S. Drissi Oudghiri, P. Perny, O. Spanjaard, M. Hachimi, Using the Sugeno integral in optimal assignment problems with qualitative utilities. ECAI 2016 (IOS Press, 2016), pp. 1406–1413

[GPS10] L. Galand, P. Perny, O. Spanjaard, Choquet-based optimisation in multiobjective shortest path and spanning tree problem. Eur. J. Oper. Res. **204**, 303–315 (2010)

[Gal+23] L. Galli, S. Martello, C. Rey, P. Toth, Lagrangian matheuristics for the quadratic multiple knapsack problem. Discrete Appl. Math. **335**, 36–51 (2023)

[GHS80] G. Gallo, P. Hammer, B. Simeone, Quadratic knapsack problems. Math. Programm. Study **12**, 132–149 (1980)

[GS94] I. Gilboa, D. Schmeidler, Additive representations of non-additive measures and the Choquet integral. Ann. Oper. Res. **52**, 43–65 (1994)

[Gra16] M. Grabisch, *Set Functions, Games and Capacities in Decision Making* (Springer, Berlin, New York, 2016)

[GL10] M. Grabisch, C. Labreuche, A decade of application of the Choquet and Sugeno integrals in multi-criteria decision aid. Ann. Oper. Res. **175**(1), 247–286 (2010)

[GR20] M. Grabisch, A. Rusinowska, k-additive upper approximation of TU-games. Oper. Res. Lett. **48**, 487–492 (2020)

[HH92] P. Hammer, R Holzman, On approximations of pseudo-Boolean functions. ZOR-Methods Models Oper. Res. **36**, 3–21 (1992)

[HT99] R. Horst, N. Thoai, DC programming: overview. J. Optim. Theory Appl. **103**, 1–43 (1999)

[Isl+19] M.A. Islam, D.T. Anderson, F. Petry, P. Elmore, An efficient evolutionary algorithm to optimize the Choquet integral. Int. J. Intell. Syst. **34**(3), 366–385 (2019)

[KPP04] H. Kellerer, U. Pferschy, D. Pisinger, *Knapsack Problems* (Springer, Berlin, Heidelberg, 2004)

[LMP13] J. Lesca, M. Minoux, P. Perny, Compact versus noncompact LP formulations for minimizing convex Choquet integrals. Discrete Appl. Math. **161**, 184–199 (2013)

[Lin98] E.Y.-H. Lin, A bibliographical survey on some well-known non-standard knapsack problems. INFOR Inf. Syst. Oper. Res. **36**(4), 274–317 (1998)

[MWM10] T. Magoc, X. Wang, F. Modave, Application of fuzzy measures and interval computation to financial portfolio selection. Int. J. Intell. Syst. **25**, 621–635 (2010)

[MM11] J.-L. Marichal, P. Mathonet, Weighted Banzhaf power and interaction indexes through weighted approximations of games. Eur. J. Oper. Res. **211**, 352–358 (2011)

[MP20] H. Martin, P. Perny, *New Computational Models for the Choquet Integral*, eds. by G.D. Giacomo et al., *ECAI 2020* (IOS Press, 2020), pp. 147–154

[Owe82] G. Owen, *Game Theory* (Academic Press, London, 1982)

[PDTHLT17] T. Pham Dinh, V. Thang Ho, H. Le Thi, DC programming and DCA for solving Brugnano–Casulli piecewise linear systems. Comput. Oper. Res. **87**, 196–204 (2017)

[Pis07] D. Pisinger, The quadratic knapsack problem—a survey. Discrete Appl. Math. **155**, 623–648 (2007)

[Pow15] M. Powell, On fast trust region methods for quadratic models with linear constraints. Math. Programm. Comput. **7**, 237–267 (2015)

[Sch+21] L.E. Schäfer et al., The binary knapsack problem with qualitative levels. Eur. J. Oper. Res. **289**(2), 508–514 (2021)

[Tib96] R. Tibshirani, Regression shrinkage and selection via the lasso. J. Roy. Stat. Soc. B (Methodol.) **58**(1), 267–288 (1996)

[Tim12] M. Timonin, Maximization of the Choquet integral over a convex set and its application to resource allocation problems. Ann. Oper. Res. **196**, 543–579 (2012)

Appendix A
Python Package pyfmtools for Handling and Fitting Fuzzy Measures

This appendix describes the programming library `pyfmtools`, Version 5, which provides various tools for handling fuzzy measures, also called *Choquet capacities*, calculating various indices, Choquet and Sugeno integrals, as well as fitting fuzzy measures to empirical data. This package is designed for Python, but it also includes the C++ source files that can be used without Python. The description of the programming library `pyfmtools` is given in Sect. A.2. Examples of its usage are provided in Sect. A.7.

Description of the Library

A.1 Installation

Installation of `pyfmtools` package can be done from sources or using `pip install pyfmtools`. This package relies on the CFFI library for Python to C interface, which needs to be installed (`pip install cffi`) if not done automatically. `Numpy` is another useful library required by `pyfmtools`. The `pyfmtools.tar.gz` contains the necessary files and should be expanded into a suitable directory.

A.2 Description of the Functions in Package pyfmtools

Scripts in the `tests` folder test the correctness of installation by running tests for some functions included in the package.

Import the library:

```
import pyfmtools as fm
```


G. Beliakov et al., *Choquet Capacities and Fuzzy Integrals*, Theory and Applications of Computability, https://doi.org/10.1007/978-3-031-97070-2

An important note: to pass Python arrays created with numpy, they need to be cast to C pointers as follows:

```
import numpy as np
a=np.array([0.2,0.1,0.6])
pa = ffi.cast("double *", a.ctypes.data)
```

In this manual it will be assumed that a) the Python arrays have been created (and sufficient memory allocated), and b) the pointers were found using the `ffi.cast` function.

`pyfmtools` allows the user to call the library functions bypassing `cffi` and numpy (the Python wrapper will convert all the arrays to the required format automatically; see `test_wrapper.py`), or without the wrapper (which avoids some overheads but allows more control over the code).

A.2.1 Using Python wrapper

Operations on Fuzzy Measures

The first operation should always be initialisation unless working with sparse representations and large n; it will be stated in the function description. The internal structures are precomputed for n variables and saved in the environment variable, which needs to be instantiated.

```
import pyfmtools as fm
fm.fm_init( n)
```

The function computes the internal structures for n variables. The output is passed to the subsequent operations. Example:

```
env=fm.fm_init(3)
env=fm.fm_free(env)
```

```
fm.fm_free( env)
```

The function frees the memory allocated by the initialisation.

The following functions involve the parameters v (the array containing the fuzzy measure in standard representation) or Mob (in Möbius representation) and env—the environment variable set to n variables. The values of the fuzzy measure almost always obey the binary ordering. It will be stated otherwise in case cardinality ordering is used.

```
fm.Choquet(x, v, env)
```

The function computes and returns the value of the Choquet integral of x, wrt fuzzy measure v. The environment variable env is precomputed in the `fm.fm_init` function. $x \in [0, 1]^n$, $v \in [0, 1]^m$. Example:

```
x = [0.2,0.1,0.6]
fm.fm_init( 3)
v = [0,0.3,0.5,0.6,0.4,0.8,0.7,1]
r= fm.Choquet(x,v,env)
```

```
env=fm.fm_free(env)
fm.ChoquetMob(x, Mob, env)
```

The function computes and returns the value of the Choquet integral of x, wrt fuzzy measure given in Möbius representation Mob (Eq. 5.7). Example:

```
x=[0.2,0.5,0.4]
Mob=[0.0,0.3,0.5,-0.2,0.4,0.1,-0.2,0.1]
r = fm.ChoquetMob(x,Mob,env)
```

```
fm.ChoquetKinter(x, v, kint, env)
```

Calculates the value of the Choquet integral of x, wrt k-interactive fuzzy measure v using compact representation of v (in cardinality ordering). Example:

```
kint=2
x=[0.2,0.5,0.4]
v=[0,0.3,0.5,0.6,0.4,0.8,0.7,1]
r=fm.ChoquetKinter(x, v, kint, env)
```

```
fm.Sugeno(x, v, env)
```

Calculates the value of the Sugeno integral of x, wrt fuzzy measure v (Eq. 1.14). Example:

```
x=[0.2,0.5,0.4]
v=[0,0.3,0.5,0.6,0.4,0.8,0.7,1]
r = fm.Sugeno(x,v,env)
```

```
fm.Orness(v, env)
```

Calculates the orness value of the Choquet integral wrt fuzzy measure v in general representation. Example:

```
v=[0,0.3,0.5,0.6,0.4,0.8,0.7,1]
r = fm.Orness(v,env)
```

```
fm.Entropy(v, env)
```

Calculates the entropy value of the Choquet integral wrt fuzzy measure v in standard representation. Example:

```
v=[0,0.3,0.5,0.6,0.4,0.8,0.7,1]
r= fm.Entropy(v,env)
```

```
fm.Mobius(v, env)
```

Calculates the Möbius representation of v. Example:

```
v=[0,0.3,0.5,0.6,0.4,0.8,0.7,1]
mob=fm.Mobius(v,env)
```

```
fm.Zeta(mob, env)
```

Calculates the inverse Möbius representation of a fuzzy measure, i.e. the standard representation. Example:

```
mob=[0.0,0.3,0.5,-0.2,0.4,0.1,-0.2,0.1]
v= fm.Zeta(mob,env)
```

```
fm.Shapley(v, env)
```

Calculates the Shapley values of v in standard representation and returns it as an array of size n. Example:

```
v=[0,0.3,0.5,0.6,0.4,0.8,0.7,1]
s= fm.Shapley(v,env)
```

`fm.ShapleyMob(Mob, env)`
Calculates the Shapley values of Mob in Möbius representation and returns it as an array of size n. Example:

```
mob=[0.0,0.3,0.5,-0.2,0.4,0.1,-0.2,0.1]
s= fm.ShapleyMob(mob,env)
```

`fm.Banzhaf(v, env)`
Calculates the Banzhaf values of v in standard representation and returns it as an array of size n. Example:

```
v=[0,0.3,0.5,0.6,0.4,0.8,0.7,1]
b= fm.Banzhaf(v,env)
```

`fm.BanzhafMob(mob, env)`
Calculates the Banzhaf values of Mob in Mobius representation and returns it as an array of size n. Example:

```
mob=[0.0,0.3,0.5,-0.2,0.4,0.1,-0.2,0.1]
b= fm.BanzhafMob(mob, env)
```

`fm.Interaction(v, env)`
Calculates all the interaction indices of fuzzy measure v in standard representation and returns it in an array of size m. Example:

```
v=[0,0.3,0.5,0.6,0.4,0.8,0.7,1]
s=fm.Interaction(v,env)
```

`fm.InteractionMob(Mob, env)`
Calculates all the interaction indices of fuzzy measure Mob in Möbius representation and returns it in an array of size m. Example:

```
mob=[0.0,0.3,0.5,-0.2,0.4,0.1,-0.2,0.1]
s=fm.InteractionMob(mob, env)
```

`fm.InteractionB(v, env)`
Calculates all the Banzhaf interaction indices of fuzzy measure v given in standard representation and returns the result in an array of size m. Example:

```
v=[0,0.3,0.5,0.6,0.4,0.8,0.7,1]
b= fm.InteractionB(v, env)
```

`fm.InteractionBMob(Mob, env)`
Calculates all the Banzhaf interaction indices of fuzzy measure Mob in Möbius representation and returns the result in an array of size m. Example:

```
mob=[0.0,0.3,0.5,-0.2,0.4,0.1,-0.2,0.1]
b=fm.InteractionBMob(mob, env)
```

`fm.BipartitionShapleyIndex(v, env)`
Calculates all the Shapley bipartition indices of fuzzy measure v given in standard representation and returns the result in an array of size m. Example:

```
v=[0,0.3,0.5,0.6,0.4,0.8,0.7,1]
```

```
s=fm.BipartitionShapleyIndex(v,env)
```

`fm.BipartitionBanzhafIndex(v, env)`
Calculates all the Banzhaf bipartition indices of fuzzy measure v given in standard representation and returns the result in an array of size m. Example:

```
v=[0,0.3,0.5,0.6,0.4,0.8,0.7,1]
b= fm.BipartitionBanzhafIndex(v,env)
```

`fm.NonadditivityIndex(v, env)`
Calculates all the nonadditivity indices of fuzzy measure v given in standard representation and returns the result in an array of size $m = 2^n$. Example:

```
v=n[0,0.3,0.5,0.6,0.4,0.8,0.7,1]
na= fm.NonadditivityIndex(v,env)
```

`fm.NonadditivityIndexMob(Mob, env)`
Calculates all the nonadditivity indices of fuzzy measure v given in Möbius representation and returns the result in an array of size $m = 2^n$. Example:

```
v=[0,0.3,0.5,0.6,0.4,0.8,0.7,1]
na= fm.py_BNonadditivityIndexMob(mob,env)
```

`fm.NonmodularityIndex(pv, env)`
Calculates all the $m = 2^n$ nonmodularity indices of fuzzy measure v given in standard representation and returns the result in an array of size m.

```
v=[0,0.3,0.5,0.6,0.4,0.8,0.7,1]
nm=fm.NonmodularityIndex(v, env)
```

`fm.NonmodularityIndexMob(pmob, env)`
Calculates all the nonmodularity indices of fuzzy measure Mob given in Möbius representation and returns the result in an array of size m. Example:

```
mob=[0.0,0.3,0.5,-0.2,0.4,0.1,-0.2,0.1],
nm-fm.NonmodularityIndexMob(mob, env)
```

`fm.NonmodularityIndexMobkadditive(pmob, k, env)`
Calculates all the $m = 2^n$ nonmodularity indices of k-additive fuzzy measure Mob given in Möbius representation (in cardinality ordering) and returns the result in an array of size m. Example:

```
mob=[0.0,0.3,0.5,-0.2,0.4,0.1,-0.2,0.1]
k=2
nm=fm.NonmodularityIndexMobkadditive(mob, k, env)
```

`fm.NonmodularityIndexKinteractive(pv, kint, env)`
Calculates all the $m = 2^n$ nonmodularity indices of k-interactive fuzzy measure v given in standard representation (in cardinality ordering) and returns the result in an array of size m.

```
v=[0,0.3,0.5,0.6,0.4,0.8,0.7,1]
kint=2
nm=fm.NonmodularityIndexKinteractive(v, kint, env))
```

`fm.dualm(v, env)`

Calculates the dual of fuzzy measure v in standard representation and returns it in an array of size m. Example:

```
v-[0,0.3,0.5,0.6,0.4,0.8,0.7,1]
d= fm.dualm(v,env)
```

```
fm.dualmMob(Mob, env)
```

Calculates the dual of fuzzy measure M in Mobius representation and returns it in an array of size m. Example as above but with Mobius input. Example:

```
mob=[0.0,0.3,0.5,-0.2,0.4,0.1,-0.2,0.1]
d=fm.dualmMob(mob, env)
```

```
fm.ConstructLambdaMeasure(singletons, env)
```

Given the values of a fuzzy measure at singletons, finds the value of λ, computes all other values of the fuzzy measure in standard representation, and returns the pair (λ, v), where v is an array of size m. The input array $singletons$ is an array of size n of the values of fuzzy measure at singletons. Example:

```
singletons=[0, 0.3, 0.5]
lambda, v=fm.ConstructLambdaMeasure(singletons, env)
```

```
fm.ConstructLambdaMeasureMob((singletons, env))
```

Given the values of a fuzzy measure at singletons, finds the value of λ, computes all other values of the fuzzy measure in Möbius representation, and returns the pair (λ, Mob), where Mob is an array of size m. The input array $singletons$ is an array of size n of the values of fuzzy measure at singletons. Example:

```
singletons=[0, 0.3, 0.5]
lambda,mob=fm.ConstructLambdaMeasureMob(singletons, env)
```

```
fm.export_maximal_chains(n, v, mc, env)
```

Returns in mc the arrays of maximal chains (there are $n!$ such arrays) of a fuzzy measure v; each array is of length n and contains the chains of discrete derivatives, which can be used to calculate the Choquet integrals for all possible orderings of the argument x, as well as serve as coefficients of the piecewise linear objective functions. That is, each maximal chain corresponds to the coefficients of a linear function on the respective simplex. All such simplices form a partition of the unit cube. Example:

```
n=3
vb=[0, 0.00224734, 0.06493396, 0.51092037, 0.00965497,
0.37405545, 0.15455057, 1]
mc=fm.export_maximal_chains(n,vb,env)
```

```
fm.ConvertCard2Bit( src, env)
```

Converts fuzzy measure src from cardinality to binary ordering. Example:

```
src=[0, 0.0340916, 0.00890771, 0.01109779, 0.07918582,
0.6305862, 0.1846705, 1]
v=fm.ConvertCard2Bit(src, env)
```

```
fm.min_subset(x, n, S)
fm.py_max_subset(x, n, S)
```

Return min or max of $x_i, i \in S$. S should be written in binary form. Example:

```
x=[0,0.3,0.5,0.6,0.7,0.8,0.85,0.9]
fm.min_subset(x,3,6)
fm.max_subset(x,3,5)
```

`fm.min_subsetC(x, n, S, env)`
Calculates the minimum of $x_i, i \in S$, in cardinality-based ordering. Example:

```
x=[0,0.3,0.5]
fm.min_subsetC(x,3,3,env)
```

`fm.max_subsetNegC(x, n, S, env)`
Calculates the maximum over the complement of subset S in cardinality ordering. Example:

```
x=[0,0.3,0.5]
fm.max_subsetNegC(x,3,3,env)
```

`fm.SizeArraykinteractive(n, k, env)`
Returns the size of the array to store k-interactive fuzzy measure. Example:

```
fm.SizeArraykinteractive(3,1,env)
```

`fm.IsSubsetC(i, j, env)`
Returns true if j is subset of i in cardinality ordering.Example:

```
fm.IsSubsetC(2,1,env)
```

`fm.IsElementC(A, j, env)`
Returns true if j is an element of A in cardinality ordering. Example:

```
fm.IsElementC(1,2,env)
```

`fm.ExpandKinteractive2Bit( src, env, kint, arraysize)`
`fm.ExpandKinteractive2Bit_m( src, env, kint, arraysize, VVC)`
Expands k-interactive fuzzy measure (from src to $dest$) given in more compact representation of size $arraysize$ into its full representation in binary ordering. The function `ExpandKinteractive2Bit` allocates working memory internally, whereas the second function expects working array VVC of length m. Example:

```
src=[0.0340916, 0.01109779, 0.07918582,0.7,1]
kint=1
arraysize=fm.SizeArraykinteractive(n, kint, env)
v=fm.ExpandKinteractive2Bit(src,env,kint,arraysize)
dest,vcc=fm.ExpandKinteractive2Bit_m(src,env,kint,arraysize)
```

`fm.fm_arraysize( n, kint, env)`
Returns the length of the array of values of k-interactive fuzzy measures. Useful for reserving memory to store them. For $kint > 5$ requires initialisation of the env variable, otherwise no. Example:

```
len=fm.fm_arraysize(n, kint, env)
```

`fm.ShowCoalitions(env)`
`fm.ShowCoalitionsCard(env)`

Return the decimal expression for the subsets A, e.g. for $A = 11$ it will show 124 for printing, in binary and in cardinality ordering, respectively. A is the array of integers to receive the decimal expressions for all 2^n coalitions. Example:
`fm.ShowCoalitions( env)`

The following routines are self-explanatory. They return 1 if yes, 0 if no. The input parameters are fuzzy measure v in standard representation and fuzzy measure Mob in Möbius representation (both in binary ordering).

```
fm.IsMeasureAdditive(v, env)
fm.IsMeasureAdditiveMob(Mob, env)
fm.IsMeasureBalanced(v, env)
fm.IsMeasureBalancedMob(Mob, env)
fm.IsMeasureSelfdual(v, env)
fm.IsMeasureSelfdualMob(Mob, env)
fm.IsMeasureSubadditive(v, env)
fm.IsMeasureSubadditiveMob(v, env)
fm.IsMeasureSubmodular(v, env)
fm.IsMeasureSubmodularMob(Mob, env)
fm.IsMeasureSuperadditive(v, env)
fm.IsMeasureSuperadditiveMob(Mob, env)
fm.IsMeasureSupermodular(v, env)
fm.IsMeasureSupermodularMob(Mob, env)
fm.IsMeasureSymmetric(v, env)
fm.IsMeasureSymmetricMob(Mob, env)
```

`fm.IsMeasureKMaxitive(v, env)` (returns k);
`fm.IsMeasureKMaxitiveMob(Mob, env)` (returns k).

A.2.2 *Without Python wrapper, Using* `numpy` *and* `cffi`

Operations on Fuzzy Measures

The first operation should always be initialisation unless working with sparse representations and large n; it will be stated in the function description. The internal structures are precomputed for n variables and saved in the environment variable, which needs to be defined.
`fm.py_fm_init( n, env)`
The function computes the internal structures for n variables. The output is passed to the subsequent operations. Example:

```
env=ffi.new("struct fm_env *")
fm.py_fm_init(3, env)
```

`fm.py_fm_free( env)`
The function frees the memory allocated by the initialisation.

The following functions involve the parameters v (the array containing the fuzzy measure in standard representation) or Mob (in Möbius representation) and env—the environment variable set to n variables. The values of the fuzzy measure almost always obey the binary ordering. it will be stated otherwise in case cardinality ordering is used.

`fm.py_Choquet(x, v, env)`

The function computes and returns the value of the Choquet integral of x, wrt fuzzy measure v. The environment variable env is precomputed in the `fm.py_fm_init` function. $x \in [0, 1]^n$, $v \in [0, 1]^m$. Example:

```
x=np.array([0.2,0.5,0.4],np.float)
px = ffi.cast("double *", x.ctypes.data)
v=np.array([0,0.3,0.5,0.6,0.4,0.8,0.7,1],np.float)
pv = ffi.cast("double *", v.ctypes.data)
r= fm.py_Choquet(px,pv,env)
```

`fm.py_ChoquetMob(x, Mob, env)`

The function computes and returns the value of the Choquet integral of x, wrt fuzzy measure given in Möbius representation Mob (Eq. 5.7). Example:

```
x=np.array([0.2,0.5,0.4],np.float)
px = ffi.cast("double *", x.ctypes.data)
Mob=np.array([0.0,0.3,0.5,-0.2,0.4,0.1,-0.2,0.1],np.float)
pMob = ffi.cast("double *", Mob.ctypes.data)
r = fm.py_ChoquetMob(px,pMob,env)
```

`fm.py_ChoquetKinter(x, v, kint, env)`

Calculates the value of the Choquet integral of x, wrt $k$interactive fuzzy measure v using compact representation of v (in cardinality ordering). Example:

```
kint=2
x=np.array([0.2,0.5,0.4],np.float)
px = ffi.cast("double *", x.ctypes.data)
v=np.array([0,0.3,0.5,0.6,0.4,0.8,0.7,1],np.float)
pv = ffi.cast("double *", v.ctypes.data)
r=lib.py_ChoquetKinter(px, pv, kint, env)
```

`fm.py_Sugeno(x, v, env)`

Calculates the value of the Sugeno integral of x, wrt fuzzy measure v (Eq. 1.14). Example:

```
x=np.array([0.2,0.5,0.4],np.float)
px = ffi.cast("double *", x.ctypes.data)
v=np.array([0,0.3,0.5,0.6,0.4,0.8,0.7,1],np.float)
pv = ffi.cast("double *", v.ctypes.data)
r = fm.py_Sugeno(px,pv,env)
```

`fm.py_Orness(v, env)`

Calculates the orness value of the Choquet integral wrt fuzzy measure v in general representation. Example:

```
v=np.array([0,0.3,0.5,0.6,0.4,0.8,0.7,1],np.float)
pv = ffi.cast("double *", v.ctypes.data)
```

```
r = fm.py_Orness(pv,env)
```

`fm.py_Entropy(v, env)`
Calculates the entropy value of the Choquet integral wrt fuzzy measure v in standard representation. Example:

```
v=np.array([0,0.3,0.5,0.6,0.4,0.8,0.7,1],np.float)
pv = ffi.cast("double *", v.ctypes.data)
r= fm.py_Entropy(pv,env)
```

`fm.py_Mobius(v, mob, env)`
Calculates the Möbius representation of v. Example:

```
v=np.array([0,0.3,0.5,0.6,0.4,0.8,0.7,1],np.float)
pv = ffi.cast("double *", v.ctypes.data)
mob=np.zeros(env.m, np.float) # note env.m=2**n
pmob = ffi.cast("double *", mob.ctypes.data)
fm.py_Mobius(pv,pmob,env)
```

`fm.py_Zeta(pmob, pv, env)`
Calculates the inverse Möbius representation of a fuzzy measure, i.e. the standard representation. Example:

```
v=np.zeros(env.m, np.float)
pv = ffi.cast("double *", v.ctypes.data)
mob=np.array([0.0,0.3,0.5,-0.2,0.4,0.1,-0.2,0.1],np.float)
pmob = ffi.cast("double *", mob.ctypes.data)
fm.py_Zeta(pmob,pv,env)
```

`fm.py_Shapley(v, s, env)`
Calculates the Shapley values of v in standard representation and returns it as an array of size n. Example:

```
v=np.array([0,0.3,0.5,0.6,0.4,0.8,0.7,1],np.float)
pv = ffi.cast("double *", v.ctypes.data)
s=np.zeros(env.n,np.float)
ps = ffi.cast("double *", s.ctypes.data)
fm.py_Shapley(pv,ps,env)
```

`fm.py_ShapleyMob(Mob, S, env)`
Calculates the Shapley values of Mob in Möbius representation and returns it as an array of size n. Example:

```
mob=np.array([0.0,0.3,0.5,-0.2,0.4,0.1,-0.2,0.1],np.float)
pmob = ffi.cast("double *", mob.ctypes.data)
s=np.zeros(env.n, np.float)
ps = ffi.cast("double *", s.ctypes.data)
fm.py_ShapleyMob(pmob,ps,env)
```

`fm.py_Banzhaf(pv, B, env)`
Calculates the Banzhaf values of v in standard representation and returns it as an array of size n. Example:

```
v=np.array([0,0.3,0.5,0.6,0.4,0.8,0.7,1],np.float)
```

```
pv = ffi.cast("double *", v.ctypes.data)
B=np.zeros(env.n, np.float)
pB = ffi.cast("double *", B.ctypes.data)
fm.py_Banzhaf(pv,pB,env)
```

`fm.py_BanzhafMob(pmob, B, env)`
Calculates the Banzhaf values of Mob in Mobius representation and returns it as an array of size n. Example:

```
mob=np.array([0.0,0.3,0.5,-0.2,0.4,0.1,-0.2,0.1],np.float)
pmob = ffi.cast("double *", mob.ctypes.data)
B=np.zeros(env.n, np.float)
pB = ffi.cast("double *", B.ctypes.data)
fm.py_BanzhafMob(pmob, pB, env)
```

`fm.py_Interaction(v, s, env)`
Calculates all the interaction indices of fuzzy measure v in standard representation and returns it in an array of size m. Example:

```
v=np.array([0,0.3,0.5,0.6,0.4,0.8,0.7,1],np.float)
pv = ffi.cast("double *", v.ctypes.data)
s=np.zeros(env.m, np.float) # size of the array is 2**n
ps = ffi.cast("double *", s.ctypes.data)
fm.py_Interaction(pv,ps,env)
```

`fm.py_InteractionMob(Mob, s, env)`
Calculates all the interaction indices of fuzzy measure Mob in Möbius representation and returns it in an array of size m. Example:

```
mob=np.array([0.0,0.3,0.5,-0.2,0.4,0.1,-0.2,0.1],np.float)
pmob = ffi.cast("double *", mob.ctypes.data)
s=np.zeros(env.m, np.float)
ps = ffi.cast("double *", s.ctypes.data)
fm.py_InteractionMob(pmob, ps, env)
```

`fm.py_InteractionB(v, b, env)`
Calculates all the Banzhaf interaction indices of fuzzy measure v given in standard representation and returns the result in an array of size m. Example:

```
v=np.array([0,0.3,0.5,0.6,0.4,0.8,0.7,1],np.float)
pv = ffi.cast("double *", v.ctypes.data)
b=np.zeros(env.m, np.float)
pb = ffi.cast("double *", b.ctypes.data)
fm.py_InteractionB(pv, pb, env)
```

`fm.py_InteractionBMob(Mob, b, env)`
Calculates all the Banzhaf interaction indices of fuzzy measure Mob in Möbius representation and returns the result in an array of size m. Example:

```
mob=np.array([0.0,0.3,0.5,-0.2,0.4,0.1,-0.2,0.1],np.float)
pmob = ffi.cast("double *", mob.ctypes.data)
b=np.zeros(env.m, np.float)
pb = ffi.cast("double *", b.ctypes.data)
```

`fm.py_InteractionBMob(pmob, pb, env)`

`fm.py_BipartitionShapleyIndex(v, s, env)`
Calculates all the Shapley bipartition indices of fuzzy measure v given in standard representation and returns the result in an array of size m. Example:

```
v=np.array([0,0.3,0.5,0.6,0.4,0.8,0.7,1],np.float)
pv = ffi.cast("double *", v.ctypes.data)
s=np.zeros(env.m, np.float)
ps = ffi.cast("double *", s.ctypes.data)
fm.py_BipartitionShapleyIndex(pv,ps,env)
```

`fm.py_BipartitionBanzhafIndex(v, b, env)`
Calculates all the Banzhaf bipartition indices of fuzzy measure v given in standard representation and returns the result in an array of size m. Example:

```
v=np.array([0,0.3,0.5,0.6,0.4,0.8,0.7,1],np.float)
pv = ffi.cast("double *", v.ctypes.data)
b=np.zeros(env.m, np.float)
pb = ffi.cast("double *", b.ctypes.data)
fm.py_BipartitionBanzhafIndex(pv,pb,env)
```

`fm.py_NonadditivityIndex(v, na, env)`
Calculates all the nonadditivity indices of fuzzy measure v given in standard representation and returns the result in an array of size $m = 2^n$. Example:

```
v=np.array([0,0.3,0.5,0.6,0.4,0.8,0.7,1], np.float)
pv = ffi.cast("double *", v.ctypes.data)
na=np.zeros(env.m, np.float)
pna = ffi.cast("double *", na.ctypes.data)
fm.py_NonadditivityIndex(pv,pna,env)
```

`fm.py_NonadditivityIndexMob(Mob, na, env)`
Calculates all the nonadditivity indices of fuzzy measure v given in Möbius representation and returns the result in an array of size $m = 2^n$. Example:

```
mob=np.array([0.0,0.3,0.5,-0.2,0.4,0.1,-0.2,0.1],np.float)
pmob = ffi.cast("double *", mob.ctypes.data)
na=np.zeros(env.m, np.float)
pna = ffi.cast("double *", na.ctypes.data)
fm.py_BNonadditivityIndexMob(pmob,pna,env)
```

`fm.py_NonmodularityIndex(pv, pnm, env)`
Calculates all the $m = 2^n$ nonmodularity indices of fuzzy measure v given in standard representation and returns the result in an array of size m.

```
v=np.array([0,0.3,0.5,0.6,0.4,0.8,0.7,1],np.float)
pv = ffi.cast("double *", v.ctypes.data)
nm=np.zeros(env.m, np.float)
pnm = ffi.cast("double *", nm.ctypes.data)
fm.py_NonmodularityIndex(pv, pnm, env)
```

`fm.py_NonmodularityIndexMob(pmob, pnm, env)`

Calculates all the nonmodularity indices of fuzzy measure Mob given in Möbius representation and returns the result in an array of size m. Example:

```
mob=np.array([0.0,0.3,0.5,-0.2,0.4,0.1,-0.2,0.1], np.float)
pmob = ffi.cast("double *", mob.ctypes.data)
nm=np.zeros(env.m, np.float)
pnm = ffi.cast("double *", nm.ctypes.data)
fm.py_NonmodularityIndexMob(pmob, pnm, env)
```

`fm.py_NonmodularityIndexMobkadditive(pmob, pnm, k, env)`
Calculates all the $m = 2^n$ nonmodularity indices of k-additive fuzzy measure Mob given in Möbius representation (in cardinality ordering) and returns the result in an array of size m. Example:

```
mob=np.array([0.0,0.3,0.5,-0.2,0.4,0.1,-0.2,0.1],np.float)
pmob = ffi.cast("double *", mob.ctypes.data)
nm=np.zeros(env.m, np.float)
pnm = ffi.cast("double *", nm.ctypes.data)
k=2
fm.py_NonmodularityIndexMobkadditive(pmob, pnm, k, env)
```

`fm.py_NonmodularityIndexKinteractive(pv, pnm, kint, env)`
Calculates all the $m = 2^n$ nonmodularity indices of k-interactive fuzzy measure v given in standard representation (in cardinality ordering) and returns the result in an array of size m.

```
v=np.array([0,0.3,0.5,0.6,0.4,0.8,0.7,1],np.float)
pv = ffi.cast("double *", v.ctypes.data)
nm=np.zeros(env.m, np.float)
pnm = ffi.cast("double *", nm.ctypes.data)
kint=2
fm.py_NonmodularityIndexKinteractive(pv, pnm,kint, env))
```

`fm.py_dualm(v, d, env)`
Calculates the dual of fuzzy measure v in standard representation and returns it in an array of size m. Example:

```
v=np.array([0,0.3,0.5,0.6,0.4,0.8,0.7,1],np.float)
pv = ffi.cast("double *", v.ctypes.data)
d=np.zeros(env.m, np.float) # note env.m=2**n
pd = ffi.cast("double *", d.ctypes.data)
fm.py_dualm(pv,pd,env)
```

`fm.py_dualmMob(Mob, d, env)`
Calculates the dual of fuzzy measure M in Mobius representation and returns it in an array of size m. Example as above but with Mobius input. Example:

```
mob=np.array([0.0,0.3,0.5,-0.2,0.4,0.1,-0.2,0.1],np.float)
pmob = ffi.cast("double *", mob.ctypes.data)
d=np.zeros(env.m, np.float) # note env.m=2**n
pd = ffi.cast("double *", d.ctypes.data)
fm.py_dualmMob(pmob, pd, env)
```

`fm.py_ConstructLambdaMeasure(singletons, lambda, v, env)`
Given the values of a fuzzy measure at singletons, finds the value of λ, computes all other values of the fuzzy measure in standard representation, and returns the pair (λ, v), where v is an array of size m. The input array $singletons$ is an array of size n of the values of fuzzy measure at singletons. Example:

```
lambd=np.array([0.0],np.float)
plambd=ffi.cast("double *", lambd.ctypes.data)
singletons=np.array(singletons=np.array([0, 0.3, 0.5],np.float)
psingletons = ffi.cast("double *", singletons.ctypes.data)
v=np.zeros(env.m, np.float) # note env.m=2**n
pv = ffi.cast("double *", v.ctypes.data)
fm.py_ConstructLambdaMeasure(psingletons, plambd, pv, env)
```

`fm.py_ConstructLambdaMeasureMob((singletons, lambda, Mob, env))`
Given the values of a fuzzy measure at singletons, finds the value of λ, computes all other values of the fuzzy measure in Möbius representation, and returns the pair (λ, Mob), where Mob is an array of size m. The input array $singletons$ is an array of size n of the values of fuzzy measure at singletons. Example:

```
lambd=np.array([0.0],np.float)
plambd=ffi.cast("double *", lambd.ctypes.data)
singletons=np.array([0, 0.3, 0.5],np.float)
psingletons = ffi.cast("double *", singletons.ctypes.data)
Mob=np.zeros(env.m, np.float) # note env.m=2**n
pMob = ffi.cast("double *", Mob.ctypes.data)
fm.py_ConstructLambdaMeasureMob(psingletons, plambd, pMob, env)
```

`fm.py_export_maximal_chains(n, v, mc, env)`
Returns in mc the arrays of maximal chains (there are $n!$ such arrays) of a fuzzy measure v; each array is of length n and contains the chains of discrete derivatives, which can be used to calculate the Choquet integrals for all possible orderings of the argument x, as well as serve as coefficients of the piecewise linear objective functions. That is, each maximal chain corresponds to the coefficients of a linear function on the respective simplex. All such simplices form a partition of the unit cube. Example:

```
n=3
mc=np.zeros(math.factorial(n)*n,np.float)
pmc = ffi.cast("double *", mc.ctypes.data)
vb=np.array([0, 0.00224734, 0.06493396, 0.51092037, 0.00965497,
0.37405545, 0.15455057, 1],np.float)
pvb = ffi.cast("double *", vb.ctypes.data)
fm.py_export_maximal_chains(n,pvb,pmc,env)
```

`fm.py_ConvertCard2Bit(dest, src, env)`
Converts fuzzy measure src from cardinality to binary ordering. Example:

```
src=np.array([0, 0.0340916, 0.00890771, 0.01109779, 0.07918582,
0.6305862, 0.1846705, 1],np.float)
psrc = ffi.cast("double *", src.ctypes.data)
```

```
dest=np.zeros(env.m,np.float)
pdest = ffi.cast("double *", dest.ctypes.data)
fm.py_ConvertCard2Bit(pdest,psrc, env)
```

```
fm.py_min_subset(x, n, S)
fm.py_max_subset(x, n, S)
```

Return min or max of x_i, $i \in S$. S should be written 5 in binary form. Example:

```
x=np.array([0,0.3,0.5,0.6,0.7,0.8,0.85,0.9],np.float)
px = ffi.cast("double *", x.ctypes.data)
fm.py_min_subset(px,3,6)
fm.py_max_subset(px,3,5)
```

```
fm.py_min_subsetC(x, n, S, env)
```

Calculates the minimum of x_i, $i \in S$, in cardinality-based ordering. Example:

```
x=np.array([0,0.3,0.5],np.float)
px = ffi.cast("double *", x.ctypes.data)
fm.py_min_subsetC(px,3,3,env)
```

```
fm.py_max_subsetNegC(x, n, S, env)
```

Calculates the maximum over the complement of subset S in cardinality ordering. Example:

```
x=np.array([0,0.3,0.5],np.float)
px = ffi.cast("double *", x.ctypes.data)
fm.py_max_subsetNegC(px,3,3,env)
```

```
fm.py_SizeArraykinteractive(n, k, env)
```

Returns the size of the array to store k-interactive fuzzy measure. Example:

```
fm.py_SizeArraykinteractive(3,1,env)
```

```
fm.py_IsSubsetC(i, j, env)
```

Returns true if j is subset of i in cardinality ordering.Example:

```
fm.py_IsSubsetC(2,1,env)
```

```
fm.py_IsElementC(A, j, env)
```

Returns true if j is an element of A in cardinality ordering. Example:

```
fm.py_IsElementC(1,2,env)
```

```
fm.py_ExpandKinteractive2Bit(dest, src, env, kint, arraysize)
fm.py_ExpandKinteractive2Bit_m(dest, src, env, kint, arraysize,
VVC)
```

Expands k-interactive fuzzy measure (from *src* to *dest*) given in more compact representation of size *arraysize* into its full representation in binary ordering. The function ExpandKinteractive2Bit allocates working memory internally, whereas the second function expects working array VVC of length m. Example:

```
src=np.array([0.0340916, 0.01109779, 0.07918582,0.7,1],
np.float)
psrc = ffi.cast("double *", src.ctypes.data)
dest=np.zeros(env.m, np.float)
pdest = ffi.cast("double *", dest.ctypes.data)
```

```
vcc=np.zeros(env.m, np.float)
pvcc = ffi.cast("double *", vcc.ctypes.data)
kint=1
arraysize=fm.py_SizeArraykinteractive(n, kint, env)
fm.py_ExpandKinteractive2Bit(pdest,psrc,env,kint,arraysize)
fm.py_ExpandKinteractive2Bit_m(pdest,psrc,env,kint,arraysize,
pvcc)
```

`fm.py_fm_arraysize( n, kint, env)`
Returns the length of the array of values of k-interactive fuzzy measures. Useful for reserving memory to store them. For $kint > 5$ requires initialisation of the env variable, otherwise no. Example:

```
len=fm.py_fm_arraysize(n, kint, env)
v=np.zeros( len, float)
```

`fm.py_ShowCoalitions(A, env)`
`fm.py_ShowCoalitionsCard(A, env)`
Return the decimal expression for the subsets A, e.g. for $A = 11$ it will show 124 for printing, in binary and in cardinality ordering, respectively. A is the array of integers to receive the decimal expressions for all 2^n coalitions. Example:

```
A=np.zeros( 2**n, np.intc)
pA = ffi.cast("int *", A.ctypes.data)
fm.py_ShowCoalitions(pA, env)
```

The following routines are self-explanatory. They return 1 if yes, 0 if no. The input parameters are fuzzy measure v in standard representation and fuzzy measure Mob in Möbius representation (both in binary ordering).

```
fm.py_IsMeasureAdditive(v, env)
fm.py_IsMeasureAdditiveMob(Mob, env)
fm.py_IsMeasureBalanced(v, env)
fm.py_IsMeasureBalancedMob(Mob, env)
fm.py_IsMeasureSelfdual(v, env)
fm.py_IsMeasureSelfdualMob(Mob, env)
fm.py_IsMeasureSubadditive(v, env)
fm.py_IsMeasureSubadditiveMob(v, env)
fm.py_IsMeasureSubmodular(v, env)
fm.py_IsMeasureSubmodularMob(Mob, env)
fm.py_IsMeasureSuperadditive(v, env)
fm.py_IsMeasureSuperadditiveMob(Mob, env)
fm.py_IsMeasureSupermodular(v, env)
fm.py_IsMeasureSupermodularMob(Mob, env)
fm.py_IsMeasureSymmetric(v, env)
fm.py_IsMeasureSymmetricMob(Mob, env)
```

`fm.py_IsMeasureKMaxitive(v, env)` (returns k);
`fm.py_IsMeasureKMaxitiveMob(Mob, env)` (returns k).

A.3 Fuzzy Measures in Compact and Sparse Representations

These routines do not require initialisation of the fuzzy measure with `fm.py_fm_init` and can hence be used with large n (limited by the available memory). Note that 2-additive fuzzy measures require $O(n^2)$ parameters. They are stored in cardinality ordering.

Sparse Structure Definition
Sparse representation is in the form of singletons, pairs, and tuples with nonzero values, stored and indexed in the respective arrays, which are part of the structure `fm_env_sparse`.
The following routines require defining a structure *envsp* of type `fm_env_sparse` to store the relevant values, but due to sparse representation only the indicated k-tuples are stored.

A.3.1 Using Python wrapper

`envsp=fm.prepare_fm_sparse(n, tupsize)`
This function initialises this structure, if the number of tuples is known, otherwise can be initialised with 0 tuple (i.e. set tup=0). Given the list of cardinalities of the nonzero tuples (cardinality, tuple composition) like this: 2 pairs 4-tuple and a triple (2,1,2, 2, 3,1, 4, 1,2,3,4, 3,3,2,1...). It is used to allocate storage and later populate these values.

```
n=3
tup=0
envsp= fm.prepare_fm_sparse(n, tup)
```

`fm.free_fm_sparse(envsp)`
Frees the memory previously allocated in *envsp*.

`fm.tuple_cardinality_sparse(i, envsp)`
Returns the cardinality of the tuple numbered i in the list of tuples.

`fm.get_num_tuples(envsp)`
Returns the number of tuples.

`fm.get_sizearray_tuples(envsp)`
Returns the length of the array of tuples.

`fm.is_inset_sparse(A, card, i, envsp)`
Checks if element i (1-based) belongs to the tuple indexed A (whose cardinality can be 1,2, *other* (automatically determined)).

`fm.is_subset_sparse(A, cardA, B, cardB, envsp)`
Checks if tuple B is a subset of tuple A. The cardinalities of both tuples need to be supplied.

```
fm.min_subset_sparse(x, n, S, cardS, envsp)
fm.max_subset_sparse(x, n, S, cardS, envsp)
```
Calculates minimum (maximum) of x_i with the indices belonging to tuple indexed as S (its cardinality cardS can be 1,2, *other* (put 3, it will be determined automatically)). Note that x is 0-based and tuples are 1-based. Example:
```
x=[0.1,0.05,0.2]
fm.min_subset_sparse(x, 3, 0, 3, envsp)
fm.max_subset_sparse(x, 3, 0, 3, envsp)
```

```
fm.ChoquetMob_sparse(x, envsp)
```
Calculates the Choquet integral in Mobius sparse representation. Example:
```
x=[0.1,0.05,0.2]
fm.ChoquetMob_sparse(x, envsp)
```

```
fm.ShapleyMob_sparse(n, envsp)
fm.BanzhafMob_sparse( n, envsp)
```
Calculate Shapley and Banzhaf value vectors (of size n) of a sparse fuzzy measure, returned in v. Example:
```
n=3
s=fm.ShapleyMob_sparse( n, envsp)
b=fm.BanzhafMob_sparse( n, envsp)
```

```
fm.Nonmodularityindex_sparse(w, n, envsp)
```
Calculates all 2^n nonmodularity indices (returned in w) using Mobius transform of a fuzzy measure (of length $2^n = m$), using sparse representation. Example:
```
n=3
nm=fm.Nonmodularityindex_sparse( n, envsp)
```

```
fm.populate_fm_2add_sparse(singletons, numpairs, pairs,
indicesp1, indicesp2, envsp)
```
Populates 2-additive sparse capacity with nonzero values using the singletons and two arrays of indices (of size $numpairs$). Indices need to be 1-based. Example:
```
singletons=[0.1,0.2,0.3]
numpairs=3
pairs = [0.4,0.5,0.6]
indicesp1 = [1,1,2]
indicesp2 = [2,3,3]
fm.populate_fm_2add_sparse(singletons, numpairs, pairs,
indicesp1, indicesp2,envsp)
```

```
fm.add_pair_sparse(i, j, v, envsp)
```
For populating capacities, adds pair (v_{ij}) to the structure. Indices are 1-based. Example:
```
fm.add_pair_sparse(1, 2, 0.4, envsp)
fm.add_pair_sparse(1, 3, 0.5, envsp)
fm.add_pair_sparse(2, 3, 0.6, envsp)
```

```
fm.add_tuple_sparse(tupsize, tuple, v, envsp)
```

For populating capacities, adds a tuple of size *tupsize*, whose 1-based indices are in *tuple*. Example:

```
tup=[1,3,4]
v=0.2
fm.add_tuple_sparse(3, tup, v, envsp)
fm.add_pair_sparse(2,5, v, envsp)
v=[0.1,0.2,0.3]
fm.add_singletons_sparse(v,envsp)
```

`fm.populate_fm_2add_sparse_from2add(n, v, envsp)`
Given 2-additive capacity (singletons+pairs in one array v), selects nonzero pairs and populates sparse capacity *envsp*. Example:

```
n=3
y,v,length=fm.py_generate_fm_2additive_convex_withsome
independent(1, n)
envsp=fm.prepare_fm_sparse(n, 0 )
fm.py_populate_fm_2add_sparse_from2add(n, v, envsp)
```

`fm.expand_2add_full( envsp)`
From sparse to full representation of 2-additive capacity (singletons and pairs, augmented with 0s). Vector v has to be allocated; its size is $n+n(n-1)/2$. Example:

```
v=fm.py_expand_2add_full(envsp)
```

`fm.expand_sparse_full(envsp)`
Exports from sparse to full capacity (vector v, size 2^n has to be preallocated). Example:

```
v=fm.expand_sparse_full(envsp)
```

```
fm.sparse_get_singletons(n, envsp)
fm.sparse_get_pairs(pairs, envsp)
fm.sparse_get_tuples(tuples, envsp)
```

Export the internal arrays of the sparse capacity as arrays of singletons, pairs, and tuples. Return the numbers of pairs and tuples. Example:

```
sin=fm.sparse_get_singletons(3, envsp)
pairs, v = fm.sparse_get_pairs(envsp)
tuples, v = fm.py_sparse_get_tuples(envsp)
```

The following functions do not use sparse representation, but they do not require initialisation by `fm_init` either; they work with 2-additive fuzzy measures stored on an array of Möbius values of singletons and pairs.

`fm.dualMobKadd(n, length, k, src, env)`
Calculates the dual of a k-additive fuzzy measure for n inputs. Generally requires initialisation but not for 2-additive fuzzy measures. Parameters: *length* is the size of the array in cardinality-based ordering (can be calculated by function `py_fm_arraysize`), k is the k-additivity, *src* is the source measure, *dest* is its dual, and *env* must be initialised for $k > 2$ only. Example:

```
k=2
n=20
```

```
Mob = [0.0,0.3,0.5,-0.2,0.4,0.1,-0.2,0.1]
dual,length= fm.dualMobKadd(k, v, env)
```

```
fm.Shapley2addMob(v, n)
fm.Banzhaf2addMob(v, n)
```

Calculate the Shapley and Banzhaf values of a 2-additive fuzzy measure for n inputs given in Mobius representation. The results are in arrays s or b. Example:

```
v = [0.0, 0.3, 0.5, -0.2, 0.4, 0.1]
s=fm.Shapley2addMob(v,3)
b=fm.Banzhaf2addMob(v,3)
```

```
fm.Choquet2addMob(x, v, n)
```

Calculates the Choquet integral value of x for a 2-additive fuzzy measure for n inputs given in Mobius representation. Example:

```
x=[0.2,0.5,0.4]
v=[0.2, 0.3, 0.5, -0.2, 0.4, 0.1]
r=fm.Choquet2addMob(x, v, 3)
```

```
fm.OWA( x, w, n)
```

Returns the OWA function of x wrt weights w. Example:

```
n=3
x=[0.2,0.5,0.4]
w=[0.3,0.3,0.4]
r=fm.OWA( x, w, n)
```

```
fm.WAM( x, w, n)
```

Returns the WAM function of x wrt weights w. Example:

```
n=3
x=[0.2,0.5,0.4]
w=[0.3,0.3,0.4]
r=fm.WAM( px, pw, n)
```

A.3.2 Without Python wrapper, Using numpy and cffi

The sparse capacity structure needs to be defined as follows:

```
envsp=ffi.new("struct fm_env_sparse *")
  fm.py_prepare_fm_sparse(n, tupsize, tuples, envsp)
```

This function initialises this structure, if the number of tuples is known, otherwise can be initialised with 0 tuple (i.e. set tup=0). Given the list of cardinalities of the nonzero tuples (cardinality, tuple composition) like this: 2 pairs 4-tuple and a triple (2,1,2, 2, 3,1, 4, 1,2,3,4, 3,3,2,1…). It is used to allocate storage and later populate these values.

```
n=3
tup=0
tuples=np.zeros( 0, np.intc)
```

```
ptuples = ffi.cast("int *", tuples.ctypes.data)
fm.py_prepare_fm_sparse(n, tup, ptuples, envsp)
```

`fm.py_free_fm_sparse(envsp)`
Frees the memory previously allocated in *envsp*.

`fm.py_tuple_cardinality_sparse(i, envsp)`
Returns the cardinality of the tuple numbered i in the list of tuples.

`fm.py_get_num_tuples(envsp)`
Returns the number of tuples.

`fm.py_get_sizearray_tuples(envsp)`
Returns the length of the array of tuples.

`fm.py_is_inset_sparse(A, card, i, envsp)`
Checks if element i (1-based) belongs to the tuple indexed A (whose cardinality can be 1,2, *other* (automatically determined)).

`fm.py_is_subset_sparse(A, cardA, B, cardB, envsp)`
Checks if tuple B is a subset of tuple A. The cardinalities of both tuples need to be supplied.

`fm.py_min_subset_sparse(x, n, S, cardS, envsp)`
`fm.py_max_subset_sparse(x, n, S, cardS, envsp)`
Calculate minimum (maximum) of x_i with the indices belonging to tuple indexed as S (its cardinality cardS can be 1,2, *other* (put 3, it will be determined automatically)). Note that x is 0-based and tuples are 1-based. Example:

```
x=np.array([0.1,0.05,0.2],np.float)
px = ffi.cast("double *", x.ctypes.data)
fm.py_min_subset_sparse(px, 3, 0, 3, envsp)
fm.py_max_subset_sparse(px, 3, 0, 3, envsp)
```

`fm.py_ChoquetMob_sparse(x, n, envsp)`
Calculates the Choquet integral in Mobius sparse representation. Example:

```
x=np.array([0.1,0.05,0.2],np.float)
px = ffi.cast("double *", x.ctypes.data)
n=3
fm.py_ChoquetMob_sparse(px, n, envsp)
```

`fm.py_ShapleyMob_sparse(v, n, envsp)`
`fm.py_BanzhafMob_sparse(v, n, envsp)`
Calculate Shapley and Banzhaf value vectors (of size n) of a sparse fuzzy measure, returned in v. Example:

```
n=3
v=np.zeros(3, np.float)
pv = ffi.cast("double *", v.ctypes.data)
fm.py_ShapleyMob_sparse(pv, n, envsp)
v_1=np.zeros(3,np.float)
pv_1 = ffi.cast("double *", v_1.ctypes.data)
```

```
fm.py_BanzhafMob_sparse(pv_1, n, envsp)
```

```
fm.py_Nonmodularityindex_sparse(w, n, envsp)
```
Calculates all 2^n nonmodularity indices (returned in w) using Mobius transform of a fuzzy measure (of length $2^n = m$), using sparse representation. Example:
```
n=3
w=np.zeros(2**n, np.float)
pw = ffi.cast("double *", w.ctypes.data)
fm.py_Nonmodularityindex_sparse(pw, n, envsp)
```

```
fm.py_populate_fm_2add_sparse(singletons, numpairs, pairs,
indicesp1, indicesp2, envsp)
```
Populates 2-additive sparse capacity with nonzero values using the singletons and two arrays of indices (of size *numpairs*). Indices need to be 1-based. Singletons 0-based. Example:
```
singletons=np.array([0.1,0.2,0.3],np.float)
psingletons = ffi.cast("double *", singletons.ctypes.data)
numpairs=3
pairs = np.array([0.4,0.5,0.6],np.float)
ppairs = ffi.cast("double *", pairs.ctypes.data)
indicesp1 = np.array([1,1,2], np.intc)
pindicesp1 = ffi.cast("int *", indicesp1.ctypes.data)
indicesp2 = np.array([2,3,3], np.intc)
pindicesp2 = ffi.cast("int *", indicesp2.ctypes.data)
fm.py_populate_fm_2add_sparse(psingletons, numpairs, ppairs,
pindicesp1, pindicesp2,envsp)
```

```
fm.py_add_pair_sparse(i, j, v, envsp)
```
For populating capacities, adds pair (v_{ij}) to the structure. Indices are 1-based. Example:
```
fm.py_add_pair_sparse(1, 2, 0.4, envsp)
fm.py_add_pair_sparse(1, 3, 0.5, envsp)
fm.py_add_pair_sparse(2, 3, 0.6, envsp)
```

```
fm.py_add_tuple_sparse(tupsize, tuple, v, envsp)
```
For populating capacities, adds a tuple of size *tupsize*, whose 1-based indices are in *tuple*. Example:
```
tup=np.array([1,3,4],np.intc)
ptup = ffi.cast("int *", tup.ctypes.data)
v=0.2
fm.py_add_tuple_sparse(3, ptup, v, envsp)
fm.py_add_pair_sparse(2,5, v, envsp)
v=np.array([0.1,0.2,0.3],np.float)
pv = ffi.cast("double *", v.ctypes.data)
fm.py_add_singletons_sparse(pv,envsp)
```

```
fm.py_populate_fm_2add_sparse_from2add(n, v, envsp)
```

Given 2-additive capacity (singletons+pairs in one array v), selects nonzero pairs and populates sparse capacity $envsp$. Example:

```
n=3
v=np.zeros(8,np.float)
pv = ffi.cast("double *", v.ctypes.data)
fm.py_generate_fm_2additive_convex_withsomeindependent(1, n,
pv)
tuples=np.zeros( 0, np.intc)
ptuples = ffi.cast("int *", tuples.ctypes.data)
fm.py_prepare_fm_sparse(n, 0 , ptuples , envsp)
fm.py_populate_fm_2add_sparse_from2add(n, pv, envsp)
```

```
fm.py_expand_2add_full(v, envsp)
```

From sparse to full representation of 2-additive capacity (singletons and pairs, augmented with 0s). Vector v has to be allocated; its size is $n+n(n-1)/2$. Example:

```
v1=np.array([0.0,0.0,0.0,0.0,0.0,0.0],np.float)
pv1 = ffi.cast("double *", v1.ctypes.data)
fm.py_expand_2add_full(pv1, envsp)
```

```
fm.py_expand_sparse_full(v, envsp)
```

Exports from sparse to full capacity (vector v, size 2^n has to be preallocated). Example:

```
v=np.zeros(8,np.float)
pv = ffi.cast("double *", v.ctypes.data)
fm.py_expand_sparse_full(pv, envsp)
```

```
fm.py_sparse_get_singletons(n, v, envsp)
fm.py_sparse_get_pairs(pairs, v, envsp)
fm.py_sparse_get_tuples(tuples, v, envsp)
```

Export the internal arrays of the sparse capacity as arrays of singletons, pairs, and tuples. Return the numbers of pairs and tuples. Example:

```
s=np.zeros(n,np.float)
ps = ffi.cast("double *", s.ctypes.data)
fm.py_sparse_get_singletons(n, ps,envsp)

pairs=np.zeros(32,np.intc)
ppairs = ffi.cast("int *", pairs.ctypes.data)
vals=np.zeros(32,np.float)
pvals = ffi.cast("double *", vals.ctypes.data)
sizepairs= fm.py_sparse_get_pairs(ppairs, pvals,envsp)

indextuples=np.zeros(32,np.intc)
pindextuples = ffi.cast("int *", indextuples.ctypes.data)
vvt=np.zeros(32,np.float)
pvvt = ffi.cast("double *", vvt.ctypes.data)
sizetuples = fm.py_sparse_get_tuples(pindextuples, pvvt,envsp)
```

The following functions do not use sparse representation, but they do not require initialisation by `py_fm_init` either; they work with 2-additive fuzzy measures stored on an array of Möbius values of singletons and pairs.

`fm.py_dualMobKadd(n, length, k, src, dest, env)`
Calculates the dual of a k-additive fuzzy measures for n inputs. Generally requires initialisation but not for 2-additive fuzzy measures. Parameters: *length* is the size of the array in cardinality-based ordering (can be calculated by function `py_fm_arraysize`), k is the k-additivity, *src* is the source measure, *dest* is its dual, and *env* must be initialised for $k > 2$ only. Example:

```
k=2
n=20
env=ffi.new("struct fm_env *") # no need to initialise, not
used
sz=fm.py_fm_arraysize(n,k,env)
v=np.zeros( n*n, np.float)
pv = ffi.cast("double *", v.ctypes.data)
d=np.zeros( n*n, np.float)
pd = ffi.cast("double *", d.ctypes.data)
lengthv=fm.py_generate_fm_2additive_convex(1, n, pv)
fm.py_dualMobKadd(n, lengthv, k, pv, pd, env)
```

`fm.py_Shapley2addMob(v, s, n)`
`fm.py_Banzhaf2addMob(v, b, n)`
Calculate the Shapley and Banzhaf values of a 2-additive fuzzy measure for n inputs given in Mobius representation. The results are in arrays s or b. Example:

```
n=20
s=np.zeros( n, np.float)
ps = ffi.cast("double *", s.ctypes.data)
b=np.zeros( n, np.float)
pb = ffi.cast("double *", b.ctypes.data)
fm.py_Shapley2addMob(pv,ps,n)
fm.py_Banzhaf2addMob(pv,pb,n)
```

`fm.py_Choquet2addMob(x, v, n)`
Calculates the Choquet integral value of x for a 2-additive fuzzy measure for n inputs given in Mobius representation. Example:

```
x=np.array([0.2,0.5,0.4],np.float)
px = ffi.cast("double *", x.ctypes.data)
v=np.array([0,0.3,0.5,0.6,0.4,0.8,0.7,1],np.float)
pv = ffi.cast("double *", v.ctypes.data)
r=fm.py_Choquet2addMob(px, pv, 3)
```

`fm.py_OWA( x, w, n)`
Returns the OWA function of x wrt weights w. Example:

```
n=3
x=np.array([0.2,0.5,0.4],np.float)
```

```
px = ffi.cast("double *", x.ctypes.data)
w=np.array([0.3,0.3,0.4],np.float)
pw = ffi.cast("double *", w.ctypes.data)
r=fm.py_OWA( px, pw, n)
```

`fm.py_WAM( x, w, n)`
Returns the WAM function of x wrt weights w. Example:

```
n=3
x=np.array([0.2,0.5,0.4],np.float)
px = ffi.cast("double *", x.ctypes.data)
w=np.array([0.3,0.3,0.4],np.float)
pw = ffi.cast("double *", w.ctypes.data)
r=fm.py_WAM( px, pw, n)
```

A.4 Fuzzy Measure Fitting Routines

The package contains six functions taking the empirical data, some options, and returning the (k-additive) fuzzy measure.

A.4.1 Using Python wrapper

Important note: after fitting, the user must re-initialise the internal structures with `fm_init`, because these functions create these structures internally and then free the memory.

`fm.FuzzyMeasureFit(datanum, kadditive, env, dataset)`
Returns (k-additive) fuzzy measure fitted to data in standard representation. Example:

```
n=3
datanum=10
kadditive=2
env=fm.fm_init(n);
dataset=np.random.rand(datanum,n+1)
v= fm.py_FuzzyMeasureFit(datanum, kadditive, env, dataset)
```

`fm.FuzzyMeasureFitMob(datanum, kadditive, env, dataset)`
Returns (k-additive) fuzzy measure fitted to data in Möbius representation (binary ordering). Example:

```
datanum=10
kadditive=3
dataset=np.random.rand(datanum,n+1)
mob=fm.FuzzyMeasureFitMob(datanum, kadditive, env, pdataset)
```

`fm.FuzzyMeasureFitLP(datanum, additive, env, options, indexlow, indexhihg, option1, orness, dataset)`
Returns fuzzy measure fitted to data in standard representation, with optional constraints on the orness measure and the upper and lower values of Shapley values and interaction indices. Example:

```
datanum=10
additive=2
dataset=np.random.rand(datanum,n+1)
options=0
indexlow = [0,0,0]
indexhihg = [1,1,1]
option1=0
orness = 0.5
v= fm.py_FuzzyMeasureFitLP(datanum, additive, env, options,
indexlow, indexhihg, option1, orness, dataset)
```

`fm.FuzzyMeasureFitLPMob(datanum, additive, env, options, indexlow, indexhihg, option1, orness, dataset)`
Returns fuzzy measure in Möbius representation fitted to data in standard representation, with optional constraints on the orness measure and the upper and lower values of Shapley values and interaction indices. Example:

```
datanum=10
additive=2
dataset=np.random.rand(datanum,n+1)
options=0
indexlow = [0,0,0]
indexhihg = [1,1,1]
option1=0
orness = 0.5
mob= fm.FuzzyMeasureFitLPMob(datanum, additive, env, mob,
options, indexlow, indexhihg, option1, orness, dataset)
```

`fm.FuzzyMeasureFitKtolerant(datanum, additive, env, dataset)`
Returns k-tolerant fuzzy measure fitted to data in standard representation, with no optional constraints. Example:

```
datanum=10
additive=2
dataset=np.random.rand(datanum,n+1)
v=fm.py_FuzzyMeasureFitKtolerant(datanum, additive, env,
dataset)
```

`fm.FuzzyMeasureFitLPKmaxitive(datanum, additive, env, dataset)`
Returns k-maxitive fuzzy measure fitted to data in standard representation, with no optional constraints. Example:

```
datanum=10
additive=2
dataset=np.random.rand(datanum,n+1)
```

```
v=fm.FuzzyMeasureFitLPKmaxitive(datanum, additive, env, dataset)
```

`fm.fittingOWA(datanum, env, dataset)`
Returns symmetric fuzzy measure fitted to data. The vector of OWA weights is returned as in this case the Choquet integral becomes the OWA function. Example:

```
datanum=10
dataset=np.random.rand(datanum,n+1)
v=fm.fittingOWA(datanum, env, dataset)
```

`fm.fittingWAM(datanum, env, dataset)`
Returns additive fuzzy measure fitted to data. The vector of WAM weights is returned as in this case the Choquet integral becomes the WAM function. Example:

```
datanum=10
dataset=np.random.rand(datanum,n+1)
v=fm.fittingWAM(datanum, env, dataset)
```

List of parameters and options of the `fm.FuzzyMeasureFitLP` function.
`fm.FuzzyMeasureFitLP(datanum, additive, env, options, indexlow, indexhihg, option1, orness, dataset)`
$data$—an array of size $K \times (n+1)$, where each row is the pair $(\mathbf{x}_k, y_k)$, $\mathbf{x}_k \in [0, 1]^n$, $y_k \in [0, 1]$, K data altogether.
$Kadd$—k in k-additive fuzzy measures, $1 < Kadd < n + 1$, if $Kdd = n$—FM is unrestricted.
$options$ (default value is 0)

- 1—lower bounds on Shapley values supplied in $indexlow$
- 2—upper bounds on Shapley values supplied in $indexhigh$
- 3—lower and upper bounds on Shapley values supplied in $indexlow$ and $indexhigh$
- 4—lower bounds on all interaction indices supplied in $indexlow$
- 5—upper bounds on all interaction indices supplied in $indexhigh$
- 6—lower and upper bounds on all interaction indices supplied in $indexlow$ and $indexhigh$

All these values will be treated as additional constraints in the LP.
$indexlow$, $indexhigh$—an array of size n (options =1,2,3) or m (options=4,5,6) containing the lower and upper bounds on the Shapley values or interaction indices.
$options1$ (default 0) is a flag whose bits indicate which additional properties are needed. If the first bit is set, then the desired interval of orness values specified in array $orness$ will be used. If the second bit is set, then the FM will be forced to be balanced (currently not implemented). If the third bit is set, then in addition to fitting the data, the order of output values will be preserved.
If the fourth bit is set, the fuzzy measure will be forced to be submodular.
Note that this constraint may lead to inconsistent set of conditions, in which case the problem will have no solution (and no output vector returned).
$orness$—an array of size 2 which contains the lower and the upper bounds on the orness value. These values should be from [0,1] and could coincide if an exact orness

value is needed. If the bounds are 0 and 1, respectively, they are ignored. Only used if the first bit of $options1$ is set.
Notes: 1. Arrays $indexlow$ and $indexhigh$ are 0-based if they contain Shapley values (i.e. the bound on Shapley value of the first input is in indexlow[0]), but when these arrays contain interaction indices, these are 1-based (since there is a nonzero value of the interaction index corresponding to empty set). In this case the bounds are arranged in cardinality order, i.e. the bounds correspond to the sets in this ordering (for n=3) Ø{1}{2}{3}{12}{13}{23}{123}.
2. If an exact value of some index or Shapley value is needed, use $indexlow[i] = indexhigh[i] = thisvalue$; if no value for some index is required, use $indexlow[i] = -1, indexhigh[i] = 1$.
3. Note that Shapley values have range [0,1], whereas interaction indices have range [− 1,1].

Note that the options as listed above are not implemented for the following methods.

`fm.FuzzyMeasureFitLPKinteractive(datanum, additive, env, v, K, dataset)`

Returns k-interactive fuzzy measure fitted to data in standard representation, with fixed parameter KC (the value $v(A)$ at sets of cardinality $k+1$). Example:

```
datanum=10
additive = 2
dataset=np.random.rand(datanum,n+1)
K=0.5
v = fm.FuzzyMeasureFitLPKinteractive(datanum, additive, env, K,
dataset)
```

`fm.FuzzyMeasureFitLPKinteractiveAutoK(datanum, additive, env, v, K, maxiters, dataset)`

Returns k-interactive fuzzy measure fitted to data in standard representation, with the parameter KC automatically determined from the data. Example:

```
datanum=10
additive = 2
dataset=np.random.rand(datanum,n+1)
K=0.5
maxiters=100
v=fm.FuzzyMeasureFitLPKinteractiveAutoK(datanum, additive, env,
K, maxiters, dataset)
```

`fm.FuzzyMeasureFitLPKinteractiveMaxChains(datanum, additive, env, K, dataset)`

Returns k-interactive fuzzy measure fitted to data in standard representation, with fixed parameter KC (the value $v(A)$ at sets of cardinality $k+1$). This method uses maximal chain approach and is more efficient for smaller datasets. Example:

```
datanum=10
additive = 2
dataset=np.random.rand(datanum,n+1)
```

```
K=0.5
v=fm.FuzzyMeasureFitLPKinteractiveMaxChains(datanum, additive,
env, K, dataset)
```

```
fm.FuzzyMeasureFitLPKinteractiveMarginal(datanum, additive,
env, K, submod, dataset)
```

Returns k-interactive fuzzy measure fitted to data in standard representation, with fixed parameter KC (the value $v(A)$ at sets of cardinality $k+1$). This method uses marginal contribution representation and may involve more variables and constraints than the other methods. It is useful when imposing sub- or supermodularity constraints, in which case parameter $submod$ should be set to -1 (supermodularity), +1 (submodularity), or 0 (none). Example:

```
datanum=10
additive = 2
dataset=np.random.rand(datanum,n+1)
K=0.5
submod=1
v=fm.FuzzyMeasureFitLPKinteractiveMarginal(datanum, additive,
env, K, submod, dataset)
```

```
fm.FuzzyMeasureFitLPKinteractiveMarginalMaxChain(datanum,
additive, env, K, maxiters, submod, dataset)
```

Returns k-interactive fuzzy measure fitted to data in standard representation, with fixed parameter KC (the value $v(A)$ at sets of cardinality $k+1$). This method uses marginal contribution representation and maximal chain approach and may involve more variables and constraints than the other methods. It is useful when imposing sub- or supermodularity constraints, in which case parameter $submod$ should be set to -1 (supermodularity), +1 (submodularity), or 0 (none). It is more efficient than $fittingKinteractiveMarginal$ in the case of a small dataset.

Note: at the moment only supermodular option works in this method. If a submodular measure is needed, fit its dual supermodular to dual data. Example:

```
datanum=10
additive = 2
dataset=np.random.rand(datanum,n+1)
K=0.5
maxiters=100
submod=1
v=fm.FuzzyMeasureFitLPKinteractiveMarginalMaxChain(datanum,
additive, env, K, maxiters, submod, dataset)
```

A.4.2 Without Python wrapper, Using numpy and cffi

Important note: after fitting, the user must re-initialise the internal structures with `py_fm_init`, because these functions create these structures internally and then free the memory.

`fm.py_FuzzyMeasureFit(datanum, kadditive, env, dataset)`
Returns (k-additive) fuzzy measure fitted to data in standard representation. Example:

```
n=3
env=ffi.new("struct fm_env *")
datanum=10
kadditive=2
fm.py_fm_init(n, env);
v=np.array([0,0,0,0,0,0,0,0],np.float)
pv = ffi.cast("double *", v.ctypes.data)
dataset=np.random.rand(datanum,n+1)
fm.py_FuzzyMeasureFit(datanum, kadditive, env, pv, pdataset)
```

`fm.py_FuzzyMeasureFitMob(datanum, kadditive, env, Mob, dataset)`
Returns (k-additive) fuzzy measure fitted to data in Möbius representation (binary ordering). Example:

```
datanum=10
kadditive=3
mob=np.array([0,0,0,0,0,0,0,0],np.float)
pmob = ffi.cast("double *", mob.ctypes.data)
dataset=np.random.rand(datanum,n+1)
pdataset = ffi.cast("double *", dataset.ctypes.data)
fm.py_FuzzyMeasureFitMob(datanum, kadditive, env, pmob,
pdataset)
```

`fm.py_FuzzyMeasureFitLP(datanum, additive, env, v, dataset, options, indexlow, indexhihg, option1, orness, dataset)`
Returns fuzzy measure fitted to data in standard representation, with optional constraints on the orness measure and the upper and lower values of Shapley values and interaction indices. Example:

```
datanum=10
additive=2
v=np.array([0,0,0,0,0,0,0,0],np.float)
pv = ffi.cast("double *", v.ctypes.data)
dataset=np.random.rand(datanum,n+1)
pdataset = ffi.cast("double *", dataset.ctypes.data)
options=np.array([0],np.intc)
poptions= ffi.cast("int *", options.ctypes.data)
indexlow = np.array([0,0,0],np.float)
pindexlow = ffi.cast("double *", indexlow.ctypes.data)
```

```
indexhihg = np.array([1,1,1],np.float)
pindexhihg = ffi.cast("double *", indexhihg.ctypes.data)
option1=np.array([0],np.intc)
poption1= ffi.cast("int *", option1.ctypes.data)
orness = np.array([0.5],np.float)
porness = ffi.cast("double *", orness.ctypes.data)
fm.py_FuzzyMeasureFitLP(datanum, additive, env, pv, pdataset,
poptions, pindexlow, pindexhihg, poption1, porness)
```

`fm.py_FuzzyMeasureFitLPMob(datanum, additive, env, Mob, dataset, options, indexlow, indexhihg, option1, orness)`
Returns fuzzy measure in Möbius representation fitted to data in standard representation, with optional constraints on the orness measure and the upper and lower values of Shapley values and interaction indices. Example:

```
datanum=10
additive=2
mob=np.array([0,0,0,0,0,0,0,0],np.float)
pmob = ffi.cast("double *", mob.ctypes.data)
dataset=np.random.rand(datanum,n+1)
pdataset = ffi.cast("double *", dataset.ctypes.data)
options=np.array([0],np.intc)
poptions= ffi.cast("int *", options.ctypes.data)
indexlow = np.array([0,0,0],np.float)
pindexlow = ffi.cast("double *", indexlow.ctypes.data)
indexhihg = np.array([1,1,1],np.float)
pindexhihg = ffi.cast("double *", indexhihg.ctypes.data)
option1=np.array([0],np.intc)
poption1= ffi.cast("int *", option1.ctypes.data)
orness = np.array([0.5],np.float)
porness = ffi.cast("double *", orness.ctypes.data)
fm.py_FuzzyMeasureFitLPMob(datanum, additive, env, pmob,
pdataset, poptions, pindexlow, pindexhihg, poption1, porness)
```

`fm.py_FuzzyMeasureFitKtolerant(datanum, additive, env, v, dataset)`
Returns k-tolerant fuzzy measure fitted to data in standard representation, with no optional constraints. Example:

```
datanum=10
additive=2
v=np.array([0,0,0,0,0,0,0,0],float)
pv = ffi.cast("double *", v.ctypes.data)
dataset=np.random.rand(datanum,n+1)
pdataset = ffi.cast("double *", dataset.ctypes.data)
fm.py_FuzzyMeasureFitKtolerant(datanum, additive, env, pv,
pdataset)
```

`fm.py_FuzzyMeasureFitLPKmaxitive(datanum, additive, env, v, dataset)`
Returns k-maxitive fuzzy measure fitted to data in standard representation, with no optional constraints. Example:

```
datanum=10
additive=2
v=np.array([0,0,0,0,0,0,0,0],np.float)
pv = ffi.cast("double *", v.ctypes.data)
dataset=np.random.rand(datanum,n+1)
pdataset = ffi.cast("double *", dataset.ctypes.data)
fm.py_FuzzyMeasureFitLPKmaxitive(datanum, additive, env, pv,
pdataset)
```

`fm.py_fittingOWA(datanum, env, v, dataset)`
Returns symmetric fuzzy measure fitted to data. The vector of OWA weights is returned as in this case the Choquet integral becomes the OWA function. Example:

```
datanum=10
v=np.array([0,0,0],np.float)
pv = ffi.cast("double *", v.ctypes.data)
dataset=np.random.rand(datanum,n+1)
pdataset = ffi.cast("double *", dataset.ctypes.data)
fm.py_fittingOWA(datanum, env, pv, pdataset)
```

`fm.py_fittingWAM(datanum, env, v, dataset)`
Returns additive fuzzy measure fitted to data. The vector of WAM weights is returned as in this case the Choquet integral becomes the WAM function. Example:

```
datanum=10
v=np.array([0,0,0],np.float)
pv = ffi.cast("double *", v.ctypes.data)
dataset=np.random.rand(datanum,n+1)
pdataset = ffi.cast("double *", dataset.ctypes.data)
fm.py_fittingWAM(datanum, env, pv, pdataset)
```

List of parameters and options of the `fm.py_FuzzyMeasureFitLP` function.
`fm.py_FuzzyMeasureFitLP(datanum, additive, env, v, dataset, options, indexlow, indexhihg, option1, orness)`
$data$—an array of size $K \times (n+1)$, where each row is the pair $(\mathbf{x}_k, y_k)$, $\mathbf{x}_k \in [0, 1]^n$, $y_k \in [0, 1]$, K data altogether.
$Kadd$—k in k-additive fuzzy measures, $1 < Kadd < n + 1$, if $Kdd = n$—FM is unrestricted.
$options$ (default value is 0)

- 1—lower bounds on Shapley values supplied in $indexlow$
- 2—upper bounds on Shapley values supplied in $indexhigh$
- 3—lower and upper bounds on Shapley values supplied in $indexlow$ and $indexhigh$
- 4—lower bounds on all interaction indices supplied in $indexlow$

- 5—upper bounds on all interaction indices supplied in *indexhigh*
- 6—lower and upper bounds on all interaction indices supplied in *indexlow* and *indexhigh*

All these values will be treated as additional constraints in the LP.
indexlow, *indexhigh*—an array of size n (options =1,2,3) or m (options=4,5,6) containing the lower and upper bounds on the Shapley values or interaction indices.
*options*1 (default 0) is a flag whose bits indicate which additional properties are needed. If the first bit is set, then the desired interval of orness values specified in array *orness* will be used. If the second bit is set, then the FM will be forced to be balanced (currently not implemented). If the third bit is set, then in addition to fitting the data, the order of output values will be preserved.
If the fourth bit is set, the fuzzy measure will be forced to be submodular.
Note that this constraint may lead to inconsistent set of conditions, in which case the problem will have no solution (and no output vector returned).
orness—an array of size 2 which contains the lower and the upper bounds on the orness value. These values should be from [0,1] and could coincide if an exact orness value is needed. If the bounds are 0 and 1, respectively, they are ignored. Only used if the first bit of *options*1 is set.
Notes: 1. Arrays *indexlow* and *indexhigh* are 0-based if they contain Shapley values (i.e. the bound on Shapley value of the first input is in indexlow[0]), but when these arrays contain interaction indices, these are 1-based (since there is a nonzero value of the interaction index corresponding to empty set). In this case the bounds are arranged in cardinality order, i.e. the bounds correspond to the sets in this ordering (for n=3) Ø{1}{2}{3}{12}{13}{23}{123}.
2. If an exact value of some index or Shapley value is needed, use $indexlow[i] = indexhigh[i] = thisvalue$; if no value for some index is required, use $indexlow[i] = -1, indexhigh[i] = 1$.
3. Note that Shapley values have range [0,1], whereas interaction indices have range [− 1,1].

Note that the options as listed above are not implemented for the following methods.

```
fm.py_FuzzyMeasureFitLPKinteractive(datanum, additive, env, v,
dataset, K)
```

Returns k-interactive fuzzy measure fitted to data in standard representation, with fixed parameter KC (the value $v(A)$ at sets of cardinality $k+1$). Example:

```
datanum=10
additive = 2
v=np.array([0,0,0,0,0,0,0,0],np.float)
pv = ffi.cast("double *", v.ctypes.data)
dataset=np.random.rand(datanum,n+1)
pdataset = ffi.cast("double *", dataset.ctypes.data)
K=np.array([0.5],np.float)
pK = ffi.cast("double *", K.ctypes.data)
```

`fm.py_FuzzyMeasureFitLPKinteractive(datanum, additive, env, pv, pdataset, pK)`

`fm.py_FuzzyMeasureFitLPKinteractiveAutoK(datanum, additive, env, v, dataset, K, maxiters)`
Returns k-interactive fuzzy measure fitted to data in standard representation, with the parameter KC automatically determined from the data. Example:

```
datanum=10
additive = 2
v=np.array([0,0,0,0,0,0,0,0],np.float)
pv = ffi.cast("double *", v.ctypes.data)
dataset=np.random.rand(datanum,n+1)
pdataset = ffi.cast("double *", dataset.ctypes.data)
K=np.array([0.5],np.float)
pK = ffi.cast("double *", K.ctypes.data)
maxiters=np.array([100],np.intc)
pmaxiters = ffi.cast("int *", maxiters.ctypes.data)
fm.py_FuzzyMeasureFitLPKinteractiveAutoK(datanum, additive,
env, pv, pdataset, pK, pmaxiters)
```

`fm.py_FuzzyMeasureFitLPKinteractiveMaxChains(datanum, additive, env, v, dataset, K)`
Returns k-interactive fuzzy measure fitted to data in standard representation, with fixed parameter KC (the value $v(A)$ at sets of cardinality $k + 1$). This method uses maximal chain approach and is more efficient for smaller datasets. Example:

```
datanum=10
additive = 2
v=np.array([0,0,0,0,0,0,0,0],np.float)
pv = ffi.cast("double *", v.ctypes.data)
dataset=np.random.rand(datanum,n+1)
pdataset = ffi.cast("double *", dataset.ctypes.data)
K=np.array([0.5],np.float)
pK = ffi.cast("double *", K.ctypes.data)
fm.py_FuzzyMeasureFitLPKinteractiveMaxChains(datanum, additive,
env, pv, pdataset, pK)
```

`fm.py_FuzzyMeasureFitLPKinteractiveMarginal(datanum, additive, env, v, dataset, K, submod)`
Returns k-interactive fuzzy measure fitted to data in standard representation, with fixed parameter KC (the value $v(A)$ at sets of cardinality $k + 1$). This method uses marginal contribution representation and may involve more variables and constraints than the other methods. It is useful when imposing sub- or supermodularity constraints, in which case parameter $submod$ should be set to -1 (supermodularity), +1 (submodularity), or 0 (none). Example:

```
datanum=10
additive = 2
v=np.array([0,0,0,0,0,0,0,0],np.float)
```

```
pv = ffi.cast("double *", v.ctypes.data)
dataset=np.random.rand(datanum,n+1)
pdataset = ffi.cast("double *", dataset.ctypes.data)
K=np.array([0.5],np.float)
pK = ffi.cast("double *", K.ctypes.data)
submod=1
fm.py_FuzzyMeasureFitLPKinteractiveMarginal(datanum, additive,
env, pv, pdataset, pK, submod)
```

`fm.py_FuzzyMeasureFitLPKinteractiveMarginalMaxChain(datanum, additive, env, v, dataset, K, maxiters, submod)`
Returns k-interactive fuzzy measure fitted to data in standard representation, with fixed parameter KC (the value $v(A)$ at sets of cardinality $k+1$). This method uses marginal contribution representation and maximal chain approach and may involve more variables and constraints than the other methods. It is useful when imposing sub- or supermodularity constraints, in which case parameter $submod$ should be set to -1 (supermodularity), +1 (submodularity), or 0 (none). It is more efficient than $fittingKinteractiveMarginal$ in the case of a small dataset.
Note: at the moment only supermodular option works in this method. If a submodular measure is needed, fit its dual supermodular to dual data.
Example:

```
datanum=10
additive = 2
v=np.array([0,0,0,0,0,0,0,0],np.float)
pv = ffi.cast("double *", v.ctypes.data)
dataset=np.random.rand(datanum,n+1)
pdataset = ffi.cast("double *", dataset.ctypes.data)
K=np.array([0.5],np.float)
pK = ffi.cast("double *", K.ctypes.data)
maxiters=np.array([100])
pmaxiters = ffi.cast("int *", maxiters.ctypes.data)
submod=1
fm.py_FuzzyMeasureFitLPKinteractiveMarginalMaxChain(datanum,
additive, env, pv, pdataset, pK, pmaxiters, submod)
```

A.5 Random Generation of Fuzzy Measures

The package `pyfmtools` provides several methods to randomly generate fuzzy measures of different types: general, k-interactive, and k-additive convex and concave (i.e. supermodular and submodular). General and k-interactive fuzzy measures require initialisation of the env structures, but sparse and k-additive (for small k) do not; see Sect. A.3.

A.5.1 Using Python wrapper

`fm.generate_fm_tsort(num, n, kint, markov, option, K, env)`
`fm.generate_fm_minplus(num, n, kint, markov, option, K, env)`
Both functions generate several random fuzzy measures (*num* is their number) stored in **cardinality** ordering in the array v (the ith fuzzy measure starts at position $i * lengthv$ and has length $length$, which is returned by this function). The tsort method is based on topological sort and minplus is based on MinimalsPlus method (see Chap. 9). Other parameters: $0 < kint \leqslant n$ is for k-interactive fuzzy measures; $markov$ is the number of Markov steps to take—the randomness increases with that number, but slows down the process; K is the constant in k-interactive fuzzy measures; and v is the array to store the generated values (preallocated of size $num * 2^n$ or less for k-interactive fuzzy measures; see `fm.py_fm_arraysize`). $option = 1$ employs internal rejection method to improve uniformity, but for $n > 5$ is not essential. Returns the size of the array for each generated fuzzy measure. Example:

```
num=10
n=6
kint=3
env=fm.fm_init(n)
lengthv, v =fm.generate_fm_tsort(num, n, kint, 1000, 0, 0.7,
env)
```

`fm.generate_fmconvex_tsort(num, n, kint, markov, option, K, env)`
Generates *num* convex random fuzzy measures stored consecutively in **cardinality** ordering in the array v (the ith fuzzy measure starts at position $i * 2^n$ and has length 2^n). The parameters are as for the previous two functions. Returns the size of the array for each generated fuzzy measure. $kint$ for the moment is not implemented. This method is quite reliable for $n < 10$ but then becomes slow. There is an internal rejection process which guarantees supermodularity for small $n < 9$, but otherwise the result must be checked as in the example below. Increase Markov chain length for $n > 6$. Example:

```
num=1
n=5
env=fm.fm_init(n)
len=fm.fm_arraysize(n, n, env)
lengthv,v=fm.py_generate_fmconvex_tsort(num, n, n, 1000, 0, 1,
env)
w=fm.ConvertCard2Bit( v, env)
super=fm.IsMeasureSupermodular(w, env)
```

The following routines do not require initialisation of the fuzzy measure with `fm_init` and can hence be used with large n (limited by the available memory).

Note that 2-additive fuzzy measures require $O(n^2)$ parameters. They are stored in cardinality ordering.

`fm.generate_fm_2additive_convex(num, n)`
`fm.generate_fm_2additive_concave(num, n)`
`fm.generate_fm_2additive_convex_withsomeindependent(num, n)`
Generates *num* 2-additive convex (supermodular) or concave (submodular) fuzzy measures for n inputs. Returns the length of the part of the array v allocated for each fuzzy measure, and the array with singletons and pairs in Mobius representation. The array needs to be reserved in the calling program of size at most $num * n^2$. The routine "withsomeindependent" means that a proportion of pairwise interaction indices will be set to 0, making these pairs independent. The different fuzzy measures are stored consecutively, so that the ith measure starts at position $i * lengthv$. Example:

```
num=10
n=20
lengthv,v=fm.py_generate_fm_2additive_convex(num, n)
```

See also `fm.generate_fm_2additive`, `fm.generate_fm_2additive_randomwalk2` in version 5 below.

`fm.generate_fm_2additive_convex_sparse( n, envsp)`
Generates a random 2-additive supermodular fuzzy measure in sparse representation. Some Möbius values will be set to 0. Example:

```
n=5
tup=0
envsp=fm.py_prepare_fm_sparse(n, tup)
fm.py_generate_fm_2additive_convex_sparse(n, envsp)
```

`fm.generate_fm_kadditive_convex_sparse( n, k, nonzero, envsp)`
Generates a random k additive belief fuzzy measure (i.e. totally monotone) in sparse representation. Some Möbius values will be set to 0. Example:

```
n=5
tup=0
envsp=fm.py_prepare_fm_sparse(n, tup)
k=4
fm.generate_fm_Kadditive_convex_sparse(n, k, 10, envsp)
```

A.5.2 *Without Python wrapper, Using* `numpy` *and* `cffi`

`fm.py_generate_fm_tsort(num, n, kint, markov, option, K, v, env)`
`fm.py_generate_fm_minplus(num, n, kint, markov, option, K, v, env)`

Both functions generate several random fuzzy measures (*num* is their number) stored in **cardinality** ordering in the array v (the ith fuzzy measure starts at position $i * lengthv$ and has length $length$, which is returned by this function). The tsort method is based on topological sort and minplus is based on MinimalsPlus method (see Chap. 9). Other parameters: $0 < kint \leqslant n$ is for k-interactive fuzzy measures; *markov* is the number of Markov steps to take—the randomness increases with that number, but slows down the process; K is the constant in k-interactive fuzzy measures; and v is the array to store the generated values (preallocated of size $num * 2^n$ or less for k-interactive fuzzy measures; see `fm.py_fm_arraysize`). $option = 1$ employs internal rejection method to improve uniformity, but for $n > 5$ is not essential. Returns the size of the array for each generated fuzzy measure. Example:

```
num=10
n=6
kint=3
env=ffi.new("struct fm_env *")
fm.py_fm_init(n, env)
len=fm.py_fm_arraysize(n, kint, env)
v=np.zeros(num * len, np.float)
pv = ffi.cast("double *", v.ctypes.data)
lengthv=fm.py_generate_fm_tsort(num, n, kint, 1000, 0, 0.7, pv,
env)
```

`fm.py_generate_fmconvex_tsort(num, n, kint, markov, option, K, v, env)`

Generates *num* convex random fuzzy measures stored consecutively in **cardinality** ordering in the array v (the ith fuzzy measure starts at position $i * 2^n$ and has length 2^n). The parameters are as for the previous two functions. Returns the size of the array for each generated fuzzy measure. *kint* for the moment is not implemented. This method is quite reliable for $n < 10$ but then becomes slow. There is an internal rejection process which guarantees supermodularity for small $n < 9$, but otherwise the result must be checked as in the example below. Increase Markov chain length for $n > 6$. Example:

```
num=1
n=5
env=ffi.new("struct fm_env *")
fm.py_fm_init(n, env)
len=fm.py_fm_arraysize(n, n, env)
v=np.zeros(num * len, np.float)
pv = ffi.cast("double *", v.ctypes.data)
lengthv=fm.py_generate_fmconvex_tsort(num, n, n, 1000, 0, 1,
pv, env)
w=np.zeros(num * len, np.float)
pw = ffi.cast("double *", w.ctypes.data)
fm.py_ConvertCard2Bit(pw, pv, env)
```

```
super=fm.py_IsMeasureSupermodular(pw, env)
```

The following routines do not require initialisation of the fuzzy measure with `py_fm_init` and can hence be used with large n (limited by the available memory). Note that 2-additive fuzzy measures require $O(n^2)$ parameters. They are stored in cardinality ordering.

```
fm.py_generate_fm_2additive_convex(num, n, v)
fm.py_generate_fm_2additive_concave(num, n, v)
fm.py_generate_fm_2additive_convex_withsomeindependent(num, n,
v)
```

Generates num 2-additive convex (supermodular) or concave (submodular) fuzzy measures for n inputs. Returns the length of the part of the array v allocated for each fuzzy measure, and the array with singletons and pairs in Mobius representation. The array needs to be reserved in the calling program of size at most $num * n^2$. The routine "withsomeindependent" means that a proportion of pairwise interaction indices will be set to 0, making these pairs independent. The different fuzzy measures are stored consecutively, so that the ith measure starts at position $i * lengthv$. Example:

```
num=10
n=20
v=np.zeros(num * n*n, np.float)
pv = ffi.cast("double *", v.ctypes.data)
lengthv=fm.py_generate_fm_2additive_convex(num, n, pv)
```

```
fm.py_generate_fm_2additive_convex_sparse( n, envsp)
```

Generates a random 2-additive supermodular fuzzy measure in sparse representation. Some Möbius values will be set to 0. Example:

```
env_sp=ffi.new("struct fm_env_sparse *")
n=5
tup=0
tuples=np.zeros( 0, np.intc)
ptuples = ffi.cast("int *", tuples.ctypes.data)
fm.py_prepare_fm_sparse(n, tup, ptuples, env_sp)
fm.py_generate_fm_2additive_convex_sparse(n, env_sp)
```

```
fm.py_generate_fm_kadditive_convex_sparse( n, k, nonzero, envsp)
```

Generates a random k-additive belief fuzzy measure (i.e. totally monotone) in sparse representation. Some Möbius values will be set to 0. Example:

```
env_sp=ffi.new("struct fm_env_sparse *")
n=5
tup=0
tuples=np.zeros( 0, np.intc)
ptuples = ffi.cast("int *", tuples.ctypes.data)
fm.py_prepare_fm_sparse(n, tup, ptuples, env_sp)
k=4
fm.py_generate_fm_Kadditive_convex_sparse(n, k, 10, env_sp)
```

A.6 New Functions in Version 5

A.6.1 Using Python wrapper

`fm.generate_fm_2additive(num, n)`
Generates num random 2-additive fuzzy measures in Mobius representation. The output contains singletons and pairs but no empty set. Example:

```
num=10
n=5
ret_code, v, len = fm.generate_fm_2additive(num, n)
print(v)
print(ret_code)
print(len)
```

`fm.CheckMonMob2additive2(v, n)`
Checks monotonicity of the 2-additive set function v in Mobius representation using fast check. The auxiliary array of length n^2 may or may not be specified (if speed matters, then preallocate it). No need for initialisation with `fm.Init`. Example:

```
ret_code, v, len =fm.generate_fm_2additive(1, 10)
n=10
check = fm.CheckMonMob2additive2(v, n)
```

`fm.generate_fm_2additive_randomwalk2(num, n, markov, option, step, Fn)`
Generates num random 2-additive fuzzy measures in Mobius representation by using random walk. The output contains singletons and pairs but no empty set. The function Fn can be used for additional checks; see below. Example:

```
num=10
n=5
ret_code, v, len = fm.generate_fm_2additive_randomwalk2(num, n,
1000, 0, 0.001, None)
print(v)
print(ret_code)
print(len)
```

`fm.generate_fm_sorting(num, n, markov, option, env)`
Generates num general random fuzzy measures in standard representation. The output contains an array of length num $*$ 2^n. It uses a number of Markov steps to improve randomness. Parameter `option` is not used (reserved). Example:

```
env=fm.fm_init(3)
markovsteps=100
fuzzymeasures = fm.generate_fm_sorting(10, 3, markovsteps, 0,
env)
```

`fm.generate_fm_balanced(num, n, env)`
Generates num balanced random fuzzy measures in standard representation. The output contains an array of length num $*$ 2^n. Example:

```
env=fm.fm_init(3)
ret_code, v, len = fm.generate_fm_balanced(10, 3, env)
print(v)
print(ret_code)
print(len)
```

`fm.generate_fm_belief(num, n, kadd, env)`
Generates num random k-additive belief measures in Mobius representation. The output contains an array of length num * len. Example:

```
env=fm.fm_init(4)
num= 2
n=4
ret_code, v, len = fm.generate_fm_belief(num, n, 2, env)
print(v)
print(ret_code)
print(len)
```

`fm.generateAntibuoyant(n, env)`
Generates one antibuoyant random fuzzy measure in standard representation. The output contains an array of length 1 * 2^n. Example:

```
env=fm.fm_init(3)
antibuoyant = fm.generateAntibuoyant(3, env)
```

`fm.generate_fm_randomwalk(num, n, kadd, markov, option, step, env, Fn)`
Generates num k-additive fuzzy measures in the standard or Mobius representation using random walk of length markov of stepsize step.
Options: 0, normal; 1, convex (supermodular); 2, antibuoyant; 3, k-additive; 4, belief measure; 5, k-additive convex.
The measure generated is in standard representation for all options except 3,5. The parameter *kadd* only matters for options 3 and 5. In that case the measure is in more compact Mobius representation.
Function Fn, if not *None*, is a callback function to perform additional check at every Markov step of the current set function, i.e. any extra conditions. The function should be implemented as in the example below.
The output is a list with the first element retcode being return code, second V being the fuzzy measure, and the third one being the length of the array containing it.
Note. The parameter option can be added 256*16 to ensure that convex fuzzy measures are less likely to be belief measures (e.g. Option=256*16 + 1). Example:

```
def Fn(n, v):
out = 0.0
for i in range(n[0]):
out += v[i]
if out > 1:
return 0
else:
return 1
```

```
env=fm.fm_init(3)
step=0.0010
Option=3
n=3
ret_code, v,len= fm.generate_fm_randomwalk(2, 3, 2, 1000,
Option, step, env, Fn)
print(v)
print(ret_code)
print(len)
```

`fm.CheckMonotonicitySortMerge(n, v, env)`
Checks monotonicity of the set function v in standard representation using merge sort. The output is a list of (*True/False, indices*). The indices and values can be used at subsequent steps of monotonicity verification (e.g. values slightly perturbed). Initially indices need not be specified. Example:

```
env=fm.fm_init(3)
n=3
v=fm.generate_fm_sorting(1, 3, 1000, 0, env)
ret_code, index = fm.CheckMonotonicitySortMerge(n, v, env)

print(ret_code)
```

`fm.CheckMonotonicitySortInsert(n, v, indices, env)`
Checks monotonicity of the set function v in standard representation using insert sort. The output is a list of (*True/False, indices*). The indices and values can be used at subsequent steps of monotonicity verification (e.g. values slightly perturbed). This function is called after merge sort, so the indices are already precomputed. Example:

```
n=3
env=fm.fm_init(n)
v=fm.generate_fm_sorting(1, n, 1000, 0, env)
ret_code, index=fm.CheckMonotonicitySortMerge(n, v, env)
v[0] *= 1.1
ret_code, index=CheckMonotonicitySortInsert(n, v, index, env)
```

`fm.CheckMonotonicitySimple(v, env)`
Checks monotonicity of the set function v in standard representation using standard check. Example:

```
env=fm.fm_init(3)
v=fm.generate_fm_sorting(1, 3, 1000, 0, env)
monotonicity=CheckMonotonicitySimple(v, env)
```

`fm.CheckMonotonicityMob(Mob, len, env)`
Checks monotonicity of the set function v in Mobius representation using standard check. `len` is the length of the array of Mobius values. It can be obtained by `fm_arraysize_kadd(N, Kadd)`. Example:

```
env=fm.fm_init(3)
step=0.001
Fn=None
```

```
Option=3
ret_code, Mob, len = fm.generate_fm_randomwalk(1, 3, 2, 1000,
Option, step, env, Fn)
check=fm.CheckMonotonicityMob(Mob, len, env)
```

`fm.CheckConvexityMonMob(Mob, len, env)`
Checks supermodularity of the set function *Mob* in Mobius representation using standard check. `len` is the length of the array of Mobius values. It can be obtained by
`fm_arraysize_kadd(N, Kadd)`. Example:

```
env=fm.fm_init(3)
step=0.001
Fn=None
Option=3
ret_code, Mob, len = fm.generate_fm_randomwalk(1, 3, 2, 1000,
Option, step, env, Fn)
check=fm.CheckConvexityMonMob(Mob, len, env)
```

`fm.generate_fm_kinteractivedualconvex(num, n, kint, markov, step, env, Fn)`
Generates `num` k-interactive dual convex fuzzy measures in Mobius representation using random walk of length `markov` of stepsize `step`. Ensure $n > kint - 1$. Example:

```
env=fm.fm_init(4)
step=0.001
Fn=None
ret_code, Mob, len = fm.generate_fm_kinteractivedualconvex(10,
4, 2, 1000, step, env, Fn)
print(Mob)
print(ret_code)
print(len)
```

`fm.generate_fm_kinteractivedualconcave(num, n, kint, markov, step, env, Fn)`
Generates `num` k-interactive dual concave fuzzy measures in Mobius representation using random walk of length `markov` of stepsize `step`. Ensure $n > kint - 1$. Example:

```
env=fm.fm_init(4)
step=0.001
Fn=None
ret_code, Mob, len = fm.generate_fm_kinteractivedualconcave(10,
4, 2, 1000, step, env, Fn)
print(Mob)
print(ret_code)
print(len)
```

```
fm.FuzzyMeasureFit2Additive(num, n, options, indexlow,
indexhigh, option1, orness, data)
```

This procedure works similar to other fitting routines but is tailored to 2-additive fuzzy measures, which simplifies the process and allows one to work with larger n. The meaning of the parameters is the same as in `FuzzyMeasureFitLP` Example:

```
env=fm.fm_init(3)
num=20
n=3
d = np.array([0.00125122, 0.563568, 0.193298, 0.164338,
0.808716, 0.584991, 0.479858, 0.544309, 0.350281, 0.895935,
0.822815, 0.625868, 0.746582, 0.174103, 0.858917, 0.480347,
0.71048, 0.513519, 0.303986, 0.387631, 0.0149841, 0.0914001,
0.364441, 0.134229, 0.147308, 0.165894, 0.988495, 0.388044,
0.445679, 0.11908, 0.00466919, 0.0897714, 0.00891113, 0.377869,
0.531647, 0.258585, 0.571167, 0.601746, 0.607147, 0.589803,
0.166229, 0.663025, 0.450775, 0.357412, 0.352112, 0.0570374,
0.607666, 0.270228, 0.783295, 0.802582, 0.519867, 0.583348,
0.301941, 0.875946, 0.726654, 0.562174, 0.955872, 0.92569,
0.539337, 0.633631, 0.142334, 0.462067, 0.235321, 0.228419,
0.862213, 0.209595, 0.779633, 0.498077, 0.843628, 0.996765,
0.999664, 0.930197, 0.611481, 0.92426, 0.266205, 0.334666,
0.297272, 0.840118, 0.0237427, 0.168081]).reshape(num, 4)
indexlow=[0.1,0.1,0.2]
indexhigh=[0.9,0.9,0.5]
options=3
option1=1
orness=[0.1,0.7]
v=fm.FuzzyMeasureFit2Additive(num, n, options, indexlow,
indexhigh, option1, orness, d)
print(v)
```

```
fm.ConvertCoMob2KinterCall(n, kadd, len, Mob, fullmu, env)
```

Converts dual k-interactive fuzzy measure from Mobius to standard representation. If $fullmu$ is 1, then all 2^n are allocated; otherwise, a more compact representation for k-interactive fuzzy measures is used. Ensures $n > kint - 1$. Example:

```
env=fm.fm_init(4)
n=4
fullmu=0
step= 0.001
Fn=None
ret_code, Mob, len = fm.generate_fm_kinteractivedualconvex(1,
4, 2, 1000, step, env, Fn)
v=fm.ConvertCoMob2Kinter(4, 2, len, Mob, fullmu, env)
```

```
fm.ChoquetCoMobKInter(x, Mob, kadd, len, env)
```

Calculates the Choquet integral of x wrt dual k-interactive fuzzy measure in Mobius representation. This is the same as k-interactive fuzzy measure in co-Mobius representation. Example:

```
env=fm.fm_init(4)
step= 0.001
Fn=None
ret_code, Mob, len= fm.generate_fm_kinteractivedualconvex(1, 4,
2, 1000, step, env, Fn)
x=[0.2,0.5,0.4,0.1]
fm.ChoquetCoMobKInter(x, Mob, 2, len, env)
```

A.7 Examples

```
import numpy as np
import math
from  pyfmtools import ffi,lib as fm

n=3
env=ffi.new("struct fm_env *")
fm.py_fm_init(n,  env);

A=np.zeros(env.m, int)
pA = ffi.cast("int *", A.ctypes.data)
fm.py_ShowCoalitions(pA, env)
print("Fuzzy measure wrt n=3 criteria has ",env.m," in binary order")
print(A)
fm.py_fm_free( env);

total=1
n=4
fm.py_fm_init(n,  env);
v=np.zeros(env.m,float);
pv = ffi.cast("double *", v.ctypes.data);
vb=np.zeros(env.m,float);
pvb = ffi.cast("double *", vb.ctypes.data);

size=fm.py_generate_fm_2additive_concave(total,n,pv)
print("2-additive concave FM in Mobius and its length (n=4)")
print(v)

print("A convex FM in cardinality ordering ")
A=np.zeros(env.m, int)
```

```
pA = ffi.cast("int *", A.ctypes.data)
fm.py_ShowCoalitionsCard(pA, env)
print(A)

size=fm.py_generate_fmconvex_tsort(total,n, n-1 , 1000, 8, 1, pv,env)
#size=fm.py_generate_fm_tsort(total,n, 2 , 10, 0, 0.1, pv,env)
print(v)

fm.py_ConvertCard2Bit(pvb,pv,env)
print("a convex FM in binary ordering ")
fm.py_ShowCoalitions(pA, env)
print(A)
print(vb)
r=fm.py_IsMeasureSupermodular(pvb,env)
print("Is it convex (test)?", r)
r=fm.py_IsMeasureAdditive(pvb,env)
print("Is it additive (test)?", r)

mc=np.zeros(math.factorial(n)*n,float)
pmc = ffi.cast("double *", mc.ctypes.data)

fm.py_export_maximal_chains(n,pvb,pmc,env)
print("Export maximal chains ")
print(mc)

S=np.zeros(n,float)
pS = ffi.cast("double *", S.ctypes.data)
fm.py_Shapley(pvb,pS,env)
print("Shapley values")
print(S)

x=np.array([0.2,0.1,0.6,0.2])
px = ffi.cast("double *", x.ctypes.data)

z=fm.py_Choquet(px,pvb,env)
print("Choquet integral of ",x, " is ",z)
z=fm.py_Sugeno(px,pvb,env)
print("Sugeno integral of ",x,  " is ",z)

fm.py_fm_free( env);
```

Index

G. Beliakov et al., *Choquet Capacities and Fuzzy Integrals*, Theory and Applications of Computability, https://doi.org/10.1007/978-3-031-97070-2

FSC
www.fsc.org
MIX
Papier | Fördert
gute Waldnutzung
FSC® C083411

Zeitfracht Medien GmbH
Ferdinand-Jühlke-Straße 7
99095 Erfurt, Deutschland
produktsicherheit@kolibri360.de